Advances in
MICROBIAL
PHYSIOLOGY

Advances in
MICROBIAL PHYSIOLOGY

edited by

A. H. ROSE

School of Biological Sciences
Bath University
England

J. GARETH MORRIS

Department of Botany and Microbiology
University College Wales
Aberystwyth

Volume 18

1978

ACADEMIC PRESS
London New York San Francisco

A Subsidiary of Harcourt Brace Jovanovich, Publishers

ACADEMIC PRESS INC. (LONDON) LTD.
24/28 Oval Road
London NW1 7DX

United States Edition published by
ACADEMIC PRESS LTD.
111 Fifth Avenue
New York, New York 10003

Copyright © 1978 by
ACADEMIC PRESS INC. (LONDON) LTD.

All Rights Reserved

No part of this book may be reproduced in any form by photostat, microfilm, or any other means, without written permission from the publishers

Library of Congress Catalog Card Number: 67-19850
ISBN: 0-12-027718-2

Printed in Great Britain by
Spottiswoode Ballantyne Ltd.
Colchester and London

Contributors to Volume 18

NANCY KLEIN AMY, *Department of Biology, University of Virginia, Charlottesville, Virginia 22901, U.S.A. (Present address: Department of Biochemistry, Duke University Medical Center, Durham, North Carolina 27710, U.S.A.)*

PETER T. BORGIA, *Department of Microbiology, Southern Illinois University, Carbondale, Illinois 62901, U.S.A.*

W. R. FINNERTY, *Department of Microbiology, University of Georgia, Athens, Georgia 30602, U.S.A.*

R. H. GARRETT, *Department of Biology, University of Virginia, Charlottesville, Virginia 22901, U.S.A.*

S. W. GLOVER, *Department of Genetics, Ridley Building, University of Newcastle, Newcastle upon Tyne NE1 7RU, England*

JOHN L. PAZNOKAS, *Department of Bacteriology and Public Health, Washington State University, Pullman, Washington 99163, U.S.A.*

MICHAEL G. SARGENT, *National Institute for Medical Research, Mill Hill, London NW7 1AA, England*

PAUL S. SYPHERD, *Department of Medical Microbiology, University of California, Irvine, California 92717, U.S.A.*

Contents

Nitrate Assimilation in Fungi by R. H. GARRETT and NANCY KLEIN AMY

I. Introduction	1
II. Enzymology of Nitrate Assimilation	2
A. Nitrate Reductase	2
B. Nitrite Reductase	20
C. Concentrations of Nitrate Reductase and Nitrite Reductase in Mycelia	35
D. Nitrate and Nitrite Uptake	37
III. Regulation of Nitrate Assimilation	38
A. Genetics and Regulation of Gene Expression	40
B. Regulation of Enzymic Activity	58
IV. Conclusions and Future Prospects	61
References	62

Biochemistry of Dimorphism in the Fungus *Mucor* by PAUL S. SYPHERD, PETER T. BORGIA, AND JOHN L. PAZNOKAS

I. Introduction	68
II. Dimorphism of *Mucor* spp. at the Cellular Level	69
A. Spore Germination	70
B. Hyphal Growth	71
C. Yeast Growth	71
D. Dimorphic Transitions	71
III. Environmental Conditions which Influence Dimorphism	71
IV. Cell-Wall Structure and Synthesis	75
V. Metabolic Responses to Morphogenic Signals	79
A. A Role for Cyclic Nucleotides in Dimorphism	79
B. Carbon and Energy Metabolism Associated with Dimorphism	84
C. Alterations in Nitrogen Metabolism	87
VI. Protein Synthesis and Control of Morphogenesis	90
VII. Genetics of Mucor	99
VIII. Concluding Remarks	100
IX. Acknowledgments	101
References	101

Surface Extension and the Cell Cycle in Prokaryotes by MICHAEL G. SARGENT

I. A Survey of the Problem	106
II. The Age Distribution in Exponential-Phase Populations	109
A. Idealized and Real Age Distribution	109
B. Uses of the Idealized Age Distribution: The Unit Cell	111

CONTENTS

- C. The Origin of the Dispersion of Generation Times 112
- III. The Size of Exponential Phase Rod-Shaped Bacteria 117
  - A. Observations on Growing Bacteria 117
  - B. Interpretation of Size Distributions 118
  - C. Factors Determining Average Cell Size 120
  - D. Experimental Tests 123
  - E. Agents Which Can Synchronize Cell Division 126
- IV. Environmental Effects on Cellular Dimensions in the Steady State . . 128
  - A. Growth Rate and Medium Composition 128
  - B. Temperature 130
  - C. Plasmids 130
  - D. Morphological Effects Produced by Specific Nutritional Factors . . 131
- V. Inferences from Synchronous Cultures and Age-Classified Cells . . 132
- VI. Topography of the Bacterial Surface 134
  - A. Cocci 135
  - B. Rod-Shaped Bacteria 138
  - C. Caulobacter and Budding Bacteria 150
- VII. Inferences from Physiological Studies 151
  - A. Inhibition of Protein Synthesis 151
  - B. Inhibition of DNA Synthesis 151
  - C. Inhibition of Peptidoglycan Synthesis 154
- VIII. Genetic Approaches to the Analysis of Surface Extension . . . 156
  - A. Rod Mutants 157
  - B. Mutations Affecting Cell Width 160
  - C. Mutants Affecting Cell Size 160
  - D. Spiral Growth Mutants 161
  - E. Septation Mutants 161
  - F. Minicell-Producing Mutants 162
  - G. Mutants Producing Large Anucleate Cells 163
  - H. Initiation Mutants 165
- IX. Concluding Remarks 166
- X. Acknowledgments 168
- References 168

Physiology and Biochemistry of Bacterial Phospholipid Metabolism
by W. R. FINNERTY

- I. Introduction 177
- II. Structural, Functional and Biological Characteristics of Microbial Phospholipids 178
  - A. General Considerations of Phospholipid Structure and Composition . 178
  - B. Asymmetry of Phospholipids in Membranes 181
  - C. The Role of Phospholipids in Enzyme Activity 183
  - D. Phospholipids and Microbial Viruses 187
  - E. Regulation of Phospholipid Composition 188
- III. Biosynthesis of Microbial Phospholipids 194
  - A. Phosphatidic Acid Biosynthesis 194
  - B. Cytidine Diphosphate Diglyceride Biosynthesis 198
  - C. Phosphatidylglycerophosphate and Phosphatidylglycerol Biosynthesis . 199
  - D. O-Amino-Acyl Phosphatidylglycerol Biosynthesis 200

E. Fatty-Acyl Phosphatidylglycerol Biosynthesis		201
F. Diphosphatidylglycerol (Cardiolipin) Biosynthesis		203
G. Phosphatidylserine and Phosphatidylethanolamine Biosynthesis		204
H. Methylated Derivatives of Phosphatidylethanolamine		207
I. Ornithine-Containing Lipids		208
IV. Catabolism of Microbial Phospholipids		210
A. Phospholipase A_1		213
B. Phospholipase A_2		219
C. Lysophospholipases		219
D. Phospholipase C		220
E. Phospholipase D		221
F. Regulation		222
V. Conclusion		225
References		226

Aspects of Genetic Engineering in Micro-Organisms by S. W. GLOVER

I. Introduction	235
II. Restriction Endonucleases	237
A. Type I Restriction Endonucleases	238
B. Type II Restriction Endonucleases	239
III. End-to-End Joining of DNA Molecules	243
IV. General Methods for Inserting Specific DNA Sequences	244
V. Plasmid Vectors	247
VI. Bacteriophage Lambda Vectors	251
VII. Cloning of Recombinant DNA	254
A. Cloning Prokaryote genes	254
B. Cloning Yeast DNA	256
C. Cloning Higher Eukaryote DNA	257
D. Screening Methods for Cloned DNA	260
VIII. Potential Biohazards	262
A. Physical Containment	263
B. Biological Containment	264
IX. Acknowledgments	265
References	265
Author Index	273
Subject Index	289

Nitrate Assimilation in Fungi

R. H. GARRETT AND NANCY KLEIN AMY*

Department of Biology, University of Virginia, Charlottesville, Virginia 22901, U.S.A.

I. Introduction	1
II. Enzymology of Nitrate Assimilation	2
A. Nitrate Reductase	2
B. Nitrite Reductase	20
C. Concentrations of Nitrate Reductase and Nitrite Reductase in Mycelia	35
D. Nitrate and Nitrite Uptake	37
III. Regulation of Nitrate Assimilation	38
A. Genetics and Regulation of Gene Expression	40
B. Regulation of Enzymic Activity	58
IV. Conclusions and Future Prospects	61
References	62

I. Introduction

Our aim in this article has not been to present a comparative review of nitrate reduction by fungi but, instead, to present a detailed analysis of the results available on the process of nitrate assimilation as it occurs in two filamentous ascomycetes, *Aspergillus nidulans* and *Neurospora crassa*. Not only are these results in general representative of observations made on other fungi and yeasts, but, in particular, they provide for the most concise and comprehensive description of nitrate assimilation and its regulation. It is a vagary of scientific fortune that the genetics of nitrate assimilation is best understood for *A. nidulans*, whereas the enzymology of nitrate assimilation is better characterized in *N. crassa*, the organism more often subjected to biochemical investigation. Nevertheless, the parallels which can be drawn between

---

Present address: Department of Biochemistry, Duke University Medical Center, Durham, North Carolina 27710, U.S.A.

the results obtained with these two fungi, and the contrast which can be cited, have advanced our knowledge of nitrate assimilation substantially; therefore, we feel that the approach we have adopted is justified. In addition, recent reviews of a broader, more comparative scope dealing with this subject are available in this series (Brown et al., 1974) and can also be found elsewhere (Payne, 1973; Hewitt, 1975; Pateman and Kinghorn, 1976).

II. Enzymology of Nitrate Assimilation

The assimilatory reduction of nitrate to ammonia is achieved by a metabolic pathway composed of just two enzymes, namely nitrate reductase and nitrite reductase, which act in this sequence. Recently, as discussed below, both have been purified to homogeneity and detailed characterizations of each are being made. The results are presented in turn.

A. NITRATE REDUCTASE

1. *General Properties of the Nitrate Reductase from* Neurospora crassa

Nitrate reductase catalyses the first step in the pathway, that is, the reduction of nitrate to nitrite. This enzyme was initially investigated and partially purified from *N. crassa* by Nason and Evans in 1953, who found it to be soluble, sulphydryl-containing metalloflavoprotein. The preferred electron donor was NADPH and FAD was an essential prosthetic group, mediating electron transport between the nicotinamide nucleotide electron donor and nitrate. The nitrate reductase was inhibited by the sulphydryl-binding agent *para*-hydroxymercuribenzoate (*p*HMB). Metal-binding agents, such as cyanide azide and thiourea, were also inhibitory, indicating that a heavy metal was a constituent of the enzyme. Calculation of the change in free energy of the nitrate reductase reaction from the redox potentials for $NADP^+/NADPH$ ($E^{\circ\prime} = -0.324$ V) and NO_3^-/NO_2^- ($E^{\circ\prime} = 0.42$ V) yields a value of -34 Kcal. Thus, Nason and Evans (1953) concluded that other electron-transferring constituents with standard potentials intermediate between those of the reactants and products of the nitrate reductase might possibly be involved.

Nicholas *et al.* (1953, 1954) investigated further the nature of the metal component of the nitrate reductase. Growth studies revealed that the nitrate reductase activity was very low in mycelia grown on molybdenum-deficient media. Increasing amounts of molybdate added to the media caused a progressive increase in nitrate reductase activity. However, addition of molybdate to extracts of molybdenum-deficient mycelia did not restore nitrate reductase activity. Further investigation (Nicholas and Nason, 1954a) revealed that, with an increase in the specific activity of the nitrate reductase, there was a concomitant increase in the molybdenum content of the protein extract. These results indicated that molybdenum was indeed a component of the enzyme and essential for enzymic activity. Nicholas and Nason (1954b) further showed that molybdenum was involved in electron transport from NADPH to nitrate, according to the electron-transport pathway:

$$NADPH \to FAD\,(or\,FMN) \to Mo \to NO_3^-$$

Further, the nitrate reductase could use NADPH to reduce trichloro-indophenol in an FAD-dependent reaction that did not require molybdenum. Other studies (Kinsky and McElroy, 1958) showed that the nitrate reductase could use NADPH to reduce cytochrome c in an FAD-dependent reaction. Kinsky and McElroy (1958) found that extracts of *N. crassa* possessed a constitutive NADPH-cytochrome c reductase activity that was present under all conditions of nitrogen nutrition, and a second NADPH-cytochrome c reductase activity, which was induced by nitrate in parallel to the nitrate reductase and was present in a constant ratio to the nitrate reductase throughout the later stages of enzyme purification. The sulphydryl-binding agent, pHMB, inhibited both the nitrate reductase and the nitrate-induced NADPH-cytochrome c reductase to the same extent. Metal-binding agents, such as cyanide or azide, did not inhibit the NADPH-cytochrome c reductase activity.

Investigations with a 500-fold purified nitrate reductase (Garrett and Nason, 1967) revealed yet another component of the enzyme, a haem moiety designated cytochrome b-557, functioning in the electron transport. Difference spectra of the reduced-minus-the oxidized form of the enzyme showed absorption maxima at 557, 528 and 423 nm (Fig. 1). These maxima are typical of a b-type cytochrome. The pyridine haemochromogen derivative of the nitrate reductase was prepared, and its spectrum after reduction by sodium dithionite corresponded to that of a pyridine derivative of protoporphyrin IX similarly reduced

(Garrett and Nason, 1969). Cytochrome b-557 was demonstrated to participate in electron transport of the nitrate reductase (Fig. 1). A slight amount of the haem was reduced when NADPH was added alone (curve A). When FAD and NADPH were both present, complete reduction of the cytochrome in the enzyme preparation resulted (curve B). When nitrate was added to the NADPH plus FAD-reduced enzyme, its cytochrome was re-oxidized (curve C). This reduction of the cytochrome by NADPH and FAD was not inhibited by cyanide. During purification of the nitrate reductase, the ratio between the concentration of cytochrome b-557 and nitrate reductase activity in the various fractions did not vary significantly after the initial step (Garrett and Nason, 1967).

Through the use of artificial electron donors and acceptors, and the differential effect of inhibitors on the various activities of nitrate

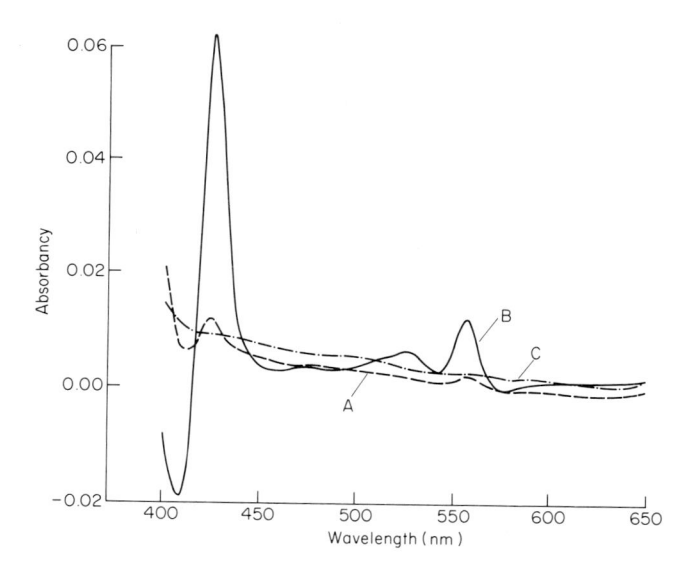

FIG. 1. Oxidized *versus* reduced difference spectra of the nitrate reductase from *Neurospora crassa*. ——— Curve A: sample cuvette contained nitrate reductase together with 0.1 mM NADPH. Reference cuvette contained the same concentration of enzyme. ——— Curve B: 0.2 μM FAD was added to both cuvettes containing enzyme and NADPH. —·— Curve C: 5 mM NaNO$_3$, 0.1 mM NADPH and enzyme added to the sample cuvette. The reference cuvette contained 5 mM NaNO$_3$ and water. Addition of a few crystals of sodium dithionite to the sample cuvette after obtaining curve C gave a spectrum essentially the same as ———. All recordings were made with samples under aerobic conditions. From Garrett and Nason (1967).

reductase, the following electron-transport sequence was proposed (Garrett and Nason, 1969):

$$\text{NADPH} \to \text{FAD} \begin{array}{c} \nearrow \text{cytochrome } b\text{-}557 \to \text{Mo} \to \text{NO}_3^- \\ \searrow \text{cytochrome } c \end{array}$$

The rationale behind using alternate electron donors and acceptors is to characterize partial functions of the overall electron-transport pathway: (a) the NADPH-nitrate reductase activity, utilizing the entire electron-transport pathway; (b) the NADPH-cytochrome c reductase (also termed the diaphorase) activity utilizing various alternate electron acceptors such as mammalian cytochrome c, ferricyanide or trichloro-indophenol to measure only the initial portion of the electron-transport sequence; (c) the reduced FAD (FADH$_2$)-nitrate reductase activity and reduced methyl viologen (MVH)-nitrate reductase activities utilizing the named artificial electron donors instead of NADPH as a measure of the terminal electron-transport process.

The overall NADPH-nitrate reductase activity required FAD and was inhibited by pHMB, cyanide and azide. The nitrate-induced NADPH-cytochrome c reductase activity (diaphorase) required FAD too, but was inhibited only by pHMB. The FADH$_2$-nitrate reductase and MVH-nitrate reductase were inhibited only by cyanide and azide. Both of the NADPH-dependent activities, namely the overall NADPH-nitrate reductase and the NADPH-cytochrome c reductase, were inactivated by a brief heat treatment at 50°C. The FADH$_2$- and MVH-nitrate reductase activities were not inactivated. Since reduction of the cytochrome b-557 by NADPH and FAD was not inhibited by cyanide, the molybdenum moiety must function in electron-transport subsequent to this haem.

The four enzymic activities were found to be associated in constant proportion throughout purification of the nitrate reductase and all were found associated with the same protein band on polyacrylamide gel electrophoresis. Thus, all four activities are attributed to a single macromolecular complex. This nitrate reductase has a molecular weight of 228,000 and a sedimentation co-efficient of 8.0 S; it has been crystallized (Garrett and Nason, 1969).

Garrett and Greenbaum (1973) have further elucidated the nature of the inhibition of the *N. crassa* nitrate reductase by metal-binding agents. These agents normally exhibit competitive inhibition with respect to nitrate. However, when nitrate reductase is pre-incubated with NADPH, FAD and cyanide, sulphide or thiourea, non-competitive inhibition results and the enzyme displays increased sensitivity to inhibition. This inhibition can be prevented if nitrate is included in the pre-incubation mixture, but nitrate will not reverse the non-competitive inhibition if added later. However, the inhibition can be reversed by addition of diaphorase electron acceptors, such as cytochrome c or ferricyanide, but the concentration of electron acceptor added must be sufficient to oxidize all of the NADPH present in the pre-incubation mixture. Thus, it appears that the redox state of the enzyme will determine its sensitivity to inhibition. It is known that cyanide displays a greater affinity for reduced molybdenum species, Mo(IV) and Mo(V), than for the oxidized Mo(VI) (Cotton and Wilkinson, 1966). The following explanation seems apt. When nitrate reductase is reduced by pre-incubation with NADPH and FAD, the molybdenum becomes reduced, the enzyme is more sensitive to inhibition and the inhibition is non-competitive. Nitrate added to the pre-incubation mixture will compete with the metal-binding agent, accept electrons from the enzyme, and keep the enzyme (and its molybdenum) in the oxidized state and thus maintain it insusceptible to the non-competitive type of inhibition. However, nitrate cannot reverse the inhibition once metal-binding agent and enzyme have interacted because the electrons from NADPH cannot flow past the block caused by the inhibitor presumably binding to molybdenum. The diaphorase electron acceptors can reverse the inhibition since they accept electrons from the transport sequence before the molybdenum and deplete the NADPH present, thereby permitting the enzyme to become oxidized once again.

By the use of FAD-agarose affinity chromatography, the nitrate reductase has recently been purified to homogeneity (Pan *et al.*, 1975). Sodium dodecyl sulphate (SDS) and 8 M urea polyacrylamide-gel electrophoresis of the enzyme revealed two protein bands with molecular weights of 130,000 and 115,000, respectively (Pan and Nason, 1976). The sum of the molecular weights of the subunits was greater than the total molecular weight, and the relative intensities of the two protein bands on the gels varied with different preparations. Gel electrofocusing exhibited multiple bands in the pH range 4.8 to

5.2. The possibility exists that there is some heterogeneity in the size of the subunits. Nevertheless the working hypothesis is that the nitrate reductase of *N. crassa* is a dimer of identical polypeptides.

The highest specific activities reported for the purified nitrate reductase were 80–100 μmoles nitrate reduced/min/mg protein (Pan *et al.*, 1975; Jacob, 1975). These data yield a turnover number for the nitrate reductase of 18–20 μmoles nitrate reduced/min/nmole of nitrate reductase.

2. *Recent Observations on the Electron-Transfer Reactions of the Nitrate Reductase from* Neurospora crassa

One of the functions of the electron transport of the nitrate reductase which has been most recently investigated is the role of the sulphydryl groups in the catalytic process. The early inhibition studies with *p*HMB indicated that sulphydryl groups were essential for activity of the enzyme (Nason and Evans, 1953) but no electron-transport role was assigned to the sulphydryl groups. Later studies implicated the sulphydryl groups in the initial portion of the electron transport since only the NADPH-dependent activities (the overall NADPH-nitrate reductase, and NADPH-cytochrome *c* reductase activities) were inhibited. The MVH- and $FADH_2$-nitrate reductase activities were not strongly affected (Nicholas and Nason, 1954b; Garrett and Nason, 1969). Amy *et al.* (1977) proposed that these *p*HMB-sensitive sulphydryl groups are directly involved in transfer of electrons from NADPH to the enzyme, rather than acting in some ancillary role. These investigators used as probes the active site-specific NAD(P) analogues, 3-aminopyridine adenine dinucleotide $(AAD)^+$ and 3-aminopyridine adenine dinucleotide phosphate $(AADP)^+$ which differ from NAD^+ and $NADP^+$ in that the carboxamide group at the 3-position of the pyridine ring is replaced by an amino group (Fig. 2; Fisher *et al.*, 1973; Anderson *et al.*, 1975). These analogues were shown to be coenzyme-competitive inhibitors of several NAD- or NADP-requiring enzymes (Chan and Anderson, 1975). Further, the amino group of the $AAD(P)^+$ can be treated with nitrous acid to yield the diazonium derivative which reacts specifically with sulphydryl groups at neutral pH values. The interesting property of these diazotized $AAD(P)^+$ derivatives is that they may act as site-labelling reagents through a covalent reaction with

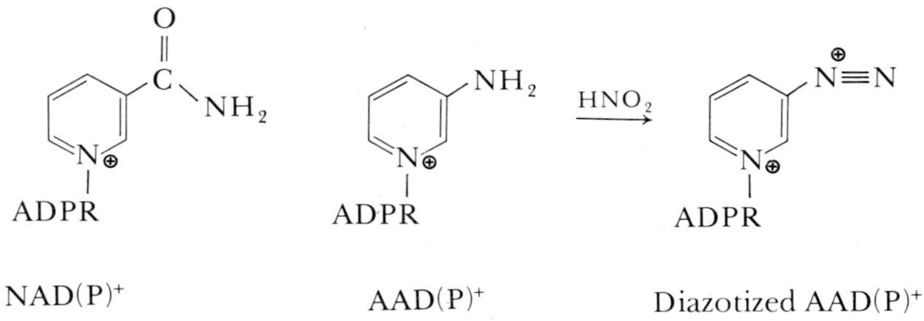

FIG. 2. Comparison of the structures of NAD(P)⁺, 3-aminopyridine adenine dinucleotide(phosphate) AAD(P)⁺ and the diazotized derivative of AAD(P)⁺. ADPR indicates adenosine diphosphoribose. From Amy et al. (1977).

sulphydryl groups at the NAD(P)-binding sites of enzymes. The nitrate reductase from *N. crassa* was competitively inhibited by the AADP⁺ with respect to NADPH (Amy et al., 1977). When the diazotized derivative of AADP⁺ was incubated with nitrate reductase, there was an irreversible, time-dependent inactivation of the enzyme, indicating that the analogue was covalently binding to enzyme sulphydryl groups in the vicinity of the NADPH-binding site. Only the two NADPH-dependent activities of the enzyme (NADPH-nitrate reductase and NADPH-cytochrome *c* reductase activities) were inactivated. The MVH- and FADH₂-nitrate reductase activities persisted. The presence of NADP⁺ or FAD protected the enzyme from inactivation by the diazo-AADP⁺. Further, when the enzyme was treated with FAD, then dialysed to remove the FAD, the enzyme was still protected against inactivation. This result was somewhat surprising. Presumably FAD treatment exerted a change in the enzyme that persisted even after removal of the FAD. When the enzyme was treated with FAD, dialysed, then treated with NADPH, and dialysed again, the nitrate reductase was again susceptible to inactivation by diazotized AAD(P)⁺. These results can be explained by postulating that the sulphydryl groups are participating in the electron transport between the electron donor, NADPH, and the enzyme-bound flavin coenzyme, FAD (Fig. 3). When the enzyme is treated with FAD, the FAD accepts electrons from the enzyme, causing an oxidation of specific sulphydryl groups. The oxidized form of the enzyme sulphydryl function would not be susceptible to the chemical attack of diazotized AADP⁺, and consequently the enzyme would be protected from inactivation. When

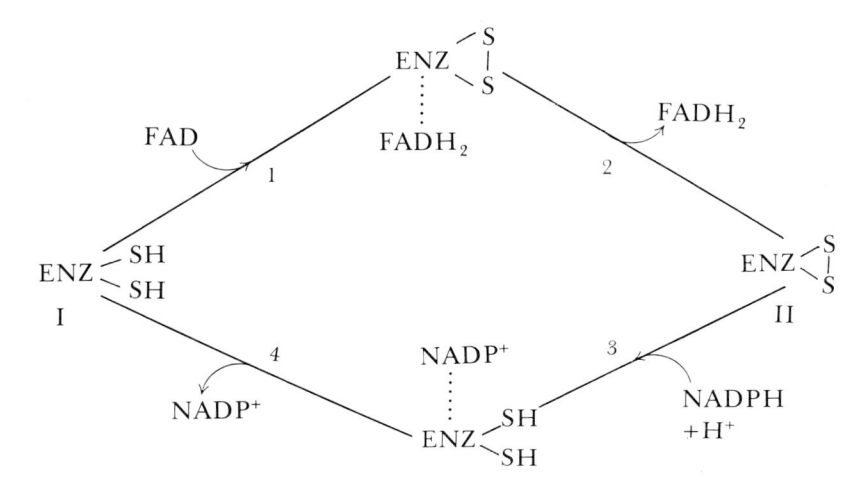

FIG. 3. Proposed mechanism of the nitrate reductase from *Neurospora crassa* in the initial electron-transfer reactions from NADPH. Form I is postulated to be susceptible to inhibition by diazotized AADP⁺, whereas form II would be insensitive to such inhibition. From Amy *et al.* (1977).

nitrate reductase is treated with NADPH, the NADPH donates electrons to the enzyme, causing a reconversion to the reduced sulphydryl form. The enzyme with free sulphydryl groups is then once again susceptible to inactivation by diazotized AAD(P)⁺ (Amy *et al.*, 1977). This interpretation also finds some support in the relative values of the S—S/(SH) and FAD/FADH$_2$ redox potentials.

Other recent studies have added significantly to knowledge about electron transfers in nitrate reductase. Metal analysis of the purified enzyme has revealed that there is one mole of haem iron and one mole of molybdenum per mole of enzyme (Jacob, 1975). Low-temperature electron paramagnetic resonance (e.p.r.) signals of this purified nitrate reductase from *N. crassa* revealed $g = 2.98$ and $g = 2.27$ resonances, which are consistent with a low-spin haem iron complexed to two imidazole residues (typical of *b*-type cytochromes). There was no resonance at $g = 2$ which could be ascribed to high-spin haem. The haem resonances were unaffected by nitrate. However, during reductive titration with NADPH and FAD, the cytochrome *b*-557 resonances disappeared, which confirmed spectral observations that the haem becomes reduced (Garrett and Nason, 1967).

Early studies by Nicholas and Stevens (1955) attempted to determine the valency changes that the molybdenum of the nitrate reductase undergoes during the catalytic process. They concluded that Mo(V) was as effective as NADPH as an electron donor for enzymic reduction of nitrate. However, Mo(III) could reduce nitrate non-enzymically. Studies with model chemical systems for biological reduction of nitrate and the valency states of molybdenum which might be involved have proposed both Mo(V) (Garner *et al.*, 1974) and Mo(III) (Ketchum *et al.*, 1976) as the reductant for nitrate.

Jacob (1975) also has used e.p.r. techniques to investigate valency changes in the molybdenum moiety of the nitrate reductase. Anaerobic samples of nitrate reductase gave e.p.r. signals in the region of $g = 2$. These spectra are ascribed to a species of Mo(V). This species is evident before titration with NADPH in the presence of FAD, rises to a maximum, and then declines as NADPH is added. These data are consistent with the existence of both Mo(V) and Mo(VI) species in the enzyme as isolated, from which Mo(V) and Mo(IV) species are produced during the reductive titration. Electron paramagnetic resonance signals with g-values at 4.15, 3.96 and 2.01 were also observed upon further reduction of the enzyme with NADPH and FAD. These signals are due to Mo(III) since the haem is reduced, and thus gives no e.p.r. signal. These signals disappeared upon re-oxidation of the enzyme with nitrate. The e.p.r. titration experiments suggest that reduction of the haem and production of the species ascribed to Mo(V) and Mo(III) occur at similar redox potentials. Therefore, both of these molecular species appear to be equally competent to be the proximal reductant of nitrate. However, it seems more likely that only one set of molybdenum (or Mo plus haem) species is active in the reduction. Kinetic experiments must be done to resolve this point. The possibility exists that the nitrate reductase may, under certain circumstances, use all of its potential electron capacity, and store at least seven electrons: two in the FAD, one in the haem, and four in the Mo transitions (Mo(VI) → Mo(II)) (Jacob, 1975). Further, the results of Amy *et al.* (1977) suggest additional functional electron capacity within the sulphydryl moieties.

3. *Properties of the Nitrate Reductase from* Aspergillus nidulans

The enzymology of the nitrate reductase from *Aspergillus nidulans* suggests that this enzyme is essentially similar to that from *Neurospora crassa*.

The nitrate reductase from *A. nidulans* was initially purified by Cove and Coddington (1965), who showed that the enzyme was specific for NADPH as the electron donor, required FAD and possessed an FAD-dependent NADPH-cytochrome *c* reductase activity which remained associated with the nitrate reductase throughout a 300-fold purification. The absorption spectrum of the purified enzyme revealed a peak at 420 nm, which can be indicative of the presence of a cytochrome. However, no spectrum of the reduced enzyme was reported. Further studies with this nitrate reductase (Downey, 1971; MacDonald and Coddington, 1974) revealed that it possesses a reduced benzylviologen(BVH)-nitrate reductase activity, has a molecular weight of 190,000–197,000, and a sedimentation co-efficient of 7.6–7.8 S. The enzyme extract is inhibited by cyanide and azide, but not by organic nitrogenous compounds such as urea, ammonia, or glutamic or aspartic acids. The NADPH-nitrate reductase and NADPH-cytochrome *c* reductase activities are heat-labile and inhibited by low concentrations of *p*HMB. This inhibition can be reversed by cysteine or dithiothreitol. If NADPH was added before *p*HMB, it protected against inhibition suggesting that, in *A. nidulans* like *N. crassa*, there is a sensitive sulphydryl residue at the NADPH binding site. Downey (1971) detected spectral changes due to oxidation and reduction of the FAD during enzymic activity. Inhibitors of enzymic activity, such as *p*HMB or cyanide, inhibited the FAD transitions. Downey (1971) also could find no evidence for the presence of a cytochrome associated with nitrate reductase. In contrast, MacDonald and Coddington (1974) published an absorption spectrum of purified nitrate reductase from *A. nidulans* which revealed absorption maxima for the reduced enzyme at 423, 524 and 554 nm, indicative of a *b*-type cytochrome and very similar to the spectrum of the enzyme from *N. crassa* (Garrett and Nason, 1967).

Kinetic studies of the effect of different concentrations of one substrate on the K_m value for another indicated that there is no interaction between the binding sites for NADPH, nitrate and cytochrome *c*. Product inhibition by $NADP^+$ and nitrite under non-saturating and saturating concentrations of the substrates NADPH and nitrate suggests a random order, rapid equilibrium kinetic mechanism for this enzyme's action (MacDonald and Coddington, 1974).

Downey (1973) analysed the nitrate reductase from *A. nidulans* and found bound FAD and molybdenum. Urea- and sodium dodecyl sulphate polyacrylamide gel electrophoresis of the purified enzyme revealed a single protein band with a subunit molecular weight of 49,000, indicating that the nitrate reductase from *A. nidulans*, molecular weight 196,000, is a tetramer of four identical subunits (Downey and Focht, 1974).

4. *Properties of Nitrate Reductase Mutants in* Neurospora crassa

Various mutations in *N. crassa* and *A. nidulans* result in loss of NADPH-nitrate reductase activity while an associated activity of the nitrate reductase is retained, e.g. MVH-nitrate reductase or NADPH-cytochrome *c* reductase activity. Characterization of the protein components displaying such activities and a comparison of their properties with the wild-type nitrate reductase has provided much insight into the structure and functions of this electron-transport enzyme.

Sorger (1966) described two such mutants in *N. crassa*, namely *nit*-1 and *nit*-3. The *nit*-1 mutant possesses the nitrate-inducible NADPH-cytochrome *c* reductase activity but no other nitrate reductase-related activity. This NADPH-cytochrome *c* reductase exhibits a sedimentation co-efficient of about 4.5–5.2 S (Nason *et al.*, 1970; Ketchum and Downey, 1975) compared with the wild-type nitrate reductase value of 8.0 S (Garrett and Nason, 1969). The molecular weight of this nitrate-induced NADPH-diaphorase activity in *nit*-1 has been estimated by various workers to be 85,000 (Coddington, 1976) to 130,000 (Ketchum and Downey, 1975). Purification of this protein has not been reported. The biochemical lesion in the *nit*-1 mutant is an inability to make an essential molybdenum-containing cofactor (see p. 13).

The *nit*-3 mutant possesses MVH- and $FADH_2$-nitrate reductase activity, but not overall NADPH-nitrate reductase or NADPH-cytochrome *c* reductase activities (Sorger, 1966). Antoine (1974) has purified and characterized the nitrate reductase from the *nit*-3 mutant from *N. crassa*. This enzyme retains the cytochrome *b*-557; it has a molecular weight of 160,000 and a sedimentation co-efficient of 6.8; thus it is somewhat smaller than the wild-type enzyme. The enzyme activities of the *nit*-3 mutant are stable to brief heat treatment at 50°C, but do not increase in activity (Antoine, 1974), whereas the wild-type enzyme demonstrates a 2–3 fold increase in MVH-nitrate reductase activity upon mild heat treatment (Garrett and Nason, 1969). Substrate affinities of the *nit*-3 nitrate reductase are slightly different from the wild-type enzyme, but are very similar to the substrate affinities of the heat-treated wild-type nitrate reductase.

These results on the *nit*-3 nitrate reductase, particularly with regard to its apparent molecular weight of 160,000 (Antoine, 1974) to 180,000 (Coddington, 1976), are not consistent with speculations about the dimer-subunit composition of the native nitrate reductase as already discussed. Presumably the *nit*-3 gene is the major structural locus for this enzyme (see Section III.A), and thus its gene product should correspond to the monomer molecular weight of 115,000–130,000 (Pan and Nason, 1976). A possible explanation for these discrepancies is as follows. The nitrate reductase in the *nit*-3 mutant is either: (a) a dimer of a cryptic gene product (deletion or nonsense mutation) of approximately 80,000–90,000 molecular weight which can still interact with the molybdenum-containing cofactor and cytochrome b-557 to give a partially functional nitrate reductase molecule (i.e. it has MVH-nitrate reductase activity); or (b) a dimer of a proteolytic fragment generated by protease action on the abnormal *nit*-3 gene product, which nevertheless retains the ability to participate in the partial functions already cited. Consistent with either of these interpretations is the measured molecular weight for the nitrate-inducible NADPH-cytochrome c reductase in *nit*-1 extracts of approximately 130,000 given by Ketchum and Downey (1975).

There are many more *A. nidulans* mutants defective in nitrate reductase activity with varied phenotypes and thus they are more appropriately discussed in the section on regulation (Section III).

5. *Molybdenum-Containing Component of Nitrate Reductase*

With the goal of elucidating the quaternary structure of the NADPH-nitrate reductase from *N. crassa*, Nason *et al.* (1970) undertook the reconstruction of active nitrate reductase molecules by judicious recombination of crude extracts of the various mutants of *N. crassa* lacking this enzymic activity. *In vitro* complementation of nitrate reductase could be obtained by mixing cell-free preparations of these mutants, and the nitrate reductase thus formed was a soluble, cytochrome b-containing molybdoflavoprotein with the same molecular weight, sedimentation co-efficient, substrate affinities, and temperature and inhibitor sensitivities as the wild-type nitrate reductase. In all successful cases, it was necessary that a crude extract of nitrate-induced mycelia from the *nit*-1 mutant be present. When extracts of nitrate-induced *nit*-1 were mixed with those from either nitrate-induced or ammonia-grown mycelia from *nit*-2 or *nit*-3 or ammonia-grown wild-type mycelia, native NADPH-nitrate reductase sedimenting at 8.0 was formed. Other combinations of extracts of mutant or ammonia-grown *nit*-1 mycelium could not replace the component contributed by the induced *nit*-1 extract. Nason *et al.* (1971) initially interpreted these results as indicating that the nitrate reductase consists of at least two subunits, namely an inducible subunit supplied by *nit*-1, and a constitutive subunit necessary for nitrate reduction to occur. Further studies (Ketchum *et al.*, 1970; Nason *et al.*, 1971) have revealed that the inducible *nit*-1 subunit could interact with a component obtainable from a variety of molybdenum-containing proteins from different animal (xanthine, aldehyde and sulphite oxidases), bacterial (nitrogenase, respiratory nitrate reductase; Ketchum and Swarin, 1973; Ketchum and Sevilla, 1973) and plant (assimilatory NADH-nitrate reductase) sources to give *in vitro* reconstitution and formation of NADPH-nitrate reductase. One essential requirement for this *in vitro* complementation was that the molybdoproteins first be dissociated by acid treatment at pH 2. Molybdate, complexes of molybdenum with cysteine or glutathione (Lee *et al.*, 1974a), or several molybdenum-amino acid complexes (Nason *et al.*, 1971), were inactive in this reconstitution. As a result, the hypothesis was modified to indicate that *in vitro* formation of nitrate reductase is due to an interaction between the nitrate-inducible, NADPH-cytochrome *c* reductase moiety supplied by *nit*-1 and a molybdenum-containing cofactor found in enzymes from diverse phylogenetic sources.

Further results (Lee *et al.*, 1974a) reported a significant enhancement of *in vitro* nitrate reductase assembly from *nit*-1 extracts and acid-treated molybdoproteins if high concentrations (10 mM) of sodium molybdate were also included in the complementation mixture. The exogenously added molybdate is then incorporated into the protein, as seen by the fact that, when radioactive (^{99}Mo) molybdate was added, the resultant nitrate reductase activity profile found after sucrose density-gradient centrifugation co-incided with the profile of ^{99}Mo radioactivity. Molybdenum was the only metal ion that enhanced the complementing activity. Brief heat treatment or seven days storage in the cold caused the acid-treated molybdoprotein preparation (xanthine oxidase) to lose complementing activity. This loss could be prevented by inclusion of molybdate with the acid-treated molybdo-enzyme. These results suggest that the molybdenum cofactor is labile and capable of losing and/or exchanging its molybdenum with molybdate in solution, and that the enhancing effect of added molybdate was rather a restoration of activity. Experiments with acid-treated molybdo-enzyme (xanthine oxidase) or cell-free extracts of mycelia of ammonia-grown wild-type *N. crassa* showed that the molybdenum in the molybdenum cofactor was exchangeable with exogenously added molybdate (^{99}Mo). In contrast, the molybdenum in the completely formed nitrate reductase was tightly bound, and there was no *in vitro* exchange or incorporation of ^{99}Mo, even after heat treatment or extended storage.

Tungsten and vanadium (added as tungstate and vanadate) were found to compete with molybdenum for incorporation *in vivo* in the wild-type enzyme (Lee *et al.*, 1974b). When these metals are incorporated into the nitrate reductase in place of molybdenum, an inactive nitrate reductase, which possesses only NADPH-cytochrome *c* reductase activity, is formed. Molybdenum, tungsten or vanadium is not incorporated by extracts of nitrate-induced *nit*-1 mycelia. Tungsten or vanadium can also be incorporated *in vitro* in the reconstitution of nitrate reductase with induced *nit*-1 extracts, but the resultant enzyme is inactive. These tungsten or vanadium analogues of nitrate reductase are considerably more labile than the enzyme containing molybdenum, as indicated by the loss of tungsten or vanadium from the reconstituted enzyme by trichloroacetic acid or heat treatment. These metals, like molybdenum, are exchangeable when incorporated into the non-induced wild-type cofactor from *N. crassa* or in acid-treated fractions of molybdo-enzyme. Consequently, high concentrations of exogenously added molybdate will prevent tungsten or vanadium incorporation into nitrate reductase. However, molybdate could not restore activity to the tungsten or vanadium analogue of the enzyme in cell-free extracts, nor would molybdate exchange with tungsten or vanadium *in vitro*. Thus, tungsten or vanadium can be incorporated into the cofactor, which can subsequently combine with induced *nit*-1 extracts to form the 8.0 S enzyme.

It is interesting to note that nitrate reductase formed by *in vitro* combination of nitrate-induced *nit*-1 extracts and the molybdenum cofactor from various sources possessed cytochrome b-557 which was contributed by the *nit*-1 component, illustrating that this haem participant in the electron-transfer sequence is associated with the *nit*-1 protein (Lee *et al.*, 1974a).

The following picture emerges. The *nit*-1 component represents the proteinaceous portion of the nitrate reductase. This component is the *nit*-3 gene product. In the presence of an unknown molybdenum-binding cofactor (obtainable from a variety of molybdoproteins), this *nit*-3 protein can self-assemble into a functional molybdenum-containing nitrate reductase. Molybdate enhances the stability of the cofactor, which is necessary for both the assembly and function of the nitrate reductase complex. Presumably, the *nit*-1 mutant cannot form a functional cofactor and hence is nitrate reductase-less.

In vitro complementation of the nitrate reductase activity has also been accomplished with extracts of *A. nidulans* (Garrett and Cove, 1976). A number of *A. nidulans* mutants lack nitrate reductase activity (Table 1). The *cnx* mutants of *A. nidulans* are analogous to the *nit*-1 mutant in *N. crassa* in that they possess only the nitrate reductase-related NADPH-cytochrome *c* reductase activity (Pateman *et al.*, 1964; MacDonald *et al.*, 1974). Loss of function in any of the *cnx* loci leads to a concomitant loss of both nitrate reductase activity and xanthine dehydrogenase activity, thus suggesting that the *cnx* genes are involved in formation of the molybdenum-containing cofactor essential to both of these molybdo-enzymes. The *niaD* gene is the presumed structural gene for the nitrate reductase polypeptide in which the NADPH-cytochrome *c* reductase activity resides.

The parallel nature of the enzymology in both *A. nidulans* and *N. crassa* has led to attempts to confirm and refine the hypothesis regarding the multimeric nature of the nitrate reductase by demonstrating biochemically that the *nit*-1 locus of *N. crassa* is equivalent to a *cnx* locus of *A. nidulans*, and the *nit*-3 locus of *N. crassa* corresponds to the *niaD* locus of *A. nidulans*. Ketchum and Downey (1975) demonstrated that extracts of ammonium-grown *A. nidulans* wild-type or *niaD* mycelia can participate in the *in vitro* formation of NADPH-nitrate reductase when combined with extracts of *N. crassa nit*-1 mycelia. Several earlier attempts to obtain *in vitro* complementation using only the *niaD* and *cnx* mutants of *A. nidulans* to form active NADPH-nitrate reductase were not successful (Ketchum and Downey, 1975; Downey, 1973). However, *in vitro* complementation with extracts of the *A.*

TABLE 1. Genetic loci involved in nitrate assimilation in *Aspergillus nidulans* and *Neurospora crassa*

Gene	Function	References
Aspergillus nidulans		
*nia*D	Nitrate reductase structural gene	Pateman et al. (1967), MacDonald and Cove (1974)
*nii*A	Nitrite reductase structural gene	Pateman et al. (1967),
cnx (five genes designated *cnx*ABC, E, F, G, and H, respectively)	Formation of the molybdenum-containing cofactor required by both nitrate reductase and xanthine dehydrogenase	Pateman et al. (1967), MacDonald and Cove (1974), Scazzocchio (1974), Garrett and Cove (1976)
*nir*A	Regulator gene necessary for nitrate induction of nitrate and nitrite reductases	Pateman et al. (1967), Cove (1970)
*are*A	Regulator gene involved in ammonia repression of nitrate assimilation and other pathways of nitrogen acquisition	Arst and Cove (1973)
*tam*A	A second regulator gene involved in ammonia repression of nitrate assimilation.	Kinghorn and Pateman (1975a)
*gdh*A	NADP$^+$-Specific glutamate dehydrogenase structural gene, also involved in ammonia repression	Arst and MacDonald (1973), Kinghorn and Pateman (1973)
Neurospora crassa		
nit-1	Formation of the molybdenum-containing cofactor required by both nitrate reductase and xanthine dehydrogenase (analogous to a *cnx* gene)	Sorger (1966), Nason et al. (1970, 1971), Ketchum et al. (1970)
nit-2	Ammonia repression of nitrate assimilation (analogous to *are*A gene)?	Arst and Cove (1973), Coddington (1976)
nit-3	Nitrate reductase structural gene (analogous to *nia*D gene)	Sorger (1966), Nason et al. (1970, 1971), Coddington (1976).
nit-4	Regulator gene necessary for nitrate induction of nitrate and nitrite reductases (analogous to *nir*A gene)?	Coddington (1976)
nit-5	A second regulator gene necessary for nitrate induction of nitrate assimilation (also analogous to *nir*A gene)?	Coddington (1976)

TABLE 1—cont.

Gene	Function	Reference
nit-6	Nitrite reductase structural gene (analogous to niiA gene)	Chang et al. (1975)

Growth characteristics of *Aspergillus nidulans* and *Neurospora crassa* mutants[a]

Mutant		Nitrogen source				
Aspergillus nidulans	Neurospora crassa	Ammonium ion	Nitrate	Nitrite	Hypoxanthine	Uric acid
niaD	nit-6	+	−	+	+	+
niiA	nit-6	+	−	−	+	+
cnx	nit-1	+	−	+	−	+
nirA	nit-4, nit-5	+	−	−	+	+
areA	nit-2	+	−	−	−	−

[a] After Pateman et al. (1967), Arst and Cove (1973), Chang et al. (1975) and Coddington (1976).

nidulans niaD and cnx mutants was recently achieved using the technique of homogenizing the two mutant mycelia together (Garrett and Cove, 1976). *In vitro* complementation also occurred when *N. crassa* nit-1 was cohomogenized with mycelia of the *A. nidulans* niaD mutant. No NADPH-nitrate reductase was found after cohomogenizing mycelia from all five *A. nidulans* cnx mutants.

Further experiments were performed to determine if the amount of functional molybdenum-containing cofactor was regulated by the nitrogen source (Garrett and Cove, 1976). Comparison of the complementing activity of the niaD mutant mycelia grown on different nitrogen sources revealed that more molybdenum cofactor was present in mycelia grown in the presence of urea, with or without nitrate, than in mycelia grown in medium containing ammonia. Comparison of the activity levels obtained using niaD mycelia grown on urea or on urea and nitrate indicates that synthesis of the molybdenum cofactor does not appear to be nitrate-inducible. Further, addition of co-inducers (uric acid or nicotinate) of xanthine dehydrogenases did not cause an increase in the amount of the molybdenum cofactor present over that found in urea-grown mycelia. Thus, the molybdenum cofactor does not appear to be inducible but, rather, much less cofactor is found when the mycelia are grown on ammonia, suggesting that synthesis of the cofactor may be ammonium-repressed.

The nature of the molybdenum cofactor remains enigmatic. In 1964, Pateman *et al.* proposed from genetic evidence that there is a "common cofactor" in *A. nidulans* which is necessary for both nitrate reductase and xanthine dehydrogenase activity. Subsequently, the molybdenum-containing substance capable of interacting with extracts from *N. crassa nit*-1 to yield active NADPH-nitrate reductase was found in a wide variety of sources. The ubiquity of this substance, presumably as a functional component of all molybdoproteins, militates against its being of a polypeptidic nature, unless it were a highly conserved amino-acid sequence, such as cytochrome *c*. Ganelin *et al.* (1972) isolated a low molecular-weight molybdenum-peptide complex from nitrogenase with a molecular weight of 1,000. Further results (McKenna *et al.*, 1974) with these preparations revealed that the complementing activity of the cofactor was stable only as long as the cofactor was protein-bound. There was a preliminary report (Lee *et al.*, 1974a) that the active cofactor had been isolated free from protein, and that it was a small, dialysable and highly labile molecule with a molecular weight of approximately 1,000 or less. It demonstrated an absolute requirement for high concentrations of molybdate in order to express its *in vitro* assembly activity. Ketchum and Swarin (1973), investigating the complementing activity from various nitrogen-fixing or nitrate-utilizing bacteria, reported that the molybdenum cofactor from these organisms was dialysable and insensitive to trypsin and protease. Yet, molybdenum-amino acid complexes could not replace the molybdenum cofactor in restoring nitrate reductase activity.

On the other hand, one observation strongly suggests that the molybdenum cofactor may include or require a protein moiety. MacDonald and Cove (1974) reported that a temperature-sensitive mutation in the *cnx*H gene of *A. nidulans* results in a temperature-sensitive NADPH-nitrate reductase. This result implies that the *cnx*H gene specifies a polypeptide which becomes part of the nitrate reductase molecule.

Lewis (1975) generated antibodies in rabbits to purified nitrate reductase and to xanthine dehydrogenase in *A. nidulans*. Both of these enzymes require the molybdenum cofactor, and are dependent on the *cnx*H gene product for enzyme activity. Enzyme extracts of nitrate reductase and xanthine dehydrogenase were tested against the two classes of antisera for antigenic determinants in common, which would be an indication of antibodies produced against the molybdenum cofactor and *cnx*H gene product. However, no antigenic determinants in common were found, indicating that either the *cnx*H gene product is very small, or that this polypeptide is completely buried (Lewis, 1975).

In summary, the data from the complementation experiments support the notion that nitrate reductase is composed of identical protein subunits (present in *nit*-1 mycelia in *N. crassa* and in *cnx* mycelia in *A. nidulans*) which are held together to form a functional enzyme in the presence of the molybdenum cofactor. Further advances must await an elucidation of the chemical and physical properties of this molybdenum cofactor and the description of its biosynthesis. In addition, the subunit structure of the nitrate reductases needs substantiation and clarification.

B. NITRITE REDUCTASE

Nitrite, the product of the nitrate reductase action, is reduced to ammonia in a six-electron transfer reaction mediated by the second and only other enzyme in the nitrate assimilatory pathway, namely nitrite reductase. Reduction of nitrite to ammonia was originally thought to proceed via discrete two-electron steps each catalysed by a unique and specific enzyme (see discussions in Nason, 1962; Losada, 1975/76). In this hypothesis, nitrite reductase referred to an enzyme concerned with reducing NO_2^- to the +1 oxidation state, e.g. (HNO) or nitroxyl. This substance or an analogue of it was then reduced to hydroxylamine (NH_2OH) which was in turn reduced to ammonia via a two-electron transfer mediated by an enzyme called hydroxylamine reductase. The postulated second step was never seriously assigned a corresponding enzyme (nitroxyl reductase?) because of the difficult technical problem in attempting experimental verification of its presence due to the instability of its presumed substrate. Enzymes designated as nitrite reductase and hydroxylamine reductase were described in a number of organisms on the basis of *in vitro* substrate disappearance measurements and/or substrate-dependent reduced nicotinamide nucleotide oxidation assays. Although this postulate of nitrite reduction to ammonia by a sequence of two-electron transfers catalysed by three distinct enzymes has been found to be incorrect, it served a heuristic purpose in establishing the ultimate proof that assimilatory reduction of nitrite to ammonia is a property residing within a single, quite specific protein rightfully termed nitrite reductase (Pateman *et al.*, 1967; Lafferty and Garrett, 1974).

The persistence of the doubt surrounding the nature of the assimilatory nitrite-reducing system into the 1960s can be attributed not only to the erroneous assumption that enzymic oxidation-reduction reactions must proceed by discrete two-electron transfer events but also to inherent difficulties in working with the system. First and foremost, unless special precautions are taken, nitrite reductase activity is very labile *in vitro* (Lafferty and Garrett, 1974) and this problem was probably a strong discouragement to many workers. Further, nitrite reductase characteristically displays hydroxylamine reductase activity, although this activity presumably has no *in vivo* significance (see p. 33). The situation is further confounded by the fact that at least two other proteins in *N. crassa* can reduce hydroxylamine *in vitro* (Siegel *et al.*, 1965). One of these proteins is, in reality, the assimilatory sulphite reductase and, in addition to reducing hydroxylamine, it can also mediate *in vitro* reduction of nitrite. Proper assignments of the physiological roles of these electron-transport enzymes have been made on genetic evidence. Mutation and loss of assimilatory sulphite reductase in *N. crassa* does not lead to a concomitant auxotrophy for ammonia (Siegel *et al.*, 1965). In both *A. nidulans* (Pateman *et al.*, 1967) and *N. crassa* (Chang *et al.*, 1975), only a single genetic locus can be ascribed with the properties of the structural gene for the nitrite reductase (see Table 1).

Published work on the biochemical characterization of the assimilatory nitrite reductase in fungi has concerned almost exclusively the enzyme from *N. crassa*. The discussion here will be restricted to ammonia-producing enzyme systems; a report of *in vitro* nitrite reduction by a low molecular-weight trypsin-resistant fraction from *N. crassa*, utilizing dithionite together with methyl viologen as electron donor, has appeared, but this reaction probably has no physiological significance in assimilation (Chang *et al.*, 1975).

1. *General Properties of Nitrite Reductase*

Nitrite reductase catalyses the stoicheiometric reduction of nitrite to ammonia using three equivalents of reduced nicotinamide nucleotide, either NADPH or NADH, to provide the necessary six electrons. *In vivo*, formation of nitrite reductase is repressed if ammonia, the end product of nitrate assimilation, is present. In the absence of ammonia, nitrite reductase, like the nitrate reductase, is synthesized if the fungus is presented with either nitrate or nitrite (Garrett, 1972). For a full discussion of these phenomena, see Section III.A. In addition to these responses to nitrogenous nutrients, expression of assimilatory nitrite reductase activity in *N. crassa* also requires the presence of ferric iron at 10 micromolar levels (Vega *et al.*, 1975b). As will become apparent later, this iron dependence is due to the involvement of a novel haem prosthetic group in the enzyme.

The molecular weight of the nitrite reductase from *N. crassa* has been determined from its behaviour upon sucrose density-gradient sedimentation and Sephadex G-200 gel filtration chromatography (Lafferty and Garrett, 1974). Centrifugation experiments according to the method of Martin and Ames (1961) gave a mean S value of 9.4 for the enzyme, and a Stokes radius (a) of 7.5 nm was obtained from gel-filtration experiments. Together, these results yielded a molecular weight of 290,000 for the nitrite reductase (Lafferty and Garrett, 1974) when inserted in the equation: mole. wt. = $6\pi\eta Nas/1 - \bar{v}\rho$, and assuming $\bar{v}$ is 0.725 cc/g (Siegel and Monty, 1966). *In vitro*, nitrite reductase activity from *N. crassa* requires exogenous FAD. Presumably, this flavin component easily dissociates from the enzyme during purification. Flavin mononucleotide cannot replace FAD in restoring activity; the K_m value for FAD is of the order of 0.1 μM (Lafferty and Garrett, 1974).

This fungal assimilatory nitrite reductase differs from the assimilatory nitrite reductase of photosynthetic organisms and tissues in its requirement for FAD, its dependency on reduced nicotinamide nucleotides as electron donors and its large molecular weight. The nitrite reductase from photosynthetic organisms is characteristically only 63,000 in molecular weight, and mediates reduction of nitrite using reduced ferredoxin formed by photosynthetic electron-transport as the electron donor. Flavins are not involved here (Hewitt, 1975). This situation contrasts with that found for nitrate reductases of fungi and higher plants which are much more similar with respect to reducing sources used, prosthetic groups involved and relative molecular weight (Hewitt, 1975). The underlying basis for the comparative differences between nitrite reductases in fungi and photosynthetic organisms is uncertain; an intriguing possibility is that fungi, because they lack ferredoxin and the photochemical means to reduce it, require a relatively larger and more complex protein in order to employ reduced nicotinamide nucleotides as electron donors.

Assimilatory nitrite reductases in general are competitively inhibited by inorganic anions which are similar to nitrite, such as sulphite, arsenite and cyanide (Lafferty and Garrett, 1974; Vega et al., 1975a). The nitrite reductase from *N. crassa* is also sensitive to inactivation by peroxide, either exogeneously added or self generated by the enzyme under aerobic conditions in the presence of NAD(P)H and FAD (Vega et al., 1975a). In fact, this conclusion grew out of attempts to show that nitrite reductase might be regulated by oxidation-reduction changes. Losada (1974) has postulated that redox changes in electron-transport enzymes and, in particular, the nitrate assimilation enzymes, are an important means to regulate their catalytic activity. The assimilatory nitrite reductase from *N. crassa* was susceptible to reductive inactivation in the presence of reduced nicotinamide nucleotides and FAD (Lafferty and Garrett, 1974). Flavin adenine dinucleotide alone had no inactivating effect, where NADPH or NADH alone inactivated to an intermediate extent. However, either NADPH or NADH together with FAD caused a 90% inactivation of the enzyme when pre-incubated with nitrite reductase for 10 min under aerobic conditions at room temperature. Several observations support the conclusion that this inactivation is a reductive phenomenon. Inactivation was not caused by NAD^+ or $NADP^+$ and FAD. Flavin adenine nucleotide, an essential cofactor for enzymic activity, is also required for full inactivation; electron acceptors of the enzyme, such as nitrite, or competitive inhibitors of the enzyme, such as sulphite, protected against inactivation. Although inactivation could be prevented, all attempts to reverse the inactivation process and restore the nitrite reductase to an active state were unsuccessful (Vega et al., 1975a).

However, it was noted that anaerobic conditions afforded significant protection against the FAD-dependent inactivation of the enzyme by NAD(P)H. Further, inclusion of catalase during the aerobic pre-incubation prevented reductive inactivation of nitrite reductase. Superoxide dismutase had no effect. Therefore, it was apparent that peroxide, generated by an FAD-mediated NAD(P)H reduction of dissolved oxygen by nitrite reductase in the absence of nitrite, was the true inactivating agent. Further, pre-incubation of the enzyme with hydrogen peroxide alone caused its inactivation, thereby confirming this notion. On the basis of the conclusion that the FAD-dependent NAD(P)H inactivation of assimilatory nitrite reductase was in reality peroxide-mediated and not a consequence of a simple reduction of the enzyme, the possibility of any redox-associated regulation of the enzyme in *N. crassa* has been rejected. The fact that such inactivation was irreversible further negates its regulatory significance (Vega *et al.*, 1975a). On the other hand, this peroxide sensitivity may have some physiological importance. Subramanian *et al.* (1968) have shown a correlated regulation of both nitrate reductase and catalase activities *in vivo* in *N. crassa*. When mycelia were exposed to nitrate, synthesis of catalase was found to be induced subsequent to induction of nitrate reductase. In addition, catalase synthesis was repressed by the same amino acids which repressed nitrate reductase synthesis. More recently, Jacob (1975) examined the catalase from *N. crassa* which appears and disappears in concert with the regulation of the nitrate assimilation enzymes, and has shown that it is an unusual type of catalase by virtue of its possession of an uncommon haem moiety. Further, Jacob (1975) has speculated that the role of this catalase may be to thwart the possibility of a peroxide-mediated inactivation of nitrite reductase, as evidenced *in vitro* in our laboratory (Vega *et al.*, 1975a).

As shown earlier by Lafferty and Garrett (1974), the partially purified (*ca.* 100-fold) nitrite reductase from *N. crassa* mediated the stoicheiometric reduction of nitrite to ammonia by NADPH. The amounts of nitrite reduced, ammonia formed and NADPH oxidized were proportional to the reaction time, and the ratios of NADPH oxidized: nitrite reduced and NADPH oxidized: ammonia formed centre about the theoretical value of three equivalents of NADPH providing the six electrons necessary for nitrite reduction to ammonia. The requirement of the enzyme for FAD was revealed by the fact that only 10 per cent of the NADPH-oxidizing, nitrite-reducing or ammonia-reducing activity resulted when FAD was omitted from reaction mixtures. Nevertheless, the stoicheiometric relationship of 3:1:1 between NADPH, nitrite and ammonia was maintained even in the absence of FAD (Lafferty and Garrett, 1974).

As already mentioned, fungal nitrite reductase has been found to be markedly unstable *in vitro* and, until recently, this problem has stymied attempts at purifying and characterizing the enzyme. Now it is understood that this instability is probably due to sensitivity of the nitrite reductase to oxidation (as seen in the peroxide-mediated inactivation) and its lability in the absence of its FAD moiety which readily dissociates *in vitro*. Excellent protection of activity has been obtained through addition of dithionite and FAD to all buffers used during enzyme purification (Vega *et al.*, 1975b). Presumably dithionite has a dual role. By virtue of its reducing power, it keeps the nitrite reductase in a reduced state; further, sulphite is generated when dithionite is added to aerobic aqueous solutions and this anion binds and protects the substrate-binding site of the enzyme. Recently, through the use of a modified version of the purification scheme reported by Lafferty and Garrett (1974), followed by affinity chromatography on blue dextran-Sepharose 4B (Thompson *et al.*, 1975), homogeneous preparations of the nitrite reductase from *N. crassa* have been obtained (Prodouz *et al.*, 1977). Only one protein band is seen upon polyacrylamide-gel electrophoresis of these preparations, and this single protein band possesses all of the enzymic functions attributed to nitrite reductase. This pure nitrite reductase exhibits a specific activity of 80.7 micromoles NADPH oxidized/min/mg protein, a value corresponding to reduction of 26.9 micromoles of nitrite reduced or ammonia formed/min/mg protein. Thus, in terms of the rate of ammonia production, the nitrite reductase has a turnover number of 7,800/min. The 3 : 1 : 1 stoicheiometry of NADPH to nitrite to ammonia has been confirmed with the homogenous nitrite reductase (Greenbaum *et al.*, 1978).

An analysis of the subunit composition of the assimilatory nitrite reductase from *N. crassa* has been performed under a variety of dissociating conditions. Treatment of the homogeneous enzyme with either 2 per cent sodium dodecyl sulphate together with 2 per cent β-mercapto-ethanol or 8 M guanidine hydrochloride and 50 mM β-mercapto-ethanol yielded a preparation showing only a single protein band after sodium dodecyl sulphate-discontinuous gel electrophoresis. From migration of the protein band relative to that of protein subunits of known molecular weights, a molecular weight of approximately 145,000 can be assigned to the nitrite reductase subunit. Thus, nitrite reductase from *N. crassa* (mol. wt. = 290,000) appears to be a dimer of identical 140–150,000 mol. wt. protomers (Prodouz *et al.*, 1977).

In general, the discussion which follows on the more detailed characterization of the nitrite reductase from *N. crassa* stems from results obtained with inhomogeneous enzyme preparations;

nevertheless the conclusions drawn have, thus far, proven to be entirely consistent with the continuing revelations regarding this enzyme.

2. Sirohaem: A Novel Prosthetic Group of Nitrite Reductase

A number of the general properties of nitrite reductase already presented are manifestations of a novel haem prosthetic group that this enzyme possesses. Visible absorption spectra of purified nitrite reductase preparations from *N. crassa* indicated the presence of haem in exhibiting absorption maxima at 390 nm and 578 nm (Lafferty and Garrett, 1974). In addition, when partially purified preparations were analysed following fractionation by sucrose density-gradient centrifugation, a close correspondence between nitrite reductase activity and absorbance in the haem Soret region (400 nm) was found (Vega *et al.*, 1975b).

Because of the consistent association of this haem-like component with the occurrence of nitrite-reducing activity, it appeared likely that this haem might be a prosthetic group of assimilatory nitrite reductase and, as such, might function in the catalysis performed by the enzyme. This suggestion was confirmed by studies of the effects of various substrates, cofactors and inhibitors on the spectral properties of nitrite reductase (Vega *et al.*, 1975b). These studies are presented in Fig. 4 where the absorbance of nitrite reductase in the 450–650 nm range is presented. Curve A is the characteristic spectrum of the enzyme; it shows the 578 nm peak and is designated as the "oxidized enzyme" spectrum. Upon addition of 2 mM NADPH and 10 μM FAD, absorbance changes occur such that maxima now are found at 588 and 556 nm, as depicted in curve B which is designated the "reduced enzyme" spectrum. A similar spectrum is obtained upon reduction of the enzyme with dithionite instead of NADPH and FAD. Nitrite when added alone causes no spectral perturbations; however when excess nitrite (2 μM) is added to the enzyme reduced by NADPH and FAD, the unique spectrum shown in curve C results, with maxima at 585 and 560 nm. This spectrum is clearly different from either the oxidized or reduced enzyme spectra, and is designated as the "reduced enzyme-nitrite complex" spectrum since it must represent formation of a complex between reduced enzyme and nitrite (or a reduction product of nitrite). Although there is presently no evidence available that this "reduced enzyme-nitrite complex" participates in the catalytic cycle of

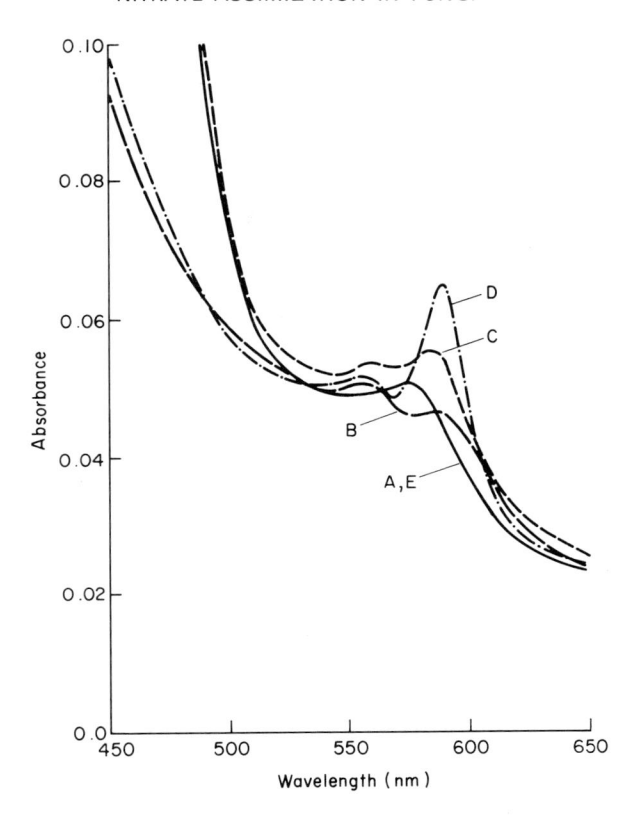

FIG. 4. Spectra of nitrite reductase from *Neurospora crassa* in the presence of reducing agents, nitrite or carbon monoxide. In all experiments, the sample cell contained nitrite reductase in 0.1 M potassium phosphate buffer (pH 7.5). Spectra were recorded *versus* a buffer blank. Additions were: A, none; B, 2 mM NADPH and 10 μM FAD; C, 2 mM NADPH, 10 μM FAD and 2 mM NaNO$_2$; D, 2 mM NADPH, 10 μM FAD and 0.9 mM carbon monoxide: E, 0.9 mM carbon monoxide. From Vega *et al.* (1975b).

the enzyme, the fact that its formation only occurs when all of the components necessary for enzymic activity (NADPH, FAD and nitrite) are present is suggestive of a functional role for this spectral species. Further, this result implies that the haem-like chromophore under observation is a critical component in the nitrite reductase reaction. When 0.9 mM carbon monoxide is added to the enzyme reduced by NADPH and FAD, a "reduced enzyme-carbon monoxide complex" was formed as judged from the absorbance changes as depicted in curve D. There is a strong α-band at 590 nm and a β-band of lesser intensity at 550 nm. Formation of this carbon monoxide complex is dependent upon reductants such as NADPH and FAD. When carbon monoxide is added to the oxidized enzyme in the absence of reducing agents (curve E), no spectral alterations were evident and, thus, curves A and E are identical. Also noteworthy in Fig. 4 is the bleaching in the various spectra in the wavelength range between 450 and 500 nm. This bleaching is evident in curves B and D where the conditions are such that any flavin components present would be in a reduced state and therefore colourless. Conversely, bleaching is absent from curves A, C and E where the flavin would be in an oxidized form and absorbing light in this wavelength range.

Additional spectral studies on the interaction of carbon monoxide with nitrite reductase have revealed that, if exogenous FAD is not added, complex formation between NADPH-reduced enzyme and carbon monoxide is significantly diminished (Vega *et al.*, 1975b). These spectral measurements were performed following the attainment of equilibrium, and thus reveal that FAD does not simply enhance the rate of reduced enzyme-carbon monoxide complex formation but actually determines the yield of this complex. This result implies that FAD functions in transfer of electrons from NADPH to the haem-like chromophore of nitrite reductase. In the absence of exogenous FAD, only those nitrite reductase molecules retaining bound flavin can interact with NADPH to promote carbon monoxide-complex formation. When FAD is added, all nitrite reductase molecules can now enter into the reduced enzyme-carbon monoxide complex form. Formation of this complex is reversible as demonstrated by the fact that, upon passage of the reduced enzyme-carbon monoxide complex over a Sephadex G-25 column, the enzyme emerging in the eluate was in the free oxidized state. The kinetics of complex formation between nitrite reductase and carbon monoxide can be determined spectrophotometrically, since the rate of reduction of the haem-like chromophore by NADPH and FAD is a rapid process in comparison to carbon monoxide-complex formation. If complex formation fits the simple relationship: $E + CO \rightarrow E \cdot CO$, a second-order rate constant of 9 M^{-1} s^{-1} can be calculated. If maximal complex formation between carbon monoxide and enzyme in the presence of NADPH and FAD is attained, and nitrite is then added, the observed spectrum changes from that typical of the reduced enzyme-carbon monoxide complex to that characteristic of the reduced enzyme-nitrite complex. Thus, nitrite competes effectively with carbon monoxide for the nitrite reductase (Vega et al., 1975b).

In addition to the spectral perturbations that carbon monoxide elicits in the haem-like chromophore of the nitrite reductase, carbon monoxide is also a potent inhibitor of nitrite reductase activity. The inhibition is maximal when nitrite reductase is pre-incubated with carbon monoxide in the presence of NADPH and FAD. When FAD is omitted, the degree of inhibition is markedly diminished. Incubation of nitrite reductase with carbon monoxide alone does not affect its activity. Once inhibition of nitrite reductase has been attained by pre-incubation with NADPH, FAD and carbon monoxide, the inhibition can be reversed by passage of the enzyme over a Sephadex G-25 column. Thus, inhibition by carbon monoxide is readily reversible. Further, the addition of nitrite to the carbon monoxide-inhibited, reduced enzyme will reverse the inhibition. As indicated earlier, sulphite, arsenite and cyanide all serve as competitive inhibitors of nitrite in the nitrite reductase. Pre-incubation of the enzyme, carbon monoxide, NADPH and FAD in the presence of any of these competitive inhibitors markedly prevented carbon monoxide inhibition of nitrite reductase activity (Vega et al., 1975b).

A comparison of the rate of reduced enzyme-carbon monoxide complex formation, as estimated from its characteristic spectrum, with the rate of onset of inhibition of nitrite reductase activity upon pre-incubation with NADPH, FAD and carbon monoxide, reveals that the two processes proceed with the same kinetics. Thus the carbon monoxide complex formation with the haem-like chromophore of the nitrite reductase and the carbon monoxide inhibition of nitrite reductase activity share many characteristics; both require the presence of the electron donor NADPH, both require FAD for maximal expression, both are reversible upon Sephadex G-25 chromatography of the enzyme or the addition of nitrite and, finally, both processes proceed with the same kinetics (Vega et al., 1975b).

The various absorption spectra of nitrite reductase from *N. crassa* and particularly that of its complex with carbon monoxide are reminiscent of those for spinach nitrite reductase and various bacterial sulphite reductases (Murphy et al., 1973a, 1974). The responsible chromophore in all of these proteins has been shown to be sirohaem (Fig. 5). Sirohaem is an iron tetrahydroporphyrin of the isobacteriochlorin type, i.e. two adjacent pyrrole rings are reduced. This novel haem is also very polar by virtue of the eight carboxylate groups it possesses (Murphy et al., 1973b). The haem-like chromophore of the nitrite reductase of *N. crassa* has been extracted from the enzyme by acetone-hydrochloric acid and compared to authentic sirohaem isolated in a similar manner from the sulphite reductase of *Escherichia coli*. By a number of criteria, the two haems appear identical. The spectra of both isolated haems in pyridine are virtually identical with absorption maxima at 401 nm and 558 nm. The spectra of the carbon monoxide complexes of dithionite-reduced sirohaem and dithionite-reduced nitrite reductase haem are very similar with both showing absorption maxima at 600 nm and 558 nm. These spectra are thus qualitatively similar to that of the "reduced enzyme-carbon monoxide" complex. Because of its polar carboxyl groups, sirohaem migrates quite differently from the more common naturally occurring exactable haem form, protohaemin. When sirohaem, the extracted nitrite reductase haem and protohaemin were subjected to thin-layer chromatography on Polyamide-6 in an acetone/water/formic acid (70:30:3) solvent system, sirohaem and the nitrite reductase haem exhibited similar R_f values (0.75) whereas the protohaemin showed minimal migration ($R_f < 0.1$). Lastly, the fluorescence excitation and emission spectra of the porphyrin derivatives of sirohaem and the nitrite reductase chromophore obtained by removal of their iron atoms were analysed. The fluorescence excitation spectra of both possess a triplet Soret band in the 400 nm region and a second major absorption at 612 nm. The fluorescence emission spectra are essentially superimposable with absorption maxima at 626 nm and shoulders at longer wavelength (Vega et al., 1975b).

The results indicate that the nitrite reductase haem-like chromophore from *N. crassa* is a highly polar substance sharing virtually identical spectral properties with sirohaem, and thus the weight of evidence forces the conclusion that sirohaem is a prosthetic group of the nitrite reductase from *N. crassa*. Its presence was suggested by the characteristic visible absorption spectra of purified enzyme preparations, and a close correspondence between this absorption and nitrite reductase activity was found following sucrose density-gradient centrifugation. The role of sirohaem as a functional prosthetic group in nitrite reduction is supported by a number of observations. Sirohaem reduction by NADPH is, like nitrite reduction, FAD-dependent *in vitro*. Addition of nitrite to the nitrite reductase reduced with

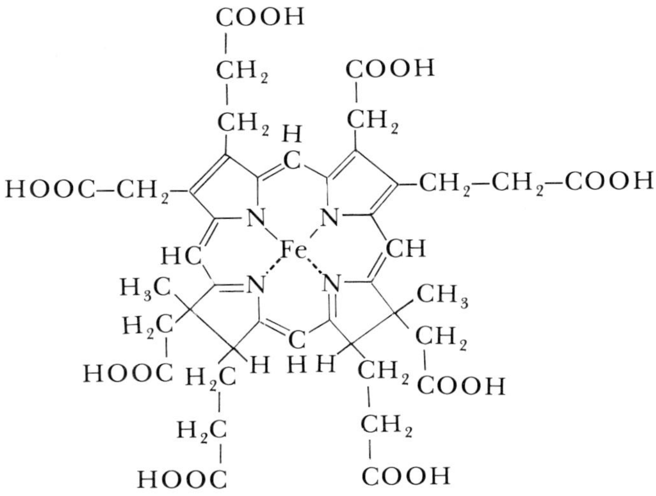

FIG. 5. Structure of sirohaem proposed by Murphy et al. (1973b).

NADPH and FAD causes unique alterations in the "reduced enzyme" sirohaem spectrum. Carbon monoxide affects both the spectral properties of the sirohaem moiety of the nitrite reductase and its enzymic activity; and both effects share a dependency for NADPH plus FAD, a reversibility upon gel filtration chromatography or addition of nitrite, and a parallel kinetic response (Vega et al., 1975b).

3. *Electron-Transfer Reactions Catalysed by Nitrite Reductase*

In addition to these studies on the complement of prosthetic groups possessed by nitrite reductase and their roles in catalysis, another avenue can be exploited to elucidate the electron-transfer capabilities of this enzyme. Nitrite reductase, besides its physiological activity of mediating the NAD(P)H-dependent reduction of nitrite to ammonia, can catalyse other electron-transfer reactions *in vitro*. These activities include:

(1) NAD(P)H-hydroxylamine reductase:

$$NAD(P)H + H^+ + NH_2OH \xrightarrow[FAD]{2e^-} NAD(P)^+ + NH_4^+ + OH^-$$

(2) NAD(P)H-diaphorase:

$$\text{NAD(P)H} + \text{H}^+ + \text{dye} \xrightarrow[\text{FAD}]{2e^-} \text{NAD(P)}^+ + \text{dye:H}_2$$

(dye indicates electron-acceptors such as cytochrome c, $Fe(CN)_6^{-3}$, menadione, 2-6-dichlorophenolindophenol)

(3) Dithionite-nitrite reductase:

$$3\,S_2O_4^{-2} + 4\,OH^- + NO_2^- \xrightarrow{6e^-} 6\,SO_3^{-2} + NH_4^+$$

In effect, these activities can be viewed as representing partial functions of the overall physiological electron transfer from NAD(P)H to nitrite performed by nitrite reductase, and thus an examination of their properties permits the experimenter to focus on a specific and restricted set of the enzyme's total catalytic capability. That all of these activities are indeed properties of the nitrite reductase from *N. crassa* has been established by: (a) their co-elution upon DEAE-cellulose chromatography, (b) their co-elution upon Sephadex G-200 gel-filtration chromatography, (c) their cosedimentation upon sucrose density-gradient centrifugation, (d) the constant proportionality of their activities during purification, and (e) their related responses to various inhibitors and inactivating agents (Lafferty and Garrett, 1974; Vega and Garrett, 1975). Further, Garrett (1972) has shown that NADPH-nitrite reductase, NADPH-hydroxylamine reductase and dithionite-nitrite reductase activities in *N. crassa* are simultaneously induced by nitrate or nitrite.

Neurospora crassa extracts possess three proteins displaying hydroxylamine reductase activity. These proteins can be resolved by sucrose density-gradient centrifugation (Leinweber *et al.*, 1965), and have been labelled peaks A, B and C in decreasing order of sedimentation rate. The peak A hydroxylamine reductase activity was NADPH-specific and was demonstrated to be a property associated with the sulphite reductase of *N. crassa*. The peak B NAD(P)H-hydroxylamine reductase activity is identical with the NAD(P)H-nitrite reductase on the basis of their similar responses to nitrate induction *in vivo*, their lack of specificity for reduced nicotinamide nucleotides and their similar sedimentation co-efficients (Lafferty and Garrett, 1974). Further, Siegel and Monty (1966) determined a molecular weight of 295,000 for the peak B hydroxylamine reductase, a value essentially identical with that determined for NAD(P)H-nitrite reductase by Lafferty and Garrett (1974). The minor peak C hydroxylamine reductase has no known physiological function (Siegel et al., 1965). The NAD(P)H-hydroxylamine reductase activity displayed by nitrite reductase, as must be expected, shares many properties with the NAD(P)H-nitrite reductase activity. It is FAD-dependent, inactivated upon pre-incubation with NAD(P)H and FAD, cyanide inhibited and sensitive to the sulphydryl agent pHMB (Lafferty and Garrett, 1974; Vega et al., 1975a). Presumably hydroxylamine and nitrite accept electrons at the same enzymic site. Hydroxylamine (K_i, 6 mM) is a competitive inhibitor of the nitrite reductase activity (Lafferty and Garrett, 1974), whereas nitrite (K_i, 9 µM) is a competitive inhibitor of the hydroxylamine reductase activity of the nitrite reductase from N. crassa (peak B NAD(P)H-hydroxylamine reductase; Siegel, 1965). A comparison of the K_m values of the nitrite reductase for nitrite (K_m, 10 µM) and hydroxylamine (K_m, 3 mM) favours assigning the physiological role of this enzyme as a nitrite reductase. Further, the large K_m value for hydroxylamine militates against this compound serving as a free intermediate in asssimilatory reduction of nitrite to ammonia, particularly in light of the fact that hydroxylamine at 0.1 mM is toxic to growth of N. crassa (McElroy and Spencer, 1956).

The NAD(P)-diaphorase activity of nitrite reductase from N. crassa is in many ways analogous to the corresponding diaphorase of nitrate reductase of this mould, as discussed earlier (p. 5). The association of a diaphorase function with nitrite reductase was reported from our laboratory (Vega and Garrett, 1975; Vega, 1976) and could be assayed by observing the FAD-dependent NAD(P)H reduction of a number of electron acceptors such as mammalian cytochrome c, ferricyanide, menadione or 2,6-dichlorophenolindophenol. Like the NAD(P)H-nitrite reductase activity, this NAD(P)H-diaphorase activity is pHMB (1 µM) sensitive but, in contrast to the nitrite reductase, this activity is not inhibited by cyanide, sulphite, arsenite or carbon monoxide nor is it inactivated upon pre-incubation with NAD(P)H and FAD.

A third activity possessed by the nitrite reductase from N. crassa is the ability to use dithionite ($S_2O_4^=$) as an electron donor to reduce nitrite in an FAD-independent reaction. This dithionite-nitrite reductase is, like the NAD(P)H-nitrite reductase, inhibited by cyanide but, in contrast to the NAD(P)H-dependent nitrite reducing activity, it is not pHMB-sensitive.

Thus, there are obvious relationships among these various activities expressed by nitrite reductase. All of the NAD(P)H-dependent activities require FAD and are *p*HMB-sensitive, whereas the dithionite-nitrite reductase is not. Conversely, those activities involving reduction of nitrite or hydroxlamine are sensitive to cyanide, carbon monoxide, and inactivation by NAD(P)H together with FAD, while the NAD(P)H-diaphorase activity is unaffected by these agents (Garrett, 1978).

4. *A Model For Nitrite Reductase*

An integration of these properties of the various activities of the nitrite reductase from *N. crassa* with the results obtained on the nature and function of this enzyme as expressed by its complement of prosthetic groups leads to the model illustrated in Fig. 6. The evidence that FAD functions to transfer electrons from NAD(P)H to sirohaem stems from the observed FAD dependency for reduction of nitrite or sirohaem by NAD(P)H and for maximal formation of the "reduced enzyme-carbon monoxide" complex. The FAD requirement expressed by the NAD(P)H-diaphorase activity indicates that this flavin is necessary for this electron-transfer reaction also. Since all of the NAD(P)H-dependent activities of nitrite reductase exhibit *p*HMB sensitivity, a functional sulphydryl moiety is implicated in the initial electron-transfer events mediated by this enzyme, but the precise location of the moiety in the sequence is unknown. Presumably, it is involved either just prior or subsequent to FAD reduction, but this problem has not yet been approached. The *p*HMB insensitivity and FAD independence of the dithionite-nitrite reductase activity suggest that dithionite can donate electrons at the nitrite-reducing terminus of this sequence. Since this dithionite-nitrite reductase is cyanide-inhibited, sirohaem is likely to be involved in this reaction. Sirohaem is placed at the end of the reaction sequence. Current interpretations of available data suggest that the peroxide-mediated inactivation of nitrite reductase, as manifested upon pre-incubation with NAD(P)H together with FAD, is due to irreversible oxidative damage to the labile sirohaem. The fact that nitrite, hydroxylamine or any of the competitive inhibitors, such as sulphite, can protect against this inactivation warrants this conclusion. Indeed, it seems likely that cyanide inhibition, peroxide inactivation and carbon monoxide poisoning of nitrite reductase are all manifestations of the enzyme's reliance on its sirohaem prosthetic group. The NAD(P)H-diaphorase is free from all

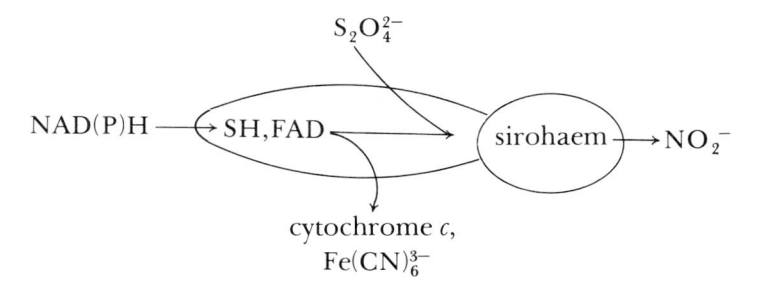

FIG. 6. Proposed model for the action of nitrite reductase from *Neurospora crassa*. From Garrett (1978).

of these inhibitions and thus sirohaem is not considered to function in this reaction.

A number of observations promote the suggestion that sirohaem may provide the site of interaction of nitrite reductase with its substrate and ultimate electron acceptor, nitrite. Nitrite can both prevent and reverse the association of enzyme-bound sirohaem with carbon monoxide, as evidenced by spectral observations. Further, nitrite can both block and alleviate inhibition of nitrite reductase activity by carbon monoxide. Sulphite, cyanide and arsenite, all competitive inhibitors of the enzyme with respect to nitrite, also decrease the ability of carbon monoxide to interact with nitrite reductase. Finally, when nitrite is added to the enzyme reduced with NADPH and FAD, a unique perturbation is observed in the sirohaem spectrum. It is worthwhile to note here also that nitrite, cyanide, sulphite, arsenite and carbon monoxide are all known ligands for haems in other systems. In summary, these results suggest that nitrite is reduced to ammonia by nitrite reductase while held directly as a ligand of the sirohaem prosthetic group. The precise nature of the six-electron transfer by which this transformation is achieved is as yet unknown. Present research is currently focused on determination of the total number and types of electron-transporting prosthetic functions that the nitrite reductase from *N. crassa* possesses, and its potential reducing equivalents which might be tapped during the catalytic cycle.

C. CONCENTRATIONS OF NITRATE REDUCTASE AND NITRITE REDUCTASE IN MYCELIA

Now that the two enzymes of nitrate assimilation have been obtained in homogenous form from *N. crassa* (Pan et al., 1975; Pan and Nason,

1976; Jacob, 1975; Greenbaum et al., 1978), calculations of their catalytic efficiency and intracellular concentrations are possible. Pure NADPH-nitrate reductase from N. crassa has a specific activity of 80–100 micromoles of nitrite produced/min/mg protein (Pan et al., 1975; Jacob, 1975) which, given the molecular weight of 228,000 for this enzyme (Garrett and Nason, 1969), reveals that the molecular activity (or turnover number) of the nitrate reductase is 18,000. Similarly the homogeneous NAD(P)H-nitrite reductase obtained by Greenbaum et al. (1978) from N. crassa had a specific activity of 80.7 micromoles NADPH oxidized in the presence of nitrite/min/mg protein. Since three equivalents of NADPH are needed to reduce nitrite to ammonia, this specific activity is equal to 26.9 micromoles of nitrite reduced to ammonia/min/mg protein. Lafferty and Garrett (1974) found the molecular weight of this N. crassa enzyme to be 290,000. Thus, one nanomole of nitrite reductase reduces 7.8 micromoles of nitrite per minute, so that the molecular activity of this enzyme is 7,800 As a point of reference, the molecular activity of succinate dehydrogenase is only 1,200 (Lehninger, 1975).

Garrett (1972) found maximal NADPH-nitrate reductase levels in N. crassa mycelia to be on the order of 30 nanomoles of nitrite produced/min/mg protein, and the NAD(P)H-nitrite reductase activity was 15 nanomoles of nitrite reduced/min/mg protein. Thus, per mg protein in the crude mycelial extracts, there were 1.7 picomoles of nitrate reductase and 1.9 picomoles of nitrite reductase. Since these extracts contained approximately 17.5 mg protein/ml and one ml extract was roughly equivalent to 0.5 g mycelia (wet weight), the levels of the nitrate reductase and nitrite reductase are approximately 0.06 nanomoles and 0.07 nanomoles per gram fresh weight of mycelia, respectively.

These results, while approximate, nevertheless dramatize the catalytic efficiency of the nitrate assimilation enzymes in fungi and also reveal a fundamental problem confronting the enzymologist who wishes to study this process. Per kilogram fresh weight of N. crassa, one can expect to find only 10 mg nitrate reductase and 20 mg nitrite reductase. Obviously, purification procedures providing significant recoveries are necessary to realize acceptable yields of pure enzyme. Further, to consider the nitrate reductase as a potential source of the molybdenum-containing component for chemical and physical characterizations is an optimistic endeavour, since this component constitutes perhaps 1% of the enzyme by weight.

D. NITRATE AND NITRITE UPTAKE

An important consideration in assimilation of nitrate by fungi is the ability of the organism to take up the substrates, nitrate and nitrite. Unfortunately, not much attention has been given to this primary process. Subramanian and Sorger (1972a) observed that ammonia did not prevent accumulation of nitrate by *N. crassa*, and mycelia exposed to ammonia and nitrate accumulated more nitrate than those exposed to nitrate alone. The lack of nitrate reductase activity in mycelia exposed to ammonia and nitrate could explain the accumulation of nitrate. The major purpose of this experiment was to show that the effects of ammonia were not simply due to exclusion of nitrate from mycelia.

In more detailed investigations, Schloemer and Garrett (1974a, b) demonstrated that *N. crassa* possesses active transport systems for nitrate and nitrite, and investigated the regulation of these systems under various conditions of nitrogen nutrition. The nitrate-uptake system (Schloemer and Garrett, 1974a) was not present in ammonia- or casamino acid-grown mycelia, but it was induced by nitrate or nitrite, even in the presence of ammonia. Inhibitors of protein or RNA synthesis prevented this induction. In the presence of cycloheximide, loss of pre-existing nitrate transport capacity occurred with a half-life of three hours, indicating that this system was undergoing metabolic turnover. Transport of nitrate is an energy-requiring process as demonstrated by the fact that it is inhibited by metabolic poisons and is capable of concentrating nitrate to concentrations 50-fold over that found in the external media. Ammonia, nitrite, or casamino acids can each affect nitrate uptake. Ammonia and nitrite are non-competitive inhibitors of nitrate uptake, while casamino acids and various amino acids (but not ammonia) repress formation of the nitrate-uptake system.

The nitrate-uptake system was shown to be distinct from nitrate reductase. Under conditions giving a non-functional nitrate reductase, such as growth of mycelia with tungsten or vanadium in place of molybdenum, or using mutants *nit*-1, *nit*-2, *nit*-3 of *N. crassa* which do not possess a functional NADPH-nitrate reductase (Table 1), the nitrate-uptake system displayed the same kinetic constants, inhibitory response to ammonia or nitrate, and repression by casamino acids, indicating that the presence of the intact nitrate reductase protein was not essential for induction or function of the nitrate-transport system.

Like the nitrate-uptake system, nitrite uptake was induced by either nitrate or nitrite (Schloemer and Garrett, 1974b). This induction was prevented by cycloheximide, puromycin or 6-methylpurine. Nitrite uptake also appeared to be an active transport process since it was inhibited by metabolic poisons such as cyanide and 2,4-dinitrophenol. In contrast to the nitrate-uptake system, nitrite uptake was not inhibited by ammonia or casamino acids nor was it inhibited by nitrate. Also, adaptive formation of the nitrite-uptake system was not prevented by casamino acids. A teleological explanation for this lack of inhibition of the nitrite-uptake system by other nitrogenous compounds is that freedom from inhibition permits immediate metabolism of the potentially toxic nitrite even in the presence of other nitrogen sources.

Mutants *nit*-1 and *nit*-3 of *N. crassa* can grow on nitrite as the sole nitrogen source, possess a functional nitrite reductase (but not nitrate reductase) and cause depletion of nitrite from the media, but do not accumulate nitrite in the mycelia (Coddington, 1976). Mutants *nit*-2, *nit*-4 and *nit*-5 cannot grow on nitrite since they lack a functional nitrite reductase; consequently, if a nitrite permease system were operable in these mutants, nitrite would accumulate in mycelia. Mutants *nit*-4 and *nit*-5 accumulated nitrite in mycelia, and *nit*-5 could concentrate nitrite in mycelia to concentrations higher than in the surrounding media (Coddington, 1976). Nitrite uptake was not found in mutant *nit*-2 under any conditions of nitrogen nutrition, even though it possesses a functional nitrate-uptake system that was induced by nitrite (Schloemer and Garrett, 1974a). Nitrite-uptake systems in mutants *nit*-1, *nit*-3, *nit*-4, and *nit*-5 appear to be regulated in a manner identical with the wild-type, i.e. the uptake system became functional upon transfer of mycelia to nitrite-containing media, and was not affected by the presence of ammonia or other nitrogenous compounds (Coddington, 1976). Since the mutants *nit*-4 and *nit*-5 lack a functional nitrite reductase yet are capable of nitrite uptake, nitrite reductase and its regulation are distinct from the nitrite-uptake system.

III. Regulation of Nitrate Assimilation

Regulation of nitrate assimilation in fungi is, in the overview, eminently simple and straightforward (Nason and Evans, 1953; Cove,

1966; Garrett, 1972). Ammonia, the end product of the nitrate assimilatory pathway, strongly represses its expression. This phenomenon is the dominant aspect in regulation of nitrate assimilation. Obviously, if ammonia is present, the organism has no need to assimilate nitrate, and indeed would find it economical to suppress its action since four equivalents of reduced nicotinamide nucleotide are consumed for each nitrate-nitrogen reduced. On the other hand, in the absence of ammonia, fungi manifest the potentiality for nitrate assimilation only in the presence of the substrates, nitrate or nitrite. Thus, absence of ammonia is insufficient to evoke this metabolic pathway; its appearance is an adaptive response to both the availability of, and necessity for, metabolizing nitrate. In general, these phenomena are discussed as the "ammonia repression" or "nitrate induction" of nitrate assimilation, and almost certainly represent dramatic changes in specific macromolecular synthesis leading to the presence or absence of nitrate assimilatory enzymes. Nevertheless, use of the terms "repression" and "induction" and, thereby, the implicit assumption of the occurrence of genetic regulation analogous to the classical operon model of Jacob and Monod, is merely a useful working supposition; the molecular biology of this regulation has not been elucidated.

The results of investigation into possible mechanisms by which ammonia repression or nitrate induction might be achieved reveal the existence of a complexity of interactions. The complexity is based on the rather large number of genes which are potentially involved, and includes the various controls exerted both over expression of these genes and the ultimate functions of the gene products. Among the latter are controls affecting enzymic activities of nitrate reductase and nitrite reductase. Thus there exists a hierarchy of regulatory influences governing expression of the metabolic potentiality inherent in the nitrate assimilatory pathway. The various levels into which this hierarchy can be divided, e.g. the genetics, regulation of gene expression and regulation of enzyme activity, will be considered in turn. One important aspect of such considerations is the likelihood that revelation of the regulatory mechanisms influencing this simple two-enzyme metabolic pathway in these lower eukaryotes might provide heuristic insights into the ultimate understanding of regulation of more intricate metabolic processes in higher eukaryotes.

A. GENETICS AND REGULATION OF GENE EXPRESSION

A summary of the principal genetic loci involved in nitrate assimilation in the fungi *A. nidulans* and *N. crassa* is presented in Table 1. In *A. nidulans*, mutations in any of more than eleven loci can lead to impaired ability to assimilate nitrate, and at least four of these loci are regulatory in nature. With regard to assimilatory enzymes, only one locus has been designated as a structural gene for nitrite reductase, the *nii*A locus (Pateman *et al.*, 1967). Nitrate reductase is encoded by the *nia*D gene, since this is the only locus specifically abolishing nitrate reductase (Pateman *et al.*, 1967). Further, *nia*D mutants have diminished levels of nitrate reductase cross-reacting material (Pateman *et al.*, 1964) and temperature-sensitive *nia*D mutants produce a nitrate reductase of greater thermolability than wild-type (MacDonald and Cove, 1974). In addition, this enzyme may include a polypeptide component that is the *cnx*H gene product. The remainder of the *cnx* genes, namely *cnx*ABC, *cnx*E, *cnx*F and *cnx*G, contribute to formation of a molybdenum-containing cofactor (Pateman *et al.*, 1964) with which the *cnx*H gene product is also certainly involved, but the precise role of these genes and their products is as yet not understood. The product of their action, the presumed molybdenum-containing component (or MCC), is also required by the other fungal molybdoenzyme, xanthine dehydrogenase (XDH), since mutation in any of the *cnx* loci leads to loss of this enzymic activity as well as nitrate reductase activity (Pateman *et al.*, 1964). The *cnx*E gene mutations are unique in that they can be repaired by growth on 0.33 M molybdate and, for this reason, it has been suggested that the *cnx*E product is responsible for insertion of molybdenum into nitrate reductase and XDH molecules, a process which can also be achieved by mass action, i.e. high concentrations of molybdate (Cove *et al.*, 1964; Arst *et al.*, 1970). An alternative explanation is that use of high concentrations of molybdate allows for sufficient inclusion of trace amounts of a different chemical form of molybdenum to meet the needs of the organism. According to this hypothesis, the normal *cnx*E gene product would catalyse formation of this molybdenum derivative using molybdate as a substrate.

The genes in *A. nidulans* specifically discussed to this point provide for the machinery necessary to assimilate nitrate; the *nia*D and presumably the *cnx*H gene products comprise the nitrate reductase molecule, the *nii*A gene product, the nitrite reductase molecule, and the *cnx*ABC, E, F and G gene products function to achieve acquisition and transformation of molybdenum into a biologically active cofactor form for inclusion in nitrate reductase. The remaining genetic loci, *nir*A, *are*A, *tam*A and *gdh*A, are involved in regulating expression of the *nia*D and *nii*A genes to lead to formation of the nitrate-assimilating enzymes.

The *nir*A gene specifies a product which acts as a positive regulator necessary for full expression of nitrate assimilation enzymes (Pateman and Cove, 1967; Cove and Pateman, 1969; Cove, 1970, 1976). Most mutants at this locus result in non-inducibility for nitrate and nitrite reductases by nitrate, i.e. the *nir*A$^-$ phenotype (Pateman *et al.*, 1967), although alleles are known which show constitutive synthesis of these enzymes in the absence of nitrate, the *nir*A^c phenotype. Cove and Pateman (1969; Cove, 1970) proposed a model for regulation of synthesis of nitrate-reducing enzymes suggesting that, in addition to the *nir*A gene product, nitrate reductase itself was involved in regulation of both its own synthesis and synthesis of nitrite reductase (Fig. 7). This suggestion was based on the observation that certain *nia*D and *cnx* mutants lacking a functional nitrate reductase exhibited constitutive synthesis of nitrite reductase in the absence of nitrate, and a proposal was made that the *nir*A gene product is an inducer substance which can be prevented from acting by nitrate reductase in the absence of nitrate. In the presence of nitrate, nitrate reductase does not block action of the *nir*A gene product, and nitrate-assimilation enzyme synthesis is induced. Therefore, nitrate reductase is itself a part of the regulatory system governing its own synthesis, i.e. an example of auto-regulation of enzyme synthesis. Further, nitrate reductase provides the site of the primary regulatory signal, the nitrate recognition site, for the control sequence. In the presence of nitrate, nitrate reductase does not affect *nir*A gene-product function; in the absence of nitrate, nitrate reductase can prevent its own induction, and induction of nitrite reductase as well, by blocking the *nir*A product's action. The *nia*D and *nii*A genes are closely linked, but do not appear to constitute an operon (Cove, 1970; Tomsett and Cove, 1976). Most *nia*D and *cnx* mutants, since they lack a normal nitrate reductase, are constitutive for nitrite reductase since the *nir*A gene product cannot be blocked in its action. It is interesting to note that a few *nia*D and *cnx* mutants retain the ability to regulate nitrite reductase activity through induction, as the wild-type *A. nidulans* does. Presumably these latter mutants

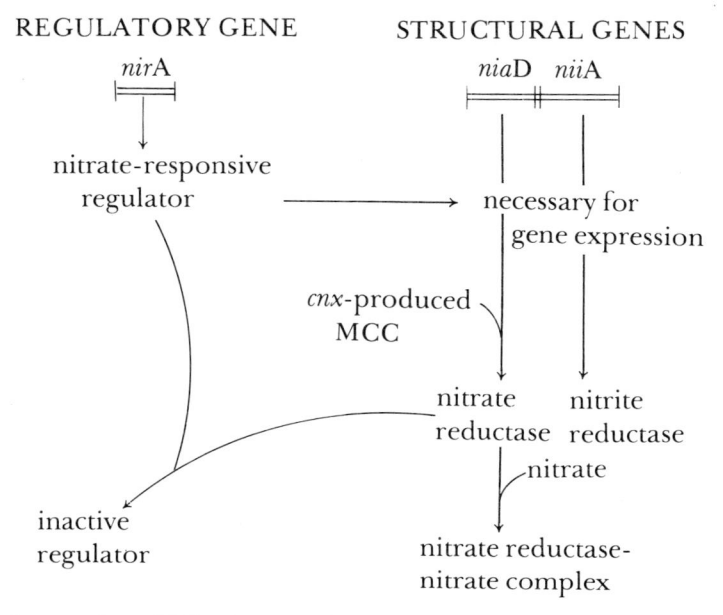

```
REGULATORY GENE           STRUCTURAL GENES
     nirA                    niaD    niiA
   ├────┤                  ├════╫════┤
      │                       │      │
      ▼                       │      │
nitrate-responsive             │      │
   regulator    ──────────▶  necessary for
         \                   │ gene expression
          \      cnx-produced\│      │
           \         MCC      \│      │
            \                  ▼      ▼
             \               nitrate  nitrite
              \              reductase reductase
               \                │  ↙nitrate
     inactive   ↙               ▼
     regulator              nitrate reductase-
                            nitrate complex
```

FIG. 7. Suggested model for the role of the *nirA*, *niaD*, *niiA* and *cnx* gene products in regulation of nitrate and nitrite reductases in *Aspergillus nidulans*. MCC stands for molybdenum-containing component. Adapted from Cove (1970).

produce a nitrate reductase which, although catalytically inactive, nevertheless, retains its regulatory properties. Thus, this first enzyme of nitrate assimilation is bifunctional; it is both catalytic and regulatory.

As discussed earlier (p. 11), wild-type nitrate reductase of both *A. nidulans* (Cove and Coddington, 1965; MacDonald et al., 1974) and *N. crassa*, (Kinsky and McElroy, 1958; Garrett and Nason, 1969) possesses an associated NADPH-cytochrome *c* reductase activity. Some of the *A. nidulans niaD* mutants and all of the *cnx* mutants retain this diaphorase activity even though they lack nitrate reductase activity. A direct correlation has been shown between nitrite reductase inducibility and cytochrome *c* reductase inducibility in these mutants; further, mutants constitutive for nitrite reductase are constitutive for NADPH-cytochrome *c* reductase (Pateman et al., 1967). These results are also consistent with the description of the *nirA* gene product as a positive regulatory element involved in induction of both nitrate and nitrite reductases, and whose action is subject to limitation through a regulatory control mediated by nitrate reductase in the absence of nitrate (Cove, 1976).

Additionally, evidence exists suggesting a more pleiotropic regulatory role for the *nir*A gene product and its interaction with nitrate reductase. Hankinson and Cove (1974) showed that growth of *A. nidulans* on nitrate causes a doubling in enzyme activities involved in the pentose phosphate pathway, and that the *nir*A gene product is responsible for mediating this nitrate induction. Also, Cove (1976) has discussed the fact that *nii*A mutants, which lack nitrite reductase activity and are thus unable to grow on nitrate, show impaired ability to grow on alternative nitrogen sources (such as L-ornithine or L-glutamate) if nitrate is present. A possible explanation is a phenomenon which has been termed nitrate repression, a cessation of various modes of catabolism, e.g. L-ornithine utilization, in the presence of nitrate. Presumably this nitrate repression is mediated by an *nir*A gene product which, in the absence of any inactivation by nitrate reductase (because of the presence of nitrate), can play a negative role and turn off processes necessary for catabolism of other nitrogen sources (Cove, 1976).

Nitrate induction as detailed above is a necessary condition for expresssion of the nitrate assimilatory pathway, but the effects of nitrate are overridden by the presence of ammonia and its repressive action (Cove, 1966). The mechanism by which ammonia repression is achieved is not at all well understood, but it appears to be quite pleiotropic in that ammonia prevents utilization of a large number of alternative nitrogen sources in addition to nitrate, such as purines, various amino acids and skimmed milk (Arst and Cove, 1973; Cove, 1976). Presumably, the fungus views ammonia as the preferred nitrogen source provided metabolizable carbohydrate is available, and thus has evolved ammonia repression as a mechanism to prevent synthesis of unneeded enzymes. Thereby, cellular economy is maximized. Genetic studies at Cambridge in England and Glasgow in Scotland have implicated at least three genetic loci in ammonia repression, namely *are*A, *tam*A and *gdh*A.

A number of mutant alleles of the *are*A gene have been obtained and fall into two general classes, designated *are*A^r and *are*A^d (Arst and Cove, 1973). The *are*A^r mutants cannot utilize a wide range of nitrogen sources including nitrate, purines, amine acids and amides, but fare very well on ammonia. This impairment in nitrogen acquisition is presumably due to an inability to synthesize the requisite enzymes. These various enzymes are normally ammonia-repressible in the wild-type, but their expression appears to be permanently repressed in these *are*A^r mutants. Because of their drastic results, the *are*A^r mutations are thought to produce an *are*A gene product whose function is grossly impaired. Thus, it is proposed that the normal *are*A gene product allows or is essential for expression of various genes whose products mediate alternate routes of nitrogen metabolism, e.g. nitrate assimilation. Consistent with this conclusion is the class of *are*A mutants termed *are*A^d. These mutants are unaffected with regard to their range of nitrogen source utilization, but instead manifest an insensitivity to ammonia repression for aspects of nitrogen metabolism. The various *are*A^d alleles, unlike the *are*A^r mutants, exhibit a highly specific phenotype. For instance, an *are*A^d allele can be specifically derepressed for ammonia repression of nitrate assimilation while retaining full ammonia repressibility of other ammonia-repressible enzymes. Presumably, the *are*A gene product of an *are*A^d mutant is altered such that its interaction with ammonia no longer limits its ability to promote expression of certain of the genes it controls, whilst others remain affected. The assumption that the *are*A gene product is acting in a positive regulatory role to foster gene expression stems from dominance studies on the various *are*A alleles in diploids and heterokaryons, and from the relative frequencies of *are*A^r compared with *are*A^d occurrence (*are*A^r mutations predominate; revertants from *are*A^r yield a very low frequency of *are*A^d phenotypes; Arst and Cove, 1973; Cove, 1976). Thus, it was concluded that ammonia repression of many nitrogen-metabolizing systems is mediated by this *are*A gene product whose action is necessary for expression of genes specifying the requisite enzymes of these systems. When ammonia is present, the *are*A gene product becomes inactive or is actively preventing expression of these same genes.

The relationship of nitrate induction to ammonia repression in control of nitrate assimilation in *A. nidulans* is mirrored in the epistasis between *nir*A^c and *are*A^r mutants. Mutant *are*A^r is completely epistatic to *nir*A^c (Arst and Cove, 1973), that is, although nitrate assimilation enzymes in an *nir*A^c strain do not require nitrate for induction but are constitutively expressed, they are not synthesized in an *are*A^r *nir*A^c double mutant and this mutant cannot grow on nitrate. This result is consistent with the observed ammonia repressibility of nitrate and nitrite reductases in *nir*A^c single mutant strains (Pateman and Cove, 1967). Further, double mutants carrying both an *are*A^d **allele giving** ammonia-insensitive synthesis of nitrate assimilation enzymes and a *nir*A⁻ allele where these enzymes are non-inducible by nitrate are also unable to grow on nitrate. Taken together, these observations substantiate the need for both nitrate induction and alleviation of ammonia repression in order to realize synthesis of nitrate-assimilation enzymes in wild-type *A. nidulans* (Arst and Cove, 1973), and further reflect the dominance of ammonia repression to nitrate induction in the hierarchy of controls governing expression of this pathway.

The *tam*A locus and the nature and action of its gene product in control of nitrogen metabolism are not yet as well understood as the *are*A gene. All mutations thus far known in the *tam*A gene are of one class, namely *tam*A^r, and show partially repressed activity for a number of ammonia-repressible systems even in the absence of ammonia. These systems include nitrate reductase, urea transport and asparaginase (Kinghorn and Pateman, 1975a). Thus, *tam*A^r mutants are somewhat similar to *are*A^r mutants, but differ in having low NADP-specific glutamate dehydrogenase (GDH) activity and the ability to excrete ammonia under certain growth conditions. The relationship between *tam*A^r and *are*A mutants was examined in double mutant strains. The *tam*^{r} *are*A^r double mutant exhibited an additive phenotype with respect to resistance to agents rendered toxic by ammonia-repressible systems (e.g. chlorate); however *tam*^{r} proved epistatic to *are*A^r with respect to NADP-GDH activity and ammonia efflux. Further, *tam*A^r was epistatic to *are*A^d allele by virtue of its ability to confer resistance to chlorate in a *tam*A^r *are*A^d double mutant (Kinghorn and Pateman, 1975a).

The remaining genetic locus thus far implicated in ammonia repression of nitrate assimilation and other systems is *gdh*A (Arst and MacDonald, 1973; Kinghorn and Pateman, 1973, 1975b). Mutations in the *gdh*A locus lead to loss of NADP-GDH activity and also result in derepression of various ammonia-repressible activities, such as nitrate assimilation and purine catabolism, whilst not affecting their inducibilities. The *gdh*A locus is presumed to be the structural gene for NADP-GDH, and thus this enzyme is also ascribed a functional role in regulatory interactions manifesting ammonia repression. The basis for the regulatory effect of NADP-GDH is not clear. Whether loss of ammonia repression in *gdh*A⁻ mutants is due to a defect in the NADP-GDH protein limiting it in a direct control function, **or whether loss in** this catalytic activity results in altered metabolite levels and thereby affects ammonia regulation, is not certain (Kinghorn and Pateman, 1975b). Further, on a genetic basis, its mode of action is unknown; *gdh*A mutations cannot suppress *are*A^r in *gdh*A *are*A^r double mutant strains (Arst and Cove, 1973). The *tam*A^r mutation is epistatic to mutations in *gdh*A (Kinghorn and Pateman, 1975a, b). Chang and Sorger (1976) have implicated glutamate dehydrogenase activity in *N. crassa* as a necessary condition for achieving ammonia repression. They concluded, on the basis of studies on repression of nitrite and nitrate reductases in both wild-type and strain *am*-1*a* (a glutamate dehydrogenaseless mutant), that ammonium must first be metabolized in order to effect repression. Their results showing that glutamate caused repression of nitrite reductase in both wild-type and *am*-1*a* mycelia led them to discount the possibility that some other component, e.g. the GDH protein itself, was lacking in *am*-1*a* mycelia and thus resulting in loss of regulation. Analogous experiments have not been done in *A. nidulans* to resolve whether the enzyme itself or its function is mediating ammonia repression.

Obviously the complexity of the overall regulatory situation, which embraces both nitrate induction and ammonia repression, stymies attempts to put forward plausible molecular biological models to describe the controls impinging just on the nitrate-assimilation system. The paucity of biochemical information is the real drawback here. An interesting beginning nevertheless is available in suggestions made by Pateman *et al.* (1973). They proposed a model in which the concentration of extracellular ammonia provides the principal regulatory signal, and that this signal is transmitted via formation of an NADP-GDH-ammonium complex within the cell membrane. This complex is designated as the first regulatory complex (Pateman *et al.*, 1973). Presumably, the *are*A, *tam*A and *nir*A gene products respond (in this sequence?) to this cue to decide induction or repression of nitrate assimilation. Elucidation of these regulatory controls and their interactions should prove most interesting.

Genetic studies on *N. crassa* reveal a strong similarity between the enzymology and regulation of nitrate assimilation in this organism as compared with *A. nidulans*. Garrett (1972) showed that, in wild-type *N. crassa*, both nitrate and nitrite reductases are induced by either nitrate or nitrite, and both are repressed by ammonia. Although far fewer alleles have been described in *N. crassa*, definite parallels can be drawn between the mutants of *N. crassa* affected in nitrate assimilation and the principal loci involved in this process in *A. nidulans* (Table 1, p. 17). Coddington (1976) has established the correlations drawn here from growth tests and biochemical studies on the *N. crassa nit* mutants. His results both confirm and extend the conclusions regarding the role of the *nit*-1 and *nit*-3 genes in contributing to the structure of nitrate reductase, as discussed in the section on nitrate reductase (p. 16). In addition, Coddington's (1976) studies provided for the assignment of functions given to the *nit*-2, *nit*-4 and *nit*-5 loci in Table 1. Sorger (1966) had earlier demonstrated that extracts of nitrate-induced *nit*-2 or *nit*-5 mycelia had very low levels of enzymic activities associated with the nitrate reductase. Coddington (1976) found absolutely no nitrate or nitrite reductase activity in *nit*-2 extracts and very low (repressed) levels of nitrate reductase activity in *nit*-4 and *nit*-5 mycelia extracts regardless of growth conditions. On the basis of these results, *nit*-2 appears analogous to the *A. nidulans are*A^r-type mutants. The repressed levels of nitrate reductase activities in *nit*-4 and *nit*-5, together with their ability to grow on purines as the sole nitrogen source, endow these mutants with properties analogous to *nir*A$^-$ mutants in *A. nidulans*. On the basis of this evidence, it would appear that *N. crassa* has two nitrate-affected regulator genes as opposed to one in *A. nidulans*. Certainly, further investigations in the genetics of *N. crassa* are warranted, to identify additional alleles in the *nit*-4 and *nit*-5 loci and, in particular, to attempt to obtain mutant alleles of these genes having the *nir*A^c phenotype (constitutive expression of the nitrate-reducing enzymes).

As indicated earlier (p. 41), in *A. nidulans*, the autogenous regulation of nitrate reductase synthesis is mediated via interactions between this enzyme and the *nir*A gene product (Fig. 7). In *N. crassa*, nitrate reductase synthesis can be considered constitutive in the *nit*-1 mutant, in that nitrate reductase-related NADPH-cytochrome *c* reductase activity is as high in mycelia deprived of a nitrogen source as in nitrate- or nitrite-induced mycelia (Coddington, 1976). This activity is still subject to ammonia repression. Thus, there is suggestive evidence for autoregulation of nitrate reductase in *N. crassa* in that, in the absence of nitrate, the *nit*-1 mutant might be incapable of preventing positive regulation of nitrate reductase synthesis by a *nir*A-type (*nit*-4 and/or *nit*-5) gene product because *nit*-1 lacks a functional nitrate reductase to abort the action of this positive regulator. Such an interpretation would at least be consistent with the interpretations given for *A. nidulans*. It should be indicated here that Coddington (1976) considered his results showed that nitrate or nitrite no longer had the ability to induce *nit*-1 NADPH-cytochrome *c* reductase, as opposed to the interpretation given here that this activity is now constitutively expressed.

However, nitrite reductase activity of the *nit*-1 mutant does not appear to be constitutively expressed (Coddington, 1976), but is instead still inducible but only by nitrite. Nitrate does not induce nitrite reductase in *nit*-1. In *A. nidulans*, if a mutant showed constitutive synthesis for cytochrome *c* reductase, it also showed constitutive synthesis for nitrite reductase (Pateman *et al.*, 1967). In contrast, the *N. crassa nit*-1 mutant, while constitutive for diaphorase activity, is inducible for nitrite reductase activity. In effect, the *nit*-3 reduced benzylviologen-nitrate reductase is analogous to the *nit*-1 diaphorase with regard to these controls, except that constitutively expressed levels of reduced benzylviologen-nitrate reductase activity in *nit*-3 are 8–10 times higher than induced levels of this activity in the wild-type and 50 times higher than constitutive levels in wild-type. It should be recalled, however, that denaturation of native NADPH-nitrate reductase by heat or *p*HMB treatment (Garrett and Nason, 1969; Antoine, 1974) causes a manifold increase in its MVH-nitrate reductase activity. Therefore, these highly elevated levels of MVH-nitrate reductase in *nit*-3 extracts, as compared to the activity in wild-type, do not necessarily mean that correspondingly greater amounts of nitrate reductase protein are present. Indeed it may only be the equivalent of induced levels in wild-type. It should also be noted that the *nit*-3 nitrite reductase is inducible by nitrite but not nitrate.

Obviously, more data are needed to reconcile the various possibilities. There is evidence for autoregulation of nitrate reductase synthesis in *N. crassa* but this fungus, in contrast to *A. nidulans*, appears to lose co-ordinate control of both nitrate reductase and nitrite reductase in mutants affecting nitrate reductase. A possibility of two distinct genetic loci, namely *nit*-4 and *nit*-5, mediating nitrate induction is suggested. Finally, on the basis of induction of nitrite reductase by nitrite but not by nitrate in *nit*-1 and *nit*-3 mutants, Coddington (1976) has suggested that nitrite might be the true co-inducer for nitrate-assimilation enzymes or, alternatively, these two enzymes might each have their own regulatory controls and perhaps their own regulatory genes. In either case, the situation differs from that described for induction of nitrate assimilation in *A. nidulans*; its resolution should prove interesting.

The genetic investigations considered above approached the regulation of nitrate assimilation by considering auxotrophy for nitrogenous compounds exhibited *in vivo* and activities of specific enzymes measurable *in vitro* among various fungal mutants unable to utilize nitrate. No direct consideration of transcriptional or translational events as sites for regulation of gene expression was made, and "induction" and "repression" were used only in an operational sense. Another approach has been to observe levels of nitrate-assimilation enzymes expressed in the wild-type and selected mutants under varying conditions of nitrogen nutrition and in the presence of either specific inhibitors of macromolecular biosynthesis (cycloheximide, actinomycin D) or specific antagonists of nitrate reductase (e.g. tungstate). By these studies it was hoped that the sites of action of nitrate and ammonia in eliciting and curtailing expression of nitrate assimilation might be elucidated. In general, these studies have favoured nitrate reductase because the results discussed above indicated that it occupied the pivotal position in these regulations.

The findings of Garrett (1972) essentially confirmed that the appearance of nitrate reductase activity in *N. crassa* mycelia depended on the presence of either nitrate or nitrite in the growth media. Ammonia prevented this "induction" of nitrate reductase activity, and essentially no enzyme activity was found in extracts of mycelia exposed to media lacking any nitrogen source. This last observation was interpreted as indicating that nitrate reductase activity was not "derepressible" in the absence of ammonia. With regard to the time-course of nitrate reductase expression, there was a rapid increase in enzyme activity in mycelia during the first six hours of exposure to nitrate, followed by a rapid decrease in activity over the ensuing six hours even though sufficient concentrations of nitrate were still present in the media. The rapidity of this loss was considered to be suggestive of an active mechanism regulating the decline in activity. Similar losses were detected when mycelia fully induced for nitrate reductase activity by prior exposure to nitrate were transferred to media containing ammonia or to media lacking any nitrogen source. These results confirmed earlier reports on nitrate reductase in the basidiomycete *Ustilago maydis* by Lewis and Fincham (1970) and in *A. nidulans* by Cove

(1966). In *U. maydis*, induced mycelia rapidly lost nitrate reductase activity upon transfer to media containing ammonia, and this loss could be prevented if the protein synthesis inhibitor cycloheximide was present. Lewis and Fincham (1970) postulated that some controlling macromolecule induced by ammonia rendered nitrate reductase unstable. In the *A. nidulans* studies, nitrate reductase activity in fully induced mycelia rapidly disappeared upon transfer to media lacking any nitrogen source, and Cove (1966) concluded that synthesis of nitrate reductase ceased upon removal of nitrate, and that the enzyme was unstable *in vivo*. Further considerations of this rapid inactivation of nitrate reductase activity in *N. crassa* and *A. nidulans* have appeared and will be discussed later.

The results obtained by Garrett (1972) for nitrite reductase from *N. crassa* stand in contrast to those of Cook and Sorger (1969) who claimed this enzyme, as measured by a reduced benzylviologen (BVH)-nitrite disappearance assay, was derepressible in the absence of ammonia, but not inducible by nitrate or nitrite. Using an assay for nitrite reductase based on its physiological function, Garrett (1972) found that nitrite reductase, like nitrate reductase, was induced in mycelia exposed to either nitrate or nitrite but was absent (repressed) in mycelia grown on ammonia. Essentially no NADPH-nitrite reductase activity was found in mycelia exposed to media lacking any nitrogen source. In support of these results, no NADPH-hydroxylamine reductase activity (an alternative assay for nitrite reductase) was found in mycelia from nitrogen-free media or in mycelia grown on ammonia. High levels of NADPH-hydroxylamine reductase activity were found in nitrate- or nitrite-induced mycelium. Further, in contrast to the conclusion of Cook and Sorger (1969), significant levels of BVH-nitrite reductase were found only in mycelia grown on nitrate or nitrite, not in mycelia grown on ammonia, or, most significantly, in mycelia exposed to media lacking any nitrogen source. The fact that the nitrite reductase from *N. crassa* is truly an enzyme inducible by nitrate or nitrite, and not an enzyme whose synthesis is derepressed in the absence of ammonia, was best shown by experiments in which the concentrations of nitrite in the media were limiting. In such instances, nitrite reductase activity increased rapidly. However, after nine hours, when all of the nitrite had been consumed, enzyme levels rapidly declined. If it were a derepressible system, nitrite reductase activity should have persisted under these conditions. In contrast to nitrate reductase, nitrite reductase activity is stable *in vivo* as long as sufficient inducer is present, that is, no evidence for the *in vivo* disappearance of this activity in the presence of nitrate or nitrite has been found. On the other hand, nitrite reductase activity is lost, albeit much less rapidly than the nitrate reductase, upon transfer of mycelia to ammonia-containing media or media lacking a nitrogen source (Garrett, 1972; R. H. Garrett, unpublished observations).

Loss of nitrate reductase activity from *N. crassa* mycelia exposed to ammonia media or to media lacking any nitrogen source, in either the presence or absence of cycloheximide, was also considered by Subramanian and Sorger (1972b, c). Mycelia from wild-type *Neurospora* and from the *nit*-1 and *nit*-3 mutants were examined in these studies. These investigators confirmed that either the removal of nitrate from, or the addition of ammonia to, cultures of wild-type mycelia fully induced for nitrate reductase resulted in the disappearance of this enzyme. However, if cycloheximide was present, this inactivation or degradation of nitrate reductase activity was significantly decreased, indicating that the inactivation mechanism may in some way be dependent on protein synthesis. In addition, this report showed that both the NADPH-cytochrome *c* reductase and BVH-nitrate reductase activities of the wild-type nitrate reductase responded in the same manner as the overall NADPH-nitrate reductase to the effects of ammonia or starvation for a nitrogen source, in the presence or absence of cycloheximide. Thus, the inactivation or degradation process was not specific for a single component function of the enzyme, but had a generalized effect on all of the catalytic activities of this enzyme. Experiments in which active extracts of nitrate reductase were mixed with inactive extracts of mycelia grown in the presence of ammonia indicated that there was neither a stable inhibitor nor a functioning destruction mechanism for nitrate reductase in the latter preparations, since no losses in enzymic activity resulted; that is, *in vitro* inactivation could not be observed. *In vivo* inactivation of nitrate reductase by ammonia was dependent on the ammonia concentration and could not be prevented by increasing nitrate concentration.

Although both the overall NADPH-nitrate reductase activity and its component NADPH-cytochrome *c* reductase and BVH-nitrate reductase activities were inactivated in wild-type *Neurospora* mycelia, neither the NADPH-cytochrome *c* reductase of *nit*-1 mycelia nor the BVH-nitrate reductase activity of *nit*-3 mycela was inactivated by exposure to ammonia or nitrogen-free media. These results suggest that the integrity of the nitrate reductase complex is required for its *in vivo* inactivation.

In order to analyse the various effects of nitrate, ammonia and nitrogen starvation on the level of nitrate reductase activity, Subramanian and Sorger (1972c) devised a procedure for experimentally separating the processes of transcription and translation in *N. crassa*. First, translation is blocked by incubation of mycelia with cycloheximide for several hours (stage I). Transcription is presumably unaffected during this time. Then, the mycelia are incubated with both cycloheximide and the transcription inhibitor actinomycin D in the presence of EDTA (stage II); EDTA increases the permeability of mycelia to actinomycin D. Finally, the mycelia are incubated with actinomycin D alone and translation of preformed mRNA is presumably observed (stage III). Nitrate and/or ammonia can be included or omitted in any or all of the stages. Following stage III, crude extracts were prepared and the nitrate reductase activity assayed to determine the extent of gene expression. Both stages II and III were necessary because EDTA prevents induction of nitrate reductase but is necessary for significant uptake of actinomycin D by mycelia. Ammonia-grown mycelia devoid of nitrate reductase activity were used as the starting material in all experiments. Although this procedure, with its prolonged exposure of mycelia to antibiotics, permitted formation of nitrate reductase to an extent of only 10 per cent of the levels seen under normal conditions, several qualitatively important observations could be made with respect to the stages at which nitrate and ammonia exert their regulatory effects.

Nitrate reductase was maximally formed in this system when nitrate was present in all stages. Omission of nitrate in stage I resulted in significant decreases (80%) in the final nitrate reductase activity, whether or not nitrate was present in stage II and/or III. Thus, the induced capacity to synthesize nitrate reductase during stage III operated efficiently only if nitrate was present in stage I. Presumably, nitrate enhanced formation of nitrate reductase-specific mRNA during this period. If nitrate was omitted from stage II, essentially no diminution was seen in the levels of enzyme synthesis from that obtained when nitrate was present throughout. Thus, the induced capacity to make nitrate reductase is stable in the absence of nitrate over the duration of stage II; or, in a dogmatic interpretation, nitrate reductase-specific mRNA is stable in the absence of nitrate. In contrast, if nitrate is omitted from stage III, the levels of enzyme activity found are from 40 to 80 per cent less than those detected when nitrate is present throughout, depending on how long stage III is allowed to proceed (the longer stage III, the less enzyme activity is found). Presumably, nitrate is not necessary for translation of nitrate reductase-specific mRNA but is required for stabilization of nitrate reductase *in vivo*, as discussed earlier (p. 51; Subramanian and Sorger, 1972b). That is, if nitrate is absent from stage III, any nitrate reductase synthesized is immediately susceptible to *in vivo* inactivation.

If ammonia is present together with nitrate during stage I, when nitrate reductase-specific mRNA would presumably be elaborated, large decreases (67%) occur in the amounts of nitrate reductase synthesis during stage III. Further, if ammonia is present during stage II, even larger diminutions (75%) in the ultimate yield of nitrate reductase are seen. Ammonia, when included with nitrate in stage III, has comparatively small effects. However, if the duration of stage III is lengthened, losses in nitrate reductase activity occur. These losses are presumably due to *in vivo* inactivation of nitrate reductase in the presence of ammonia, (Subramanian and Sorger, 1972b). Consequently, ammonia appears to act both by antagonizing the appearance of induced capacity to synthesize nitrate reductase which occurs in stage I (i.e. synthesis of nitrate reductase mRNA), and by rendering this capacity unstable during stage II incubation. The latter effect seems to be quantitatively greater. The conclusions drawn by Subramanian and Sorger were as follows. Nitrate seems to induce the appearance of nitrate reductase activity in *N. crassa* presumably by enhancing synthesis of nitrate reductase-specific mRNA. This presumed message can then be translated to give active enzyme in the absence or presence of inducer. However, if nitrate is present, the enzyme is stabilized *in vivo* and accumulates. In the absence of nitrate or the presence of ammonia, nitrate reductase is inactivated. Ammonia, on the other hand, represses the appearance of nitrate reductase activity by: (a) antagonizing formation of nitrate reductase mRNA; (b) rendering this mRNA species labile; and/or (c) leading to inactivation of nitrate reductase after it is formed. The effect cited in (a) is consistent with a classic repressor action according to the operon theory of Jacob and Monod; effects (b) and (c) are more difficult to reconcile and their causes are unknown. However, effect (c) is prevented by cycloheximide. An important aspect is that removal of nitrate or addition of ammonia not only appears to cause a cessation in synthesis of nitrate reductase mRNA but also results in the disappearance of existing nitrate reductase activity.

Sorger's laboratory subsequently re-examined the above procedures and results, and revised some of the conclusions. Sorger and Davies (1973) undertook studies quantitatively to determine the effects of cycloheximide and actinomycin D during stages I, II and III, and to re-assess the specific effects of nitrate and ammonia on formation and maintenance of nitrate reductase activity. Cycloheximide was shown to inhibit protein synthesis during stage I by more than 80 per cent as measured by incorporation of [^{14}C]leucine into acid-precipitable material. On the other hand, actinomycin D caused only a 50% inhibition of RNA synthesis under stage III conditions as determined from the amount of [^{3}H]uridine incorporated into acid-precipitable material. Thus, the results indicated that nitrate reductase mRNA should accumulate during stage I without being translated, but translation of this mRNA during stage III is complicated by new mRNA synthesis despite the presence of actinomycin D. Nevertheless, Sorger and Davies (1973) proceeded to examine the effects of the various nitrogenous nutrients on the appearance of nitrate reductase activity in both wild-type and *nit*-3 mutant mycelia using the three-stage method.

In contrast to the earlier work (Subramanian and Sorger, 1972b, c), it was found that the "induced" capacity to synthesize nitrate reductase during step I (i.e. presumably synthesis of nitrate reductase-specific mRNA) was not enhanced by nitrate since no greater levels of induced capacity were found in the presence of nitrate than in its absence (no nitrogen source). Thus, the "accumulation of the potential" to synthesize nitrate reductase was no longer "induced capacity", since nitrate did not enhance such accumulation in stage I in mycelia of either wild-type or the *nit*-3 mutant. On the other hand, ammonia antagonized accumulation of the potential to synthesize nitrate reductase during stage I in both wild-type and mutant. Presumably, ammonia repressed synthesis of nitrate reductase mRNA but nitrate did not induce it in either wild-type or *nit*-3 mycelia.

However, nitrate caused a three-fold stimulation in the rate of accumulation of nitrate reductase activity during stage III in both mutant and wild-type mycelia. This stimulation was an increase relative to the rates seen in the presence of ammonia or in the absence of any nitrogen source during the stage III incubation. The observation that the presence of ammonia in stage II labilizes the capacity to synthesize nitrate reductase during stage III (Subramanian and Sorger, 1972b) was confirmed.

The conclusions drawn by Sorger and Davies (1973) suggest that nitrate is not essential for transcription of nitrate reductase mRNA but that nitrate is important for translation of this message. Ammonia is presumed to act by repressing nitrate reductase mRNA formation and by causing inactivation of preformed nitrate reductase mRNA species. No mechanism is suggested for the latter process. The effect of nitrate on translation of nitrate reductase mRNA is suggested to occur by either influencing the mechanism of translation itself to enhance nitrate reductase production, or stabilizing the enzyme once it is formed.

The question persists as to whether nitrate induces transcription of nitrate reductase-specific mRNA or promotes its translation. Sorger *et al.* (1974) appear to favour a post-transcriptional role where nitrate enhances either (a) translation, (b) enzyme maturation or (c) stabilization against inactivation *in vivo*. The conclusions regarding the effects of ammonia are more concise (Subramanian and Sorger, 1972c; Sorger and Davies, 1973) in indicating that ammonia both antagonizes formation and/or stability of nitrate reductase mRNA and leads to *in vivo* inactivation of pre-existing nitrate reductase.

Subramanian and Sorger (1972a) have also examined the role of molybdenum in synthesis of nitrate reductase by *N. crassa*. If mycelia were induced by nitrate in molybdenum-deficient media, little NADPH-nitrate reductase or BVH-nitrate reductase activity was found, but the level of nitrate-inducible NADPH-cytochrome *c* reductase was high. This last activity exhibited a sedimentation co-efficient characteristic of the normal wild-type nitrate reductase molecule (8 S) instead of the lower sedimentation co-efficient of 4.3 S typical of the nitrate-inducible NADPH-cytochrome *c* reductase of the *nit*-1 mutant. Similar results were obtained when mycelia grown in media in which molybdate was replaced by tungstate were used as a source of nitrate reductase. Tungstate is compctititive with molybdate in interactions with nitrate reductase, but the tungstate-containing enzyme is inactive in nitrate reduction. Subramanian and Sorger (1972a) concluded that molybdenum deficiency or tungstate treatment did not affect synthesis of any of the protein components of nitrate reductase, but resulted in the production of inactive enzyme molecules. These inactive molecules could not be activated by treatment with molybdate *in vitro*. The enzyme, although inactive in nitrate reduction, was functional in the FAD-dependent NADPH-cytochrome *c* reductase assay. These results led to the development of a method to examine turnover of nitrate reductase by the use of tungsten (Sorger *et al.*, 1974).

Information regarding turnover of nitrate reductase in *N. crassa* was essential for a determination of the effects of nitrate on *in vivo* stability of this enzyme. The question asked (Sorger *et al.*, 1974) was whether the effect of nitrate on the adaptive appearance of nitrate reductase activity is solely explainable by the fact that nitrate protects this enzyme from inactivation. To examine this question, nitrate reductase was fully induced in mycelia by growth on nitrate and then the mycelium was transferred to nitrate-containing medium lacking molybdenum but containing tungstate. Newly formed enzyme would thus be catalytically inactive in the NADPH-dependent reduction of nitrate, and the fate of the pre-existing enzyme could be followed using routine assays for nitrate reductase activity. Both wild-type and *nit*-3 mycelia were used. As a control, Sorger *et al.* (1974) established that protein synthesis was required for expression of nitrate reductase activity. Mycelia fully induced for nitrate reductase were incubated in ammonia-containing media for several hours to inactivate the enzyme. When these mycelia were subsequently exposed for two hours to nitrate- and cycloheximide-containing media, nitrate reductase activity was not restored. Thus, ammonia-induced inactivation of nitrate reductase did not appear to be a reversible phenomenon in the absence of protein synthesis.

Experiments to corroborate further the notion that nitrate reductase is inactivated by degradation, and is therefore unrecoverable, were carried out using antibodies prepared against minimally purified fractions of nitrate reductase from induced wild-type mycelia. These authors (Sorger *et al.*, 1974) concluded that loss of nitrate reductase activity in mycelia under non-inducing conditions is due to disappearance of cross-reactive enzyme protein and not simply due to losses in the catalytic power of the enzyme.

Experiments involving transfer of fully induced mycelia to media containing tungstate indicated that the presence of nitrate in such media stabilized pre-existing nitrate reductase activity. Under these conditions, enzyme activity decayed at only one-third the rate of that detected in mycelia in nitrogen-free media containing tungstate, and one-half the rate observed in mycelia in tungstate- and ammonia-containing media. The half-life of pre-existing nitrate reductase activity in mycelia transferred to nitrate-containing media containing tungstate was approximately two hours. Thus, stabilization of nitrate reductase *in vivo* by nitrate is not absolute. Sorger *et al.* (1974) concluded that a major effect of nitrate on induction of nitrate reductase in *N. crassa* is its stabilization of the enzyme. However, they defer from arguing that this partial protection afforded to the enzyme by its substrate accounts for all of the effects of nitrate on nitrate reductase activity levels in mycelia. Indeed, if this were its only effect, it is difficult to understand why the rates of decay of nitrate reductase activity in mycelia exposed to ammonia-containing or nitrogen-free media are substantially decreased when these media also contain cycloheximide.

Amy and Garrett (1977) investigated the *nit* mutants of *N. crassa* using immunological techniques with antibody prepared against purified nitrate reductase. The goal was to detect whether inactive nitrate reductase protein was present in wild-type or mutant mycelia grown under various conditions of nitrogen nutrition. No cross-reacting material was detected in the wild-type or mutant mycelia grown on ammonia. When the *nit* mutants were grown on ammonia, then exposed to nitrate for four hours to induce enzyme formation, cross-reacting material was found in extracts of *nit*-1, *nit*-3 and *nit*-6. However, the *nit*-1 extract was only weakly antigenic, suggesting significant conformational differences between the enzyme in *nit*-1 and the wild-type to which the antibodies had been raised. On the other hand, the *nit*-3 extract contained large amounts of cross-reacting material implying the presence of a protein with marked identity to the wild-type enzyme. The *nit*-6 mutant lacks nitrite reductase (Table 1), but possesses NADPH-nitrate reductase activity. Extracts of *nit*-6 were very antigenic and its nitrate reductase appears identical to the wild-type enzyme. Cross-reacting material was not found in any of the mutants that lacked all of the NADPH-nitrate reductase-associated activities (*nit*-2, *nit*-4 and *nit*-5). The absence of cross-reacting material from ammonia-grown mycelia is further evidence that nitrate reductase protein is absent, and that nitrate induction involves *de novo* nitrate reductase synthesis rather than activation of a precursor.

Several points suggest that perhaps a protease is responsible for loss in nitrate reductase activity in mycelia under non-inducing conditions: (a) the inability to recover nitrate reductase activity *in vivo* in inactivated mycelia in the absence of protein synthesis; and (b) the purported proportionality between nitrate reductase activity and enzyme titrable by anti-nitrate reductase antibody in extracts from mycelia exposed to different nutritional conditions. The fact that cycloheximide strongly prevents decay of nitrate reductase activity under non-inducing conditions (Subramanian and Sorger, 1972b; R. H. Garrett, unpublished observations) might mean that the hypothetical protease is not constitutively expressed but is nitrate-repressible, and therefore derepressible in media containing ammonia or no source of nitrogen. Wallace (1974) has reported the presence of a nitrate reductase-inactivating enzyme in maize roots which also shows proteolytic activity against casein. The activity against casein raises some concern regarding a specific metabolic function for this inactivating enzyme. The maize-root protease has a molecular weight of 44,000 and is presumably a serine-type protease since it is inhibited by phenylmethylsulphonylfluoride (PMSF). Wallace (1975) has shown that this maize enzyme will inactivate activity of the nitrate reductase from *N. crassa*. However, only the overall NADPH-nitrate reductase activity and the NADPH-cytochrome *c* reductase activity (diaphorase moiety) were lost; the BVH-nitrate reductase activity was unaffected. In contrast, it is noteworthy to recall that all component activities of the nitrate reductase from *N. crassa* were lost upon inactivation *in vivo* (Subramanian and Sorger, 1972b). If an analogous protease exists in *N. crassa*, it is effective against all catalytic functions of nitrate reductase. Further, it would be anticipated that the protease from *N. crassa* would be fairly specific for nitrate reductase because the cycloheximide experiments suggest that it might be nitrate-repressible. In addition, nitrite reductase activity in *N. crassa* is relatively stable to inactivation (Garrett, 1972) despite its apparent similarities to nitrate reductase (Garrett and Nason, 1969; Lafferty and Garrett, 1974; see Section II.B).

B. REGULATION OF ENZYMIC ACTIVITY

Evidence for regulation of nitrate assimilation through direct changes in enzymic activity (as opposed to changes in enzyme concentration) has been sought on a number of occasions and, although some successes have been reported, the results are not yet conclusive. Garrett and Nason (1969) tested a number of organic nitrogenous compounds for possible feedback inhibition of the nitrate reductase from *N. crassa*. Ammonium chloride, urea, glycine, DL-alanine, DL-aspartic acid, L-asparagine, L-glutamic acid, L-glutamine, L-histidine, L-phenylalanine and L-valine (each at 10 mM) had no effect.

In a preliminary report, Ketchum *et al.* (1975) have suggested that nitrate reductase activity in *N. crassa* might be subject to activation or inhibition by certain key metabolites in nitrogen metabolism. These authors observed that nitrate reductase activity in *N. crassa* mutant am_2 was only half that found in wild-type. The am_2 mutant lacks NADP-specific glutamate dehydrogenase activity. However, nitrate reductase activity in am_2 mycelia could be increased two-fold, to wild-type levels, by transferring mycelia to media containing either glutamine or ammonia and glutamate. Further, this activity increase was not prevented by cycloheximide. This last result prompted Ketchum *et al.* (1975) to investigate the effects of various substrates and products of glutamine metabolism on the wild-type nitrate reductase activity *in vitro*. Nitrate reductase preparations used in these experiments were extensively dialysed to remove EDTA employed in the purification procedure. In the absence of EDTA, nitrate reductase was virtually inactive; however, addition of 5 mM histidine, tryptophan, glycine, arginine, glutamate or glutamine stimulated nitrate reductase activity up to 50-fold. Carbamoyl phosphate inhibited the activation provided by these amino acids. Ketchum *et al.* (1975) suggest that the nitrate reductase of *N. crassa* might be an allosteric enzyme which would be regulated by fluctuations in the intracellular concentration of certain amino acids. The advantage in these reduced nitrogen compounds activating a nitrogen-reducing enzyme is unclear. The dependency of nitrate reductase activity on the presence of substantial concentrations of specific amino acids is unanticipated.

An interesting hypothesis with respect to regulation of enzyme activity, and particularly that of oxidation-reduction enzymes, has come from Losada (1974) in his work in nitrate assimilation in photosynthetic organisms. Losada has postulated that, because plant nitrate reductases can be inactivated *in vitro* upon reduction with NAD(P)H in the absence of nitrate and re-activated by oxidation with ferricyanide, such a redox cycle might be of physiological significance in regulation of nitrate assimilation. Although *in vivo* studies with algae and higher plants substantiate Losada's ideas, the situation does not appear to hold in fungi. The nitrate reductase of *N. crassa* could be reductively inactivated *in vitro* in the presence of cyanide (Garrett and Greenbaum, 1973; see Section II.A), but no regulatory significance was ascribed to this phenomenon since evidence for reductive inactivation of this enzyme *in vivo* could not be found, nor could any potential nitrate reductase activity be restored by oxidation with ferricyanide in extracts physiologically inactivated by incubation in media containing ammonia or no nitrogen source (R. H. Garrett, unpublished observations). Further, although Vega *et al.* (1975a) could demonstrate a "reductive inactivation" of the nitrite reductase from *N. crassa*, the basis for the effect was a peroxide-mediated reaction irreversibly affecting nitrite reductase activity and, thus, this effect appeared to offer little physiological utility (see Section II.B).

Nevertheless, NADPH has been implicated as an important factor in determining the level of nitrate reductase activity in fungi. Dunn-Coleman and Pateman (1975) showed that incubation of cell-free extracts of *A. nidulans* at 25°C led to a progressive decline in their nitrate reductase activity. When added to the incubation mixture, NADPH, but not NADH, NADP$^+$ or NAD$^+$, prevented this activity decline. This effect then is directly opposite to the reductive inactivation of the photosynthetic nitrate reductases by NAD(P)H (Losada, 1974). Further, addition of NADPH to an extract whose nitrate reductase activity had significantly declined led to a progressive recovery of nitrate reductase activity to the point that it became fully restored. Therefore, this NADPH effect on the *A. nidulans* enzyme is not only protective but restorative as well.

Hankinson (Hankinson, 1974; Hankinson and Cove, 1974) showed that growth of *A. nidulans* on nitrate affected several enzymes in the pentose-phosphate pathway, the major source of NADPH in this organism. Since each nitrate ion reduced to ammonia requires four equivalents of NADPH, the correlation between nitrate metabolism and the pentose-phosphate pathway is anticipated. In addition, Hankinson (1974) showed that two classes of mutants defective in metabolism of pentoses, designated *ppp*A and *ppp*B, were characterized by poor growth on nitrate or nitrite. Such mutants would be expected to have low NADPH levels. Hynes (1973) demonstrated that nitrate reductase activity disappeared rapidly in *A. nidulans* mycelia exposed to media lacking a carbon source. In this instance, a decline in NADPH concentration is also expected. Therefore, these are *in vivo* phenomena explainable on the basis of the observed effects of NADPH on the nitrate reductase *in vitro*. Consequently, Dunn-Coleman and Pateman (1975) have hypothesized that the NADPH/NADP$^+$ ratio can determine the catalytic activity of the nitrate reductase in *A. nidulans* and changes in this ratio could regulate nitrate assimilation.

IV. Conclusions and Future Prospects

The present status of knowledge regarding nitrate assimilation in fungi as presented here reveals that an excellent operational framework now exists for understanding the enzymology and the genetics of the process. But regulation of the expression of this metabolic pathway is only superficially understood, and it is here that research interests are currently focused. Nevertheless, it is worthwhile to reflect specifically on the various accomplishments and prospects within the aspects of nitrate assimilation outlined in this review.

The fungal nitrate reductase has been well characterized at the enzymological level, and is now amenable to more sophisticated analysis of its molecular mechanism of action in electron-transfer catalysis. In particular, detailed studies on the precise roles of its sulphydryl, flavin, haem and molybdenum moieties are within the realm of experimentation. The precise nature of the subunit composition of the nitrate reductase should soon be revealed, and lead to elucidation of the assembly of this enzyme complex from its protein subunits and molybdenum-containing component.

Knowledge about the nitrite reductase is less advanced, but the recent success in achieving homogeneous preparations of this protein should provide for rapid improvement in this area. Initial goals are an analysis of the prosthetic groups involved and complete characterization of the several partial electron-transfer reactions that this enzyme can achieve. A significant mechanistic consideration is the modelling and testing of the six-electron-transfer problem posed by the reduction of nitrite to ammonia. The relationship of this reaction to the complement of prosthetic groups and subunit structure and assembly of nitrite reductase are intriguing challenges to the biochemist that are now amenable to investigation.

Further research is needed on the uptake systems by which fungi transport nitrate and nitrite. These systems have been biochemically characterized to some extent, but further evidence for active transport (e.g. using permease mutants) is desirable. In addition, the extent to which regulation of the expression of uptake at the genetic and biochemical level can influence assimilation of nitrate has received scant attention.

Regulation of nitrate assimilation is, in general, well understood only at a phenomenological level, and mechanisms underlying this regulation need to be determined not only to reveal more about the process of nitrate assimilation but to provide descriptions of the hierarchy of controls operating to govern the expression of metabolic potentiality in a simple eukaryote. Genetic studies have implicated a complexity of regulatory interactions leading to nitrate induction or ammonia repression, but the stage of gene expression where these interactions occur can only be guessed. Obviously *in vitro* transcription and translation systems would be most helpful in unravelling the complexities manifested in the action of genes such as *nir*A, *are*A and *tam*A, and their responses to metabolic signals and to one another. Also, the fate of nitrate-assimilation enzymes, their regulation by inactivation or degradation, and their modulation by effectors demand further investigation and integration into the overall mosaic of controls. The unveiling of this mosaic will have general significance beyond its importance to nitrate assimilation in that it will provide a picture encompassing regulation from gene expression to enzyme activity within organisms which can be viewed as model eukaryotic systems. It is therefore an important endeavour.

In reviewing the research on nitrate assimilation, one immediately encounters the pioneering contributions of Alvin Nason. Not only the establishment but the continuing encouragement and advancement of this field is more a result of his efforts than those of any other person. Thus, this review is a tribute to him.

REFERENCES

Amy, N. K., Garrett, R. H. and Anderson, B. M. (1977). *Biochimica et Biophysica Acta* **480**, 83.

Amy, N. K. and Garrett, R. H. (1977). *Abstracts of the Annual Meeting of the American Society for Microbiology* p. 85.

Anderson, B. M., Yuan, J. H. and Vercellotti, S. V. (1975). *Molecular and Cellular Biochemistry* **8**, 89.

Antoine, A. D. (1974). *Biochemistry, New York* **13**, 2289.

Arst, H. N., Jr., MacDonald, D. W. and Cove, D. J. (1970). *Molecular and General Genetics* **108**, 129.
Arst, H. N., Jr. and Cove, D. J. (1973). *Molecular and General Genetics* **126**, 111.
Arst, H. N., Jr. and MacDonald, D. W. (1973). *Molecular and General Genetics* **122**, 261.
Brown, C. M., MacDonald-Brown, D. S. and Meers, J. L. (1974). *Advances in Microbial Physiology* **11**, 1.
Chan, J. K. and Anderson, B. M. (1975). *Journal of Biological Chemistry* **250**, 67.
Chang, H. C. P., Mulkins, G. J., Dyer, J. C. and Sorger, G. J. (1975). *Journal of Bacteriology* **123**, 755.
Chang, H. C. P. and Sorger, G. J. (1976). *Journal of Bacteriology* **126**, 1002.
Coddington, A. (1976). *Molecular and General Genetics* **145**, 195.
Cook, K. A. and Sorger, G. J. (1969). *Biochimica et Biophysica Acta* **177**, 412.
Cotton, F. A. and Wilkinson, G. (1966). *In* "Advanced Inorganic Chemistry", (C. G. Brown, ed.), 2nd edition. pp. 930–960. Interscience, New York.
Cove, D. J. (1966). *Biochimica et Biophysica Acta* **113**, 51.
Cove, D. J. (1970). *Proceedings of the Royal Society B* **176**, 267.
Cove, D. J. (1976). *In* "Second International Symposium on the Genetics of Industrial Microorganisms", (K. D. McDonald, ed.), pp. 407–418. Academic Press, New York.
Cove, D. J., Pateman, J. A. and Rever, B. M. (1964). *Heredity* **19**, 529.
Cove, D. J. and Coddington, A. (1965). *Biochimica et Biophysica Acta* **110**, 312.
Cove, D. J. and Pateman, J. A. (1969). *Journal of Bacteriology* **97**, 1374.
Downey, R. J. (1971). *Journal of Bacteriology* **105**, 759.
Downey, R. J. (1973). *Biochemical and Biophysical Research Communications* **50**, 920.
Downey, R. J. and Focht, W. J. (1974). *Microbios* **11**, 61.
Dunn-Coleman, N. S. and Pateman, J. A. (1975). *Biochemical Society Transactions* **3**, 531.
Fisher, T. L., Vercellotti, S. V. and Anderson, B. M. (1973). *Journal of Biological Chemistry* **248**, 4293.
Ganelin, V. L., L'vov, N. P., Sergeev, N. S., Shaposhnikov, G. L. and Kretovich, V. L. (1972). *Academy of Sciences of the Union of Soviet Socialist Republics* **206**, 1236.
Garner, C. D., Hyde, M. R., Mabbs, F. E. and Routledge, V. I. (1974). *Nature, London* **252**, 579.
Garrett, R. H. (1972). *Biochimica et Biophysica Acta* **264**, 481.
Garrett, R. H. (1978). *In* "Microbiology, 1978", (D. Schlessinger, ed.), pp. 324–329. American Society of Microbiology Publication, Washington.
Garrett, R. H. and Nason, A. (1967). *Proceedings of the National Academy of Sciences of the United States of America* **58**, 1603.
Garrett, R. H. and Nason, A. (1969). *Journal of Biological Chemistry* **244**, 2870.
Garrett, R. H. and Greenbaum, P. (1973). *Biochimica et Biophysica Acta* **302**, 24.
Garrett, R. H. and Cove, D. J. (1976). *Molecular and General Genetics* **149**, 179.
Greenbaum, P., Prodouz, K. N. and Garrett, R. H. (1978). *Biochimica et Biophysica Acta* in press.
Hankinson, O. (1974). *Journal of Bacteriology* **117**, 1121.
Hankinson, O. and Cove, D. J. (1974). *Journal of Biological Chemistry* **249**, 2344.
Hewitt, E. J. (1975). *Annual Review of Plant Physiology* **26**, 73.
Hynes, M. J. (1973). *Journal of General Microbiology* **79**, 155.
Jacob, G. S. (1975). Ph.D. Thesis: University of Wisconsin, Madison.
Ketchum, P. A., Cambier, H. Y., Frazier, W. A., Madansky, C. H. and Nason, A. (1970). *Proceedings of the National Academy of Sciences of the United States of America* **66**, 1016.
Ketchum, P. A. and Sevilla, C. L. (1973). *Journal of Bacteriology* **116**, 600.

Ketchum, P. A. and Swarin, R. S. (1973). *Biochemical and Biophysical Research Communications* **52**, 1450.
Ketchum, P. A., Somerville, M. and Zeeb, D. D. (1975). *Abstracts of the Annual Meeting of the American Society for Microbiology* p. 111.
Ketchum, P. A. and Downey, R. J. (1975). *Biochimica et Biophysica Acta* **385**, 354.
Ketchum, P. A., Taylor, R. C. and Young, D. C. (1976). *Nature, London* **259**, 202.
Kinghorn, J. R. and Pateman, J. A. (1973). *Journal of General Microbiology* **78**, 39.
Kinghorn, J. R. and Pateman, J. A. (1975a). *Molecular and General Genetics* **140**, 137.
Kinghorn, J. R. and Pateman, J. A. (1975b). *Journal of General Microbiology* **86**, 294.
Kinsky, S. C. and McElroy, W. D. (1958). *Archives of Biochemistry and Biophysics* **73**, 466.
Lafferty, M. A. and Garrett, R. H. (1974). *Journal of Biological Chemistry* **249**, 7555.
Lee, K-Y., Pan, S-S., Erickson, R. and Nason, A. (1974a). *Journal of Biological Chemistry* **249**, 3941.
Lee, K-Y., Erickson, R., Pan, S-S., Jones, G., May, F. and Nason, A. (1974b). *Journal of Biological Chemistry* **249**, 3953.
Lehninger, A. L. (1975). "Biochemistry", 2nd edition, p. 208. Worth Publishers, New York.
Leinweber, F.-J., Siegel, L. M. and Monty, K. J. (1965). *Journal of Biological Chemistry* **240**, 2699.
Lewis, C. M. and Fincham, J. R. S. (1970). *Journal of Bacteriology* **103**, 55.
Lewis, N. J. (1975). Ph.D. Dissertation: University of Cambridge, England.
Losada, M. (1974). *In* "Metabolic Interconversion of Enzymes 1973", (E. H. Fischer, E. G. Krebs, H. Neurath and E. R. Stadtman, eds.), p. 257. Springer-Verlag, Berlin.
Losada, M. (1975/76). *Journal of Molecular Catalysis* **1**, 245.
McDonald, D. W. and Coddington, A. (1974). *European Journal of Biochemistry* **46**, 169.
MacDonald, D. W. and Cove, D. J. (1974). *European Journal of Biochemistry* **47**, 107.
MacDonald, D. W., Cove, D. J. and Coddington, A. (1974). *Molecular and General Genetics* **128**, 187.
McElroy, W. D. and Spencer, D. A. (1956). *In* "Symposium on Inorganic Nitrogen Metabolism", (W. D. McElroy and B. H. Glass, eds.), pp. 137–152. Johns Hopkins Press, Baltimore.
McKenna, C., L'vov, N. P., Ganelin, V. L., Sergeev, N. S. and Kretovich, V. L. (1974). *Doklady Biochem.* (English translation: *Dokl Akad. Nauk SSSR. Ser. Biokhim*) **217**, 1.
Martin, R. G. and Ames, B. N. (1961). *Journal of Biological Chemistry* **236**, 1372.
Murphy, M. J., Siegel, L. M., Kamin, H., Der Vartanian, D., Lee, J.-P., LeGall, J. and Peck, H. D. Jr. (1973a). *Biochemical and Biophysical Research Communications* **54**, 82.
Murphy, M. J., Siegel, L. M., Kamin, H. and Rosenthal, D. (1973b). *Journal of Biological Chemistry* **248**, 2801.
Murphy, M. J., Siegel, L. M., Tove, S. R. and Kamin, H. (1974). *Proceedings of the National Academy of Sciences of the United States of America* **71**, 612.
Nason, A. (1962). *Bacteriological Reviews* **26**, 16.
Nason, A. and Evans, H. J. (1953). *Journal of Biological Chemistry* **202**, 655.
Nason, A., Antoine, A. D., Ketchum, P. A., Frazier, W. A. and Lee, D. K. (1970). *Proceedings of the National Academy of Sciences of the United States of America* **65**, 137.
Nason, A., Lee, K-Y., Pan, S-S., Ketchum, P. A., Lamberti, A. and DeVries, J. (1971). *Proceedings of the National Academy of Sciences of the United States of America* **68**, 3242.
Nicholas, D. J. D., Nason, A. and McElroy, W. D. (1953). *Nature, London* **172**, 34.
Nicholas, D. J. D., Nason, A. and McElroy, W. D. (1954). *Journal of Biological Chemistry* **207**, 341.
Nicholas, D. J. D. and Nason, A. (1954a). *Journal of Biological Chemistry* **207**, 353.
Nicholas, D. J. D. and Nason, A. (1954b). *Journal of Biological Chemistry* **211**, 183.

Nicholas, D. J. D. and Stevens, H. M. (1955). *Nature, London* **176**, 1066.
Pan, S-S., Erickson, R. H. and Nason, A. (1975). *Federation Proceedings. Federation of American Societies of Experimental Biology* **34**, 682.
Pan, S-S. and Nason, A. (1976). *Federation Proceedings. Federation of American Societies of Experimental Biology* **35**, 1530.
Pateman, J. A., Cove, D. J., Rever, B. M. and Roberts, D. B. (1964). *Nature, London* **201**, 58.
Pateman, J. A. and Cove, D. J. (1967). *Nature, London* **215**, 1234.
Pateman, J. A., Rever, B. M. and Cove, D. J. (1967). *Biochemical Journal* **104**, 103.
Pateman, J. A., Kinghorn, J. R., Dunn, E. and Forbes, E. (1973). *Journal of Bacteriology* **114**, 943.
Pateman, J. A. and Kinghorn, J. R. (1976). *In* "The Filamentous Fungi", (J. E. Smith and D. R. Berry, eds.), vol. 2, pp. 159–237. John Wiley and Sons, New York.
Payne, W. J. (1973). *Bacteriological Reviews,* **37**, 409.
Prodouz, K. N., Greenbaum, P. and Garrett, R. H. (1977). *Abstracts of the Annual Meeting of the American Society for Microbiology* p. 204.
Scazzocchio, C. (1974). *Journal of Less-Common Metals* **36**, 461.
Schloemer, R. H. and Garrrett, R. H. (1974a). *Journal of Bacteriology* **118**, 259.
Schloemer, R. H. and Garrett, R. H. (1974b). *Journal of Bacteriology* **118**, 270.
Siegel, L. M. (1965). Doctoral thesis: The Johns Hopkins University, Baltimore.
Siegel, L. M., Leinweber, F.-J. and Monty, K. J. (1965). *Journal of Biological Chemistry* **240**, 2705.
Siegel, L. M. and Monty, K. J. (1966). *Biochimica et Biophysica Acta* **39**, 423.
Sorger, G. J. and Davies, J. (1973) *Biochemical Journal* **134**, 673.
Sorger, G. J., Debanne, M. T. and Davies, J. (1974). *Biochemical Journal* **140**, 395.
Sorger, G. J. (1966). *Biochimica et Biophysica Acta* **118**, 484.
Subramanian, K. N., Padmanaban, G. and Sarma, P. S. (1968). *Biochimica et Biophysica Acta* **151**, 20.
Subramanian, K. N. and Sorger, G. J. (1972a). *Biochimica et Biophysica Acta* **256**, 533.
Subramanian, K. N. and Sorger, G. J. (1972b). *Journal of Bacteriology* **110**, 538.
Subramanian, K. N. and Sorger, G. J. (1972c). *Journal of Bacteriology* **110**, 547.
Thompson, S. T., Cass, K. H. and Stellwagen, E. (1975). *Proceedings of the National Academy of Sciences of the United States of America* **72**, 669.
Tomsett, A. B. and Cove, D. J. (1976). *Heredity* **37**, 153.
Vega, J. M. (1976). *Archives of Microbiology* **109**, 237.
Vega, J. M., Greenbaum, P. and Garrett, R. H. (1975a). *Biochimica et Biophysica Acta* **377**, 251.
Vega, J. M., Garrett, R. H. and Siegel, L. M. (1975b). *Journal of Biological Chemistry* **250**, 7980.
Vega, J. M. and Garrett, R. H. (1975). *Proceedings of the Sixth Congress of the Spanish Society of Biochemistry*, p. 108.
Wallace, W. (1974). *Biochimica et Biophysica Acta* **341**, 265.
Wallace, W. (1975). *Biochimica et Biophysica Acta* **377**, 239.

Biochemistry of Dimorphism in the Fungus *Mucor*

PAUL S. SYPHERD, PETER T. BORGIA, AND JOHN L. PAZNOKAS

Respectively of the
Department of Medical Microbiology, University of California,
Irvine, California 92717, U.S.A.,

Department of Microbiology, Southern Illinois University,
Carbondale, Illinois 62901, U.S.A.,

Department of Bacteriology and Public Health,
Washington State University, Pullman, Washington 99163, U.S.A.

I. Introduction	68
II. Dimorphism of *Mucor* spp. at the Cellular Level	69
A. Spore Germination	70
B. Hyphal Growth	71
C. Yeast Growth	71
D. Dimorphic Transitions	71
III. Environmental Conditions which Influence Dimorphism	71
IV. Cell-Wall Structure and Synthesis	75
V. Metabolic Responses to Morphogenic Signals	79
A. A Role for Cyclic Nucleotides in Dimorphism	79
B. Carbon and Energy Metabolism Associated with Dimorphism	84
C. Alterations in Nitrogen Metabolism	87
VI. Protein Synthesis and Control of Morphogenesis	90
VII. Genetics of Mucor	99
VIII. Concluding Remarks	100
IX. Acknowledgements	101
References	101

I. Introduction

The appearance of reviews on species of *Mucor* and the Mucorales over the past 50 years stems more from an interest in the biology of this unusual group of organisms than from the volume of research which has been conducted on them. The purpose of this present review is more to summarize the recent biochemical studies on various aspects of development in *Mucor* spp. than to provide a comprehensive review of the entire group of organisms.

Given the overwhelming success in understanding intermediary metabolism and the molecular biology of transcription and translation, molecular biologists and biochemists have increasingly turned their attention to understanding the biochemical mechanisms of development and differentiation. The higher protists are a rich source of material for such studies since they offer many of the advantages of bacteria (rapid generation times, growth in simple media, and often the availability of a genetic system). What is it about morphogenesis that is so intriguing to the biochemist and the molecular biologist? Many facets captivate the imagination, but the differential expression of genetic information at different phases of growth seems to hold the greatest fascination. Although such studies on *Mucor* spp. are still in their elementary stages, the thrust of this review, where possible, will be to emphasize those processes which may be involved in the selective expression of information encoded in the organisms genome.

Dimorphism has intrigued microbiologists since Pasteur's time, when the transition between yeast and mycelia was used as a basis for the theory of transmutation (the conversion of one species to another). It was Pasteur who showed that the morphological form of *Mucor* sp. was dependent upon the environmental conditions under which the organism was cultured. More recently, interest in fungal dimorphism has centred on its being a model as a morphogenetic system. Dimorphism in *Mucor* spp. presents some advantages for study which do not exist in the more popular eukaryotic systems (e.g. species of *Dictyostelium* and *Neurospora*) or in prokaryotes (such as endospore formation in the bacilli). Developmental studies in micro-organisms have largely been limited to production of spores which is usually in response to nutritional deprivation. In contrast, yeast-hyphal dimorphism in *Mucor* spp. is growth-dependent and in this respect may more closely parallel development in higher eukaryotic cells. Once

BIOCHEMISTRY OF DIMORPHISM IN THE FUNGUS *MUCOR* 69

again, however, more complex developmental processes, such as embryogenesis, are unidirectional, whereas dimorphism of *Mucors* spp. is freely reversible, dependent on a variety of growth conditions.

II. Dimorphism in *Mucor* spp. at the Cellular Level

The asexual life cycle of *Mucor* spp. can be divided into three stages: sporangiospore germination, vegetative growth, and sporangiospore

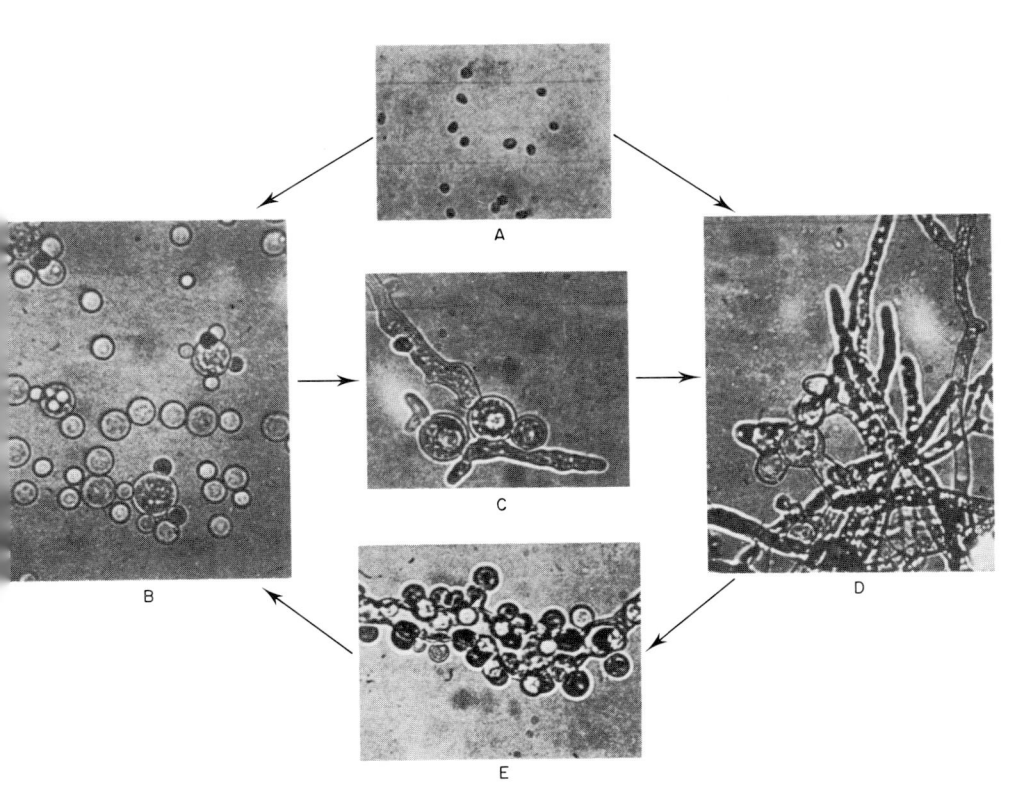

FIG. 1. Dimorphic cycle of *Mucor racemosus*. A sporangiospores; B, yeast; C, yeast-to-mycelial transition; D, mycelia; E, mycelia-to-yeast transition. Magnification ×850.

formation. Figure 1 summarizes some of the morphogenic transitions in *M. racemosus*.

A. SPORE GERMINATION

Sporangiospores of *Mucor* spp. are ellipsoid cells of about 6 × 5 μm (Fig. 1A). In most *Mucor* spp., the spores contain one or two nuclei when examined microscopically (Robinow, 1957). This conclusion has been confirmed by analysis of the phenotypes of spores arising from heterokaryons formed after protoplast fusion (P. Borgia, unpublished results). Germinating spores give rise to either the vegetative yeast cells (Fig. 1B). or hyphal cells (Fig. 1D) depending on environmental conditions during germination (see Section III).

Irrespective of the ultimate morphology, spore germination can be divided into two phases (Bartnicki-Garcia *et al.*, 1968); I, growth of the spore into a larger spherical cell (about 10–18 μm diameter); II, emergence of a germ tube (in the case of hyphal development) or a spherical bud (in the case of yeast development). Phase I represents true growth in that dry weight and protein content increase several-fold during this time (Bartnicki-Garcia *et al.*, 1968). During this phase, a physically and chemically distinct (Bartnicki-Garcia and Reyes, 1964) vegetative cell wall is synthesized *de novo* between the protoplast and the original spore wall. In this period, the surface area of the spore increases considerably with a concomitant decrease in the spore wall thickness. It is not known whether the plasticity of the spore wall is due simply to the expansion force of the vegetative cell or whether enzymic degradation is necessary. If the spore is eventually to develop as a yeast, the new vegetative cell wall is considerably thicker than if the spore is destined to give rise to a germ tube, and differs in its microfibrillar structure.

The events of phase II differ depending upon the ultimate cell morphology. In the case of mycelial development, there is an increasing polarization of cell-wall synthesis which results in the rupture of the spore wall and the emission of an apically growing germ tube (Barnicki-Garcia *et al.*, 1968). The germ-tube wall is contiguous with the vegetative wall under the spore wall. If, on the other hand, the spore is to give rise to a yeast cell, vegetative wall synthesis is not polarized and rupture of the spore wall is followed by emergence of a spherical bud in which synthesis of vegetative cell wall occurs uniformly over the cell surface.

B. HYPHAL GROWTH

Hyphal growth in *Mucor racemosus* (Fig. 1D), as in the other filamentous fungi, is characterized by apical growth of the cell wall (Bartnicki-Garcia and Lippman, 1969). Bartnicki-Garcia (1973) has written an excellent review concerning apical growth of hyphae and has proposed a model to explain the mechanisms involved.

The length, width and degree of branching of Mucor hyphae vary depending upon cultural conditions. In *M. racemosus*, long, narrow hyphae are produced in media containing carbon sources such as xylose, while wider hyphae result from growth on glucose-containing media (P. Borgia, unpublished observations). During hyphal growth, it is common for arthrospores to be produced by septation of the hyphal tube. The arthrospores can sometimes be mistaken for yeast cells since they are spherical However, arthrospores do not form buds.

C. YEAST GROWTH

Yeast growth in *M. racemosus* is characterized by spherical multipolar budding cells with cell-wall synthesis uniformly distributed over the cell surface (Bartnicki-Garcia and Lippman, 1969). Presumably, buds may originate at any location on the mother cell. In fact, it is common to observe more than ten buds arising from a single parental cell.

D. DIMORPHIC TRANSITIONS

Since dimorphism in *Mucor* spp. is freely reversible, morphological shifts in either direction may be performed. When yeast cells are induced to form hyphae, buds are often shed and one or more germ tubes are formed by polarization of cell-wall synthesis (Fig. 1C). When hyphal cells are stimulated to form yeast, spherical cells bud from the sides and tips of the hyphae (Fig. 1E). In morphological shifts in this direction, some hyphal elements always remain in the culture even after long periods, thus making quantitation more difficult.

III. Environmental Conditions which Influence Dimorphism

Several environmental factors influence dimorphism in *Mucor* spp. In general, each factor acts equally well on sporangiospores or vegetative cells. Thus, growth in an atmosphere of carbon dioxide stimulates yeast development from sporangiospores or from hyphal cells. In addition, each of the factors usually acts regardless of whether the cells are grown in liquid or on solid media. Not all dimorphic species of *Mucor* are equally sensitive to every morphopoeitic agent. Consequently, the results of experiments with one species cannot always be extrapolated to another species.

The tendency of a given species of *Mucor* to undergo either hyphal or yeast morphogenesis depends upon a complex interplay of environmental factors including carbon dioxide, oxygen and fermentable hexoses. The importance of the kind of carbon source to morphological transitions in *Mucor* spp. has been repeatedly observed. In general, the yeast morphology requires high concentrations of hexose carbon source, while hyphal development can occur on a variety of substrates including hexoses, disaccharides, pentoses and other C_2 and C_3 compounds. Bartnicki-Garcia and Nickerson (1962b) first noted that yeast cells of *M. rouxii* can be generated only on a hexose (carbon dioxide atmosphere). In *M. racemosus*, hexoses are required for any anaerobic growth. The tendency toward yeast development increases with the concentration of hexose. In a 30% carbon dioxide–70% dinitrogen atmosphere, growth of *M. rouxii* is hyphal at 0.01% glucose, and yeast-like at 1% glucose. In a 100% dinitrogen atmosphere, 8% glucose is necessary for yeast development, while lower concentrations yield a hyphal morphology (Barnicki-Garcia, 1968b). In *M. genevensis*, increased hexose concentrations induce yeast development even under aerobic conditions (Rogers *et al.*, 1974).

The importance of hexoses is strengthened by the fact that various morphopoeitic agents which induce yeast development are active only in the presence of a hexose. Thus, aerobic growth of *M. rouxii* in the presence of phenethyl alcohol (PEA; Terenzi and Storck, 1969) leads to production of yeast morphology provided that glucose is present at 2–5%. The morphology is filamentous when xylose, maltose, sucrose or amino acids are the source of carbon. A similar situation exists when chloramphenicol induces yeast development aerobically in *M. rouxii*, since the effect of the antibiotic is more pronounced if the glucose concentration is raised to 5% (Zorzopulos *et al.*, 1973). *Mucor genevensis* is induced to form aerobic yeast in the presence of chloramphenicol (Rogers *et al.*, 1974; Clark-Walker, 1973) and non-limiting glucose concentrations. When glucose is limiting in continuous culture, mycelial development occurs (Rogers *et al.*, 1974). Both *M. rouxii* and

M. racemosus can be induced to form aerobic yeast in the presence of dibutyryl 3′,5′-cyclic adenosine monophosphate and glucose. Replacement of glucose with a non-hexose carbon source results in the hyphal morphology (Larsen and Sypherd, 1974; Paveto *et al.*, 1975).

Carbon dioxide also plays a role in yeast development in *Mucor* spp., its effects often being complementary with hexoses in anaerobic cultures. At low hexose concentrations, a high partial pressure of carbon dioxide (pCO_2) is needed for complete yeast development and *vice versa* (Barnicki-Garcia, 1968b), while, in media containing 2% glucose, a pCO_2 of 0.3 ($pN_2 = 0.7$) is sufficient to give complete yeast development in *M. rouxii* (Bartnicki-Garcia and Nickerson, 1962a).

Unlike hexose, exogenous carbon dioxide is not required for yeast development. *Mucor rouxii* IM-80 was grown in the yeast form anaerobically (100% dinitrogen atmosphere) in the absence of exogenous carbon dioxide provided that 8–10% glucose was present (Bartnicki-Garcia, 1968b). Under the same gaseous environment, *M. rouxii* N.R.R.L. 1894 required 2% glucose for yeast development (Haidle and Storck, 1966). In media containing 2% glucose and in a dinitrogen atmosphere, *M. racemosus* exhibited either yeast or mycelial morphology which was completely dependent upon the flow rate of dinitrogen through the culture (Mooney and Sypherd, 1976). The stimulatory effect of carbon dioxide on yeast development is evident only in anaerobic cultures; the presence of oxygen overrides the carbon dioxide effect and mycelial development occurs (Bartnicki-Garcia and Nickerson, 1962a).

Stimulation of yeast development by carbon dioxide raises the possibility that carbon dioxide fixation may be important in yeast morphogenesis. *Mucor rouxii* can assimilate $^{14}CO_2$ into a non-dialysable cell component (Bartnicki-Garcia and Nickerson, 1962b). It could be that either exogenous or metabolically generated carbon dioxide can suffice for this purpose, thus explaining the lack of an absolute requirement for this gas. However, it has been reported that the rates of carbon dioxide evolution from cells grown anaerobically in low (hyphal development) and high (yeast development) glucose-containing media are essentially identical (Bartnicki-Garcia, 1968b).

Evidence implicating a volatile factor involved in control of dimorphism in *M. racemosus* has recently been reported (Mooney and Sypherd, 1976). When cultures were grown in glucose-containing liquid media sparged with oxygen-free dinitrogen, the morphology was dependent upon the flow rate of gas through the culture. At flow rates greater than 4 ml/min/ml culture, complete yeast development occurred, while, at flow rates of less than 1 ml/min/ml, mycelia resulted (Fig. 2). The flow-rate effect was independent of glucose concentration in complex media and dinitrogen could be replaced by argon. These results were interpreted as indicating that a volatile compound was produced by cells which either stimulated hyphal development or inhibited yeast development. The higher flow rates of gas resulted in the factor being removed by evaporation. That the compound was simply a component of uninoculated media could be discounted by preflushing experiments. Further evidence for the existence of a volatile factor came from the observation that it can be transferred from one culture to another. Passing the effluent gas through a series of connected cultures caused an accumulation of the compound throughout the series and promoted hyphal development in cultures near the end of the series. Earlier investigators may not have recognized the flow effect since none of the previous studies used a range of rigidly calibrated flow rates. In addition, it was common for dinitrogen to be prepurified through alkaline pyrogallol and/or chromous sulphate (Barnicki-Garcia, 1968b; Elmer and

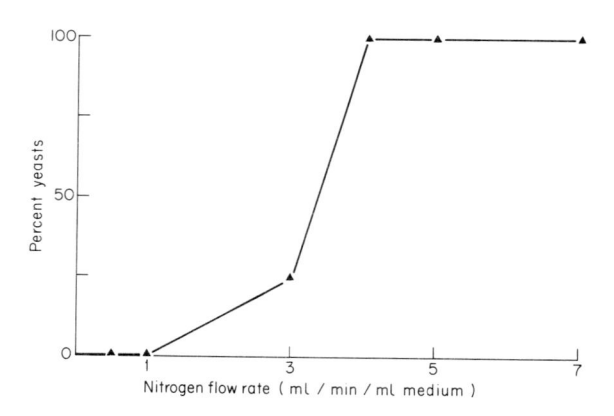

FIG. 2. Effect of nitrogen flow rate on morphogenesis of *Mucor racemosus*. Cultures in 20 ml yeast extract-peptone-2% glucose medium were incubated at 22°C with constant shaking. Gas was bubbled through the medium after being passed over copper filings at 450°C. Morphology was determined microscopically after 24 h growth. From Mooney and Sypherd (1976).

Nickerson, 1970). Since heavy metals are known to influence dimorphism (Barnicki-Garcia, 1968b), contamination may have obscured the flow effect.

The use of different dinitrogen flow rates to elicit dimorphic behaviour promises to be a very useful tool in development studies. In this manner, both morphological forms of the fungus can be generated under almost identical cultural conditions. This is especially true in studies involving differential gene expression, since aerobic-anaerobic shifts to elicit dimorphism would be expected to cause changes in numerous gene products associated with mitochrondriogenesis, the majority of which would not be developmentally significant.

Chelating agents can induce the mycelial morphology in *M. rouxii* grown in glucose-containing media under a carbon dioxide atmosphere (Barnicki-Garcia and Nickerson, 1962b). This effect can be overcome by transition group metals, especially Zn^{2+}. A report (Zorzopulos *et al.*, 1973) claiming that the effect of ethylenediaminetetraacetate (EDTA) is not due to its chelation properties can probably be discounted since the media used in those experiments (supposedly metal deficient) incorporated commercial casein hydrolysate which is known to contain considerable amounts of metal salts. It was also observed that anaerobically-grown cells (100% carbon dioxide atmosphere) were sensitive to EDTA only for the first 20 h following sporangiospore inoculation. Addition of EDTA after this time failed to induce hyphal development (Zorzopulos *et al.*, 1973).

IV. CELL-WALL STRUCTURE AND SYNTHESIS

The prominent role of the cell wall in growth and morphogenesis of fungi needs little elaboration since these areas have been the subject of several reviews in the past (Bartnicki-Garcia, 1968a, 1973; Brody, 1972). Reduced to its simplest form, the focus of the dimorphic process is in the cell wall. Any model for dimorphism must account for regulation in the vectorial orientation of wall synthesis. It is clear in both *Mucor* and *Neurospora* species that a wide variety of environmental factors, and mutations at several loci, cause alterations in the patterns of cell-wall synthesis (Brody, 1972). Consequently, elucidation of the machinery of wall synthesis and of its equally important regulatory mechanisms will be a formidable task.

Bartnicki-Garcia and his coworkers have systematically studied the composition of the cell wall of *M. rouxii* during development. Comparative studies of cell-wall composition include the two vegetative states (yeast and hyphae) and two stages of asexual reproduction (sporangiophores and spores). Table 1 shows the comparative composition of the walls of the four cell types. The qualitative and quantitative composition of the walls of yeasts and hyphae are similar, the predominant polysaccharides being chitin, chitosan and polyuronides. Significant quantitative differences between yeasts and hyphae occur only in the mannose and protein contents. The significance of these changes in dimorphism is now known. *Mucor rouxii* possesses a mannose-containing heteropolysaccharide called mucoran (Bartnicki-Garcia, 1971). Vegetative cell walls from *Mucor racemosus* contain a mannose homopolysaccharide (P. Borgia and P. S. Sypherd, unpublished) whose structure resembles that of mannan from *Saccharomyces cerevisiae* (Ballou, 1974). The presence of chitosan, a polycationic de-acetylated form of chitin, is probably neutralized by the large amounts of polyuronides and phosphate present. Chitin forms a microfibrillar skeleton in the cell wall (Bartnicki-Garcia and Nickerson, 1962c).

While the qualitative composition of the sporangiophore wall is similar to that of vegetative cells, there are significant quantitative differences. The content of polyuronides is twice as high in sporangio-

TABLE 1. Chemical differentiation of the cell wall in the life cycle of *Mucor rouxii*

Wall component	Yeasts	Hyphae	Sporangiophores	Spores
Chitin	8.4[a]	9.4	18.0	2.1[b]
Chitosan	27.9	32.7	20.6	9.5[b]
Mannose	8.9	1.6	0.9	4.8
Fucose	3.2	3.8	2.1	0.0
Galactose	1.1	1.6	0.8	0.0
Glucuronic acid	12.2	11.8	25.0	1.9
Glucose	0.0	0.0	0.1	42.6
Protein	10.3	6.3	9.2	16.1
Lipid	5.7	7.8	4.8	9.8
Phosphate	22.1	23.3	0.8	2.6
Melanin	0.0	0.0	0.0	10.3

[a]Values are percent dry weight of the cell wall. [b]Not confirmed by X-ray analysis. Value for spore chitin represents *N*-acetylated glucosamine; chitosan is nonacetylated glucosamine. From Bartnicki-Garcia (1968b).

phores, while the phosphate content is lower and the chitin/chitosan ratio is altered. The spore wall of *M. rouxii* has a qualitatively unique composition. It is rich in glucan, protein and melanin, and contains chitin and chitosan in lesser amounts than vegetative cells.

Bartnicki-Garcia and his coworkers have studied the pathway, localization and possible regulation of chitin synthesis in *M. rouxii*. Soluble mycelial cell-free extracts catalyse synthesis of the chitin precursor uridine diphosphate *N*-acetylglucosamine (UDP-GlcNAc; McMurrough *et al.*, 1971). Chitin synthetase activity is located predominantly in the cell-wall fraction (about 85% of the activity) with a lesser amount in the "microsomal" fraction. The microsomal enzyme may represent nascent enzyme synthesized prior to deposition in the wall fraction. Disrupted and washed hyphal walls incorporate UDP-GlcNAc into chitin in an apical pattern similar to that *in vivo*. The chitin synthetase is strongly activated by free *N*-acetylglucosamine (GlcNAc). Polyoxin D is a competitive inhibitor of chitin synthetase, and growth of *M. rouxii* in the presence of the antibiotic leads to osmotic sensitivity (Bartnicki-Garcia and Lippman, 1972). Chitosan is enzymically de-acetylated chitin (Araki and Oto, 1974).

The properties of the crude hyphal "microsomal (100,000 g particles) chitin synthetase have been examined (McMurrough and Bartnicki-Garcia, 1971). The enzyme was stimulated by Mg^{2+}, ATP, CTP and GTP. The enzyme shows sigmoidal saturation kinetics with respect to UDP-GlcNAc concentration. The sigmoidal character is modified by the presence of free GlcNAc, indicating a possible role for GlcNAc as a positive allosteric effector.

The chitin synthetase from membrane fractions of *M. rouxii* yeast form can be "solubilized" by incubation with UDP-GlcNAc (Ruiz-Herrera and Bartnicki-Garcia, 1974; Ruiz-Herrera *et al.*, 1975). In the absence of membranes, the enzyme was capable of forming chitin microfibrils resembling those made *in vivo*. This represents the first reported synthesis of microfibrils *in vitro*. When such preparations were examined with the electron microscope, enzyme granules 35–100 nm in diameter (multiple subunits?), with one or two chitin microfibrils originating from the terminal granule, were observed. These observations support the concept that, rather than being synthesized independently, the individual chitin chains are made simultaneously by a multimeric enzyme and assemble while still associated with the enzyme granule. In addition, they indicate that neither membranes nor other amorphous cell-wall components are necessary for microfibril synthesis.

Recent experiments by Bartnicki-Garcia and his colleagues support a regulatory mechanism for chitin synthetase involving proteolytic activation and inactivation (McMurrough and Bartnicki-Garcia, 1973; Ruiz-Herrera and Bartnicki-Garcia, 1976). The activity of chitin synthetase from the "microsomal" fraction of *M. rouxii* increases with storage at low temperature. The activation is blocked by a heat-stable, dialysable inhibitor present in the soluble fraction of the cytoplasm. The sensitivity of chitin synthetase to the inhibitor decreases with the age of the enzyme preparation. The enzyme activation can be mimicked by exogenous protease. Thus, a regulatory mechanism similar to that proposed for chitin synthetase from *Sacch. cerevisiae* (Ulane and Cabib, 1976) may be operative. Chitin synthetase may be present as a "zymogen" which can be activated proteolytically. The inhibitor could serve to block protease action, thus retaining chitin synthetase in the "zymogenic" form. Prolonged proteolysis would lead to chitin synthetase inactivation. Additional supporting evidence for such a model comes from the fact that treatments of crude chitin synthetase which remove endogenous protease activity (detergent treatment) stabilize chitin synthetase activity (Ruiz-Herrera and Bartnicki-Garcia, 1976). Proof for this model will require isolation of the components (chitin synthetase zymogen and active enzyme, protease, and inhibitor). Mutants lacking active protease or protease inhibitor, if they can be isolated, could establish cause and effect relationships.

The relative amounts of activating protease and inhibitor in yeast and hyphal cells differ (Ruiz-Herrera and Bartnicki-Garcia, 1976). In the hyphae, protease levels appear to be high while inhibitor levels are low. The situation in yeast is reversed where activity of the protease is low and of the inhibitor is high. The net result may be to produce a relatively short-lived enzyme in the hyphae and a long-lived enzyme in yeast. Ruiz-Herrera and Bartnicki-Garcia (1976) have proposed that this has morphogenetic implications. Thus, enzyme present in the apical dome of hyphal cells would be rapidly inactivated during the transition from tip wall to lateral wall. In yeast cells, inactivation would be slower and the longer lived enzyme could account for the thicker yeast cell walls (Lara and Bartnicki-Garcia, 1974). This model does not deal with questions of chitin synthetase localization (uniform or apical) nor does it say anything about regulation of the amount of chitin synthesized. There are no substantial differences in the amount of chitin in each unit dry weight between yeast and hyphae (Bartnicki-Garcia and Nickerson, 1962c). It is possible that the long- and short-lived enzymes may be a reflection of controls imposed by the long generation time of anaerobic yeast compared with the shorter generation time of aerobic hyphae used in the studies. This possibility should be investigated.

V. Metabolic Responses to Morphogenic Signals

For the purposes of discussion, one might divide the various aspects of morphogenesis into several parts, such as "signal," "perception of the signal" and "response to the signal". However, the organism may not avail itself of such simplistic divisions, and might well incorporate elements of signal perception into its response to the signal. We have chosen, therefore, to discuss a variety of metabolic activities which are correlated with the morphogenic response, and which may be involved in perceiving a signal, or which may be elementary in leading to the vast series of reactions which must take place in order for the response to be realized by a change in morphology.

A. A ROLE FOR CYCLIC NUCLEOTIDES IN DIMORPHISM

In consideration of the many nutritional and environmental factors which influence morphogenesis in *Mucor* spp., Larsen and Sypherd (1974) saw a parallel with catabolite repression in *Escherichia coli*. The involvement of cyclic 3′,5′-adenosine monophosphate (cAMP) in release of catabolite repression in *E. coli* suggested to them that this nucleotide might play a role in the morphogenic response of *Mucor* spp. By measuring intracellular cAMP levels, Larsen and Sypherd (1974) demonstrated that, very shortly after a cabon dioxide-to-air shift, the cAMP levels decreased 3–4 fold (Fig. 3). This lowering in intracellular cAMP content was followed by a decrease in the fraction of yeast forms and an increase in the fraction of hyphal forms in the population. The implication that cyclic AMP may be involved in morphogenesis was strengthened by experiments in which a lipophilic cAMP derivative (dibutyryl cAMP) was added to hyphal cultures growing in air. When dibutyryl cAMP was added to aerobic cultures,

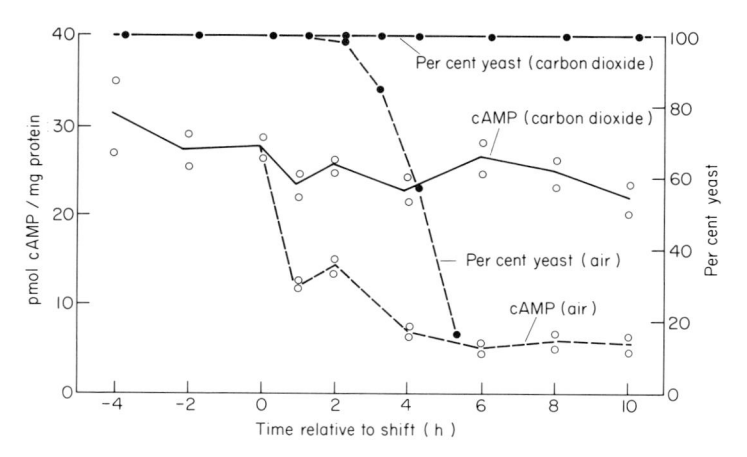

FIG. 3. Intracellular concentration of cAMP and morphology of *Mucor racemosus* during growth in 100% carbon dioxide and after a shift to air. At zero time, the culture was divided and one portion was placed in air. Symbols: ●——●, % yeast cells in carbon dioxide; ●---●, percent yeast cells after being shifted to air; ○——○, cAMP content of cells in carbon dioxide; ○---○, cAMP content of cells shifted to air. From Larsen and Sypherd (1974).

yeast growth was induced and, after several doublings, nearly all of the culture existed in the yeast form. This response to dibutyryl cAMP was found in both *M. racemosus* and *M. rouxii*.

Seiffert and Rudland (1974) proposed that cGMP acts in opposing the regulatory effects of cAMP, and some data have accumulated to support that view (Seiffert and Rudland, 1974; Steiner et al., 1972; Watson, 1975). Orlowski and Sypherd (1976) found that *M. racemosus* also contained cGMP, but there were no major differences in the intracellular levels of cGMP in ungerminated spores, vegetative yeast cells or vegetative hyphae under a variety of conditions. Spores undergoing germination had transiently elevated levels of cGMP which paralleled the changes in intracellular cAMP during this period of growth. Neither exogenously added cGMP nor dibutyryl cGMP had any effect on the morphology of *M. racemosus* in air or under conditions of anaerobic growth. Furthermore, cGMP did not antagonize the cAMP effect on morphogenesis. These results have therefore failed to implicate cGMP in controlling dimorphism of *M. racemosus*. However, the observations of Orlowski and Sypherd (1976) suggest that cGMP may be an important regulatory molecule in spore germination.

Although dibutyryl cAMP can induce yeast formation in air, and yeasts growing in the presence of 100% carbon dioxide have high levels of intracellular cAMP, the crucial question is whether or not high cAMP levels are *required* for yeast morphogenesis. By using the nitrogen gas system which, as pointed out previously, permits growth of *Mucor* spp. in either the yeast or hyphal form dependent upon the flow rate of gas, Paznokas and Sypherd (1975) examined both the intracellular cAMP levels and the effect of added cAMP on yeast morphogenesis. It was found that, under nitrogen gas, intracellular cAMP concentrations are approximately the same in cultures growing in the yeast or hyphal form. Moreover, addition of dibutyryl cAMP to dinitrogen-grown hyphae did not result in a transition to the yeast form. These studies show that morphogenesis is not obligatorily linked to high intracellular levels of cAMP. Moreover, as has been shown previously (Larsen and Sypherd, 1974; Paveto *et al.*, 1975), induction of yeast formation in aerobic cultures of *M. rouxii* or *M. racemosus* is dependent upon adequate concentrations of a metabolizable hexose in the medium. Cyclic AMP does not induce yeast formation in air if the carbon source is glycerol, succinate or maltose (A. Larsen, unpublished results). These experiments suggest that regulation of dimorphism in *Mucor* spp. is due to a variety of physiological processes. Thus, although cAMP induces yeast formation in air and is found in high concentrations in carbon dioxide-induced yeast forms, its precise role in the control of dimorphism is still uncertain. Because of the key role this important nucleotide plays in metabolism of both eukaryotic and prokaryotic cells, account will have to be made of its role in any eventual model of regulation of dimorphism in *Mucor* spp., and perhaps in other dimorphic fungi.

Cyclic AMP is synthesized by the enzyme adenyl cyclase in a reaction which converts ATP to cAMP, and is degraded to AMP by cyclic AMP phosphodiesterase. The intracellular levels of cAMP could theoretically be controlled by altering the activity of either adenyl cyclase or phosphodiesterase. In addition, intracellular levels of cAMP could be regulated by excretion of the nucleotide. Paveto *et al.* (1975) demonstrated the presence of both the cyclase and the phosphodiesterase in extracts of *M. rouxii*, and adenyl cyclase has been found in *M. racemosus* (J. L. Paznokas, unpublished data). The adenyl cyclase of *M. rouxii* is very similar to the enzyme found in *Neurospora crassa* by Flawia and Torres (1972). The enzyme from both species requires Mn^{2+} for activity, while ATP, theophylline and sodium fluoride were without effect on enzyme activity. The adenyl cyclase exhibits sigmoidal saturation kinetics for ATP (J. L. Paznokas, unpublished data), but the specific activity of the enzyme does not vary with fungal morphology. Therefore, modulation of internal cyclic AMP levels does not appear to be brought about by changes in adenyl cyclase activity.

Cyclic AMP phosphodiesterase activity was found in the soluble fraction of extracts of both yeast and mycelia of *M. rouxii* (Paveto et al., 1975). The specific activity of the enzyme was higher in extracts of mycelia (grown with or without glucose) than in extracts of yeast. Although there is a possible correlation between the intracellular levels of cAMP and phosphodiesterase activity, excretion of the nucleotide was also considered as a mechanism for controlling the intracellular levels. Paznokas and Sypherd (1975) examined the supernatant liquid of cultures that had been shifted from carbon dioxide to air for the presence of cAMP, but found none. All of the preformed cAMP could be accounted for inside the cells and there did not appear to be any degradation of pre-existing nucleotide. The decrease in cAMP in *M. racemosus*, therefore, appeared to be due to dilution. Paveto et al. (1975) also did not detect cAMP in the medium from cultures of *M. rouxii*.

The extent to which cAMP is involved in the regulation of dimorphism raises the question of how the nucleotide exerts an effect. One such mechanism could be through cAMP-dependent protein kinases which are known to function in other systems (Walsh and Krebs, 1973; Rubin and Rosen, 1975). Along these lines, Moreno *et al.* (1977) have demonstrated the presence of two protein kinases in *M. rouxii*, a particulate activity which did not respond to the presence of cAMP, and a soluble enzyme which was stimulated by the nucleotide. Moreno *et al.* (1977) have also presented evidence that the soluble protein kinase may have a regulatory subunit which binds cAMP, similar to the situation in some mammalian systems (Rubin and Rosen, 1975). However, further purification of the *M. rouxii* enzyme will be necessary to provide definitive proof that the cAMP-binding protein is the regulatory moiety of the cAMP-dependent protein kinase.

In addition to the findings that *M. rouxii* contains a cyclic AMP-dependent protein kinase, other studies show that cAMP regulates defined biochemical processes in *Mucor* spp. For example, Borgia and Sypherd (1977) reported that the *alpha*-and-*beta*-glucosidases of *M.*

racemosus were inactivated *in vivo* when 7.5 m*M* cAMP was included in the culture medium. Although Sorentino *et al.* (1977) and Flores-Carreon *et al.* (1969) claimed that these enzymes were inducible, Borgia and Sypherd (1977) demonstrated that, in *M. racemosus*, the enzymes are synthesized in the absence of glucose whether or not the specific substrate (inducer) was present. With the *M. racemosus* enzymes, cyclic AMP does not overcome the glucose repression, as it does the glucose repression of β-galactosidase synthesis in *E. coli*, but enhances the glucose repression. Cyclic-AMP alone also inhibits and inactivates preformed α- and β-glucosidase. Borgia and Sypherd (1977) further demonstrated that, when cAMP was removed from cultures whose α- and β-glucosidase had been inactivated, enzyme activity was restored to its pre-inhibited level. It is possible, therefore, that inactivation of glucosidase activity upon addition of cAMP is mediated by a protein kinase. The effect of cAMP on α- and β-glucosidase also explains the inability of *Mucor* spp. to utilize maltose or cellobiose as carbon sources under carbon dioxide since, under these conditions, cAMP levels are at their maximum level. However, the effect of cAMP on glucosidase synthesis and activity does not explain why other carbon sources cannot be utilized under anaerobic conditions.

Another example of a specific effect of cAMP on enzyme synthesis was described by Peters and Sypherd (1978). *Mucor racemosus* produces an NAD⁺-dependent and an NADP⁺-dependent glutamate dehydro-

TABLE 2. Effect of growth medium on synthesis of pyruvate kinase isozymes by *Mucor racemosus*

Atmosphere	Form	Additions[a]	Isozyme activity[b] A	Isozyme activity[b] B
Air	Mycelium	YP	66	990
Air	Mycelium	YP + G	2,950	0
Air	Mycelium	YP + cAMP	82	840
Air	Yeast	YP + G + cAMP	2,580	0
	Yeast	YP + G	4,380	0
Air	Spores	YP + G	1,001	0
Air	Spores	NA	0	1,130
Air	Mycelium	DM	0	578
Air	Mycelium	DM + G	908	0

[a] YP indicates extract-peptone, G glucose, NA nutrient agar, cAMP cyclic adenosine 3′,5′-monophosphate, and DM defined medium. [b] Isozyme activity is expressed as nanomoles of pyruvate formed per minute per milligram of protein. From Paznokas and Sypherd (1977).

genase. The NAD⁺-dependent enzyme is greatly decreased in activity when the organism is grown under carbon dioxide. When the organism is growing aerobically, enzyme synthesis is derepressed. However, synthesis of the NAD⁺-dependent enzyme is repressed by glucose or cAMP. In this case, there is no evidence to suggest that post-translational modification or allosteric modulation is the means for the decrease in enzyme synthesis (Table 3).

Other effects of cAMP have been found by Paveto *et al.* (1975) who showed that the phosphofructokinase of *M. rouxii* is strongly inhibited *in vitro* by ATP and citrate (similar to the effects seen in other tissues by Bloxham and Lardy, 1973) and that the inhibitory effect is reversed by cAMP.

B. CARBON AND ENERGY METABOLISM ASSOCIATED WITH DIMORPHISM

The hexose requirement for yeast development implies that there is a relationship between catabolism, high fermentation rates, and yeast morphology. The relationship is underscored by the isolation of a respiratory-deficient mutant of *M. bacilliformis* which does not form hyphae in air (Storck and Morrill, 1971). Clark-Walker (1972) and Paznokas and Sypherd (1975) established that, unlike the case in *Sacch. cerevisiae*, there is no glucose repression of mitochondriogenesis in *Mucor* sp. Although respiratory *capacity* is high in *M. genevensis* grown aerobically in the presence of excess glucose, respiration rate is low and the cells are essentially fermentative (Rogers *et al.*, 1974). Under these conditions, the morphology is uniformly yeast-like.

In an attempt to determine whether or not there was a change in the major pathways for dissimilation of glucose as a consequence of morphogenesis, C. Inderlied and P. S. Sypherd (unpublished observations) determined the distribution of carbon catabolized in yeast and mycelium. It was found that, although the products of glucose catabolism, namely ethanol, carbon dioxide and glycerol, were produced largely through the Embden–Meyerhof pathway, about 14% of the carbon was metabolized via the pentose phosphate pathway in yeast cells, while about 28% of the catabolized glucose was processed via the pentose phosphate pathway in air-grown mycelium. Thus, it appeared that there was a shift in the route by which glucose was catabolized, dependent upon the morphological form of the organism.

However, when mycelia were produced under anaerobic conditions (i.e. the low-flow dinitrogen system), the distribution of carbon flux in the two pathways resembled that of dinitrogen-grown yeasts. The conclusion from such studies is that, while fermentative activity may be necessary for yeast development, mycelia can also develop without the same general type of catabolic metabolism.

Carbohydrates such as maltose and cellobiose may be cleaved to yield hexoses, yet these carbon sources will not permit growth of *Mucor* sp. under anaerobic conditions. The enzymes which cleave maltose and cellobiose, α- and β-glucosidases, are readily demonstrable under aerobic conditions in the absence of glucose. However, it seems only co-incidental that these enzymes appear in mycelium, and that the enzymes are not linked to the morphogenic process.

Given the absolute requirement for a hexose in the development of the yeast phase of *Mucor* sp., it seems that an analysis of the key enzymes in glucose catabolism might provide useful information about the control of carbon flux and, at the very least, provide some enzymic correlations of the dimorphic process. Pyruvate kinase is known to be a key branch-point enzyme in glycolysis in eukaryotic organisms. Isozymes of pyruvate kinase have been demonstrated in *M. rouxii* by Terenzi *et al.* (1971). Passeron and Roselino (1971) showed that form 1 of pyruvate kinase in *M. rouxii* is synthesized in the presence of glucose, while form 3 is formed only in cells grown on non-hexose carbon sources. While these two isozymes were similar physically, they were immunologically distinct and had quite different kinetic properties. Form 1 was modulated by fructose 1-6-diphosphate and magnesium ions (Passeron and Roselino, 1971; Passeron and Terenzi, 1970) while form 3 was unaffected by fructose 1,6-diphosphate. Form 2 in *M. rouxii* may be a hybrid of forms 1 and 3. Paznokas and Sypherd (1977) demonstrated the existence of two isozymes of pyruvate kinase in *M. racemosus*. These two forms, A and B, were separable by ion-exchange chromatography on DEAE cellulose. Glucose played a role in regulation of the *M. racemosus* enzymes since form A was produced in the presence of glucose, while synthesis of form B was repressed by glucose. If cycloheximide was added simultaneously, the A enzyme was not formed but synthesis of the B enzyme was rapidly repressed. Cycloheximide also blocked induction of form B enzyme. These results were taken to indicate that there was no precursor-product relationship between the two isozymes which, therefore, represented the products of two different genes. The activity of pyruvate kinase may be controlled by enzyme levels in the cell and also by allosteric modulation. Paznokas and Sypherd (1977) examined a variety of cultural conditions to determine if there was a relationship between activity of the pyruvate kinase isozymes and the morphological form of the organism. These results showed that there was not a direct relationship between morphology and the species of pyruvate kinase present. These results are summarized in Table 2 which shows that isozyme A is present in hyphae growing in glucose and air and also in cyclic AMP-induced yeast growing in air when glucose is present in both media. As the table shows, synthesis of form A is dependent upon the presence of glucose. Synthesis of form B is dependent upon the absence of glucose irrespective of the morphology of the organism. Although regulation of pyruvate kinase isozymes is an interesting problem, in this case it appears not to be involved in the morphogenic process.

Despite the fact that pyruvate kinase, one of the key enzymes in glucose catabolism, did not prove to be correlated with morphology, it still seems worthwhile to pursue the study of various enzymes in glucose assimilation based on the observations of Inderlied and Sypherd (1978) and the unpublished observations of J. R. Garcia and V. D. Villa that there is a relationship between yeast development and high fermentative activity. In the latter case, Garcia and Villa found that, when *M. rouxii* is shifted from carbon dioxide to air to induce hyphal development, progressively higher glucose concentrations

TABLE 3. NAD^+ and $NADP^+$-Dependent glutamate dehydrogenase (GDH) activities in *Mucor* sp.

Culture conditions[a]	Morphological form	Specific activity[b] NAD-GDH	Specific activity[b] NADP-GDH
Carbon dioxide, 2% glucose	Yeast	4	1.7
Carbon dioxide, 10% glucose	Yeast	4	0.8
Air, no glucose	Hyphae	390	0.6
Air, 2% glucose	Hyphae	60	0.35
Air, 10% glucose	Hyphae	50	0.35

[a]Medium contained 0.3% yeast extract, 1.0% peptone and glucose at various concentrations.

[b]Enzyme activity was measured by the reductive amination of α-oxoglutarate by following the reduction of absorbance at 340 nm

delayed onset of germ-tube formation and, when once formed, the hyphae had lower degrees of branching with higher concentrations of glucose. These workers have also shown that, as the organism grows in air and progressively higher concentrations of glucose, it accumulates more ethanol and shows a linear increase in the internal concentration of fructose 1,6-diphosphate. It would seem, therefore, that a study of the enzymes glucose 6-phosphate dehydrogenase, 6-phosphogluconate dehydrogenase, phosphofructokinase, and diphosphofructose phosphatase, might provide useful biochemical data on the dimorphic process.

C. ALTERATIONS IN NITROGEN METABOLISM

We have referred previously (p. 72) to the fact that the growth requirements for *Mucor* spp. are considerably more stringent in the yeast than in the hyphal phase (Bartnicki-Garcia and Nickerson, 1961). Cells growing as yeasts require hexose as a carbon source along with a complex organic nitrogen source. On the other hand, hyphal growth occurs on a wide range of carbon sources as well as on ammonium salts as the sole nitrogen source. Dr. Julius Peters in our laboratory devised a defined medium containing 2% glucose with ammonium sulphate, alanine, aspartate and glutamate, each at 10 mM, which supports growth of both yeast and hyphae. Omission of glutamate results in a dramatic retardation of growth rate of the yeast form. Omission of either alanine or aspartate results in a small decrease of growth rate. These results suggest that the control of assimilation of inorganic ammonia is dramatically different in the two forms of the fungus. Peters and Sypherd (1978) studied synthesis of glutamate dehydrogenase, an enzyme which catalyses a key reaction at the juncture of carbon and nitrogen metabolism. In the biosynthetic direction, the enzyme catalyses reductive amination of α-oxoglutarate to form glutamate, and in the catabolic reaction generates ammonia and permits entry of glutamate carbon into the tricarboxylic acid cycle. In species of *Saccharomyces*, *Aspergillus* and *Neurospora*, the anabolic and catabolic reactions are catalysed *in vivo* by distinct enzymes, an NAD^+-dependent one which has primarily a catabolic role and an $NADP^+$-dependent form which is responsible for the biosynthetic reaction. Peters and Sypherd (1978) found that *M. racemosus* contains both the NAD^+- and $NADP^+$-dependent glutamate dehydrogenases. Since the yeast form of *M. racemosus* requires glutamate, it was postulated that the NADP$^+$-dependent enzyme would be repressed in the yeast form (assuming it to be the one responsible for synthesis of glutamate). However, this was not so. As Table 3 shows, the NADP$^+$-dependent enzyme is formed in yeasts growing in carbon dioxide and in mycelia growing in air. The NAD$^+$-dependent enzyme, however, is formed in high amounts in air-grown mycelia but is repressed in carbon dioxide-grown yeasts. Table 3 (p. 86) also shows that synthesis of the NAD$^+$-dependent enzyme is strongly repressed by glucose. Under all conditions examined, the NAD$^+$-dependent glutamate dehydrogenase is derepressed in hyphal-phase cells, while the NADP$^+$-dependent enzyme is found in both cell types. Although these results were somewhat unexpected, the correlation between mycelial development and high rates of synthesis of the NAD$^+$-dependent glutamate dehydrogenase has provided an excellent biochemical correlation with hyphal morphogenesis. Moreover, these results also gave a clue as to the function of these two glutamate dehydrogenases, since it appears that the NAD$^+$-dependent glutamate dehydrogenase is not the major enzyme of ammonia assimilation in *M. racemosus*. These considerations prompted Peters to look for alternative pathways, and it was soon discovered that *M. racemosus* fixes ammonia by the two-step pathway involving glutamine synthetase and glutamate synthase (J. Peters, unpublished data). Peters has also concluded that the NADP$^+$- and NAD$^+$-dependent glutamate dehydrogenases function in a catabolic, rather than anabolic, role since the activities of both enzymes generally are higher in cells growing on organic nitrogen sources than in cells growing on inorganic nitrogen salts. Through an analysis of growth responses in various media, and the properties of several classes of mutants, Peters has also developed an overall scheme for nitrogen metabolism in *M. racemosus*. This scheme is based upon the observation that the NAD$^+$-dependent glutamate dehydrogenase is associated with hyphal morphology under most conditions. The exception to this generalization is that the enzyme is absent from cells growing on either arginine or urea as the sole nitrogen source. Because the NAD$^+$-dependent glutamate dehydrogenase and NAD$^+$-dependent glutamate synthase are present in relatively high amounts in cells growing on ammonium chloride, it seems reasonable that the glutamate synthase-glutamine synthetase system is present in the cytosol, and that the NAD$^+$-dependent glutamate dehydrogenase is a mitochondrial enzyme to provide intramitochondrial ammonia for synthesis of carbamoyl phosphate. Thus, when cells are presented with either arginine or urea, ammonia can be provided by arginase or urease, eliminating the physiological requirements for glutamate dehydrogenase. On the other hand, cells growing on nitrogen-salts and assimilating ammonia through the glutamine synthetase-glutamate synthase system would transport glutamate into the mitochondrion and provide a physiological requirement for glutamate dehydrogenase as the mechanism for generating ammonia intramitochondrially. Although this scheme is speculative, it does provide a reasonable framework for developing a full picture of the interactions between the various nitrogen metabolites and enzymes responsible for nitrogen metabolism, on the one hand, and the morphogenetic transitions on the other.

The analysis of various morphological mutants, such as those which provided information about metabolism of inorganic and organic nitrogen sources, has also brought to light another class of compounds which appear to be involved either as a response to, or as a prelude to, morphogenesis. J. Peters and P. S. Sypherd (unpublished observations) have isolated conditional morphology mutants. One class of such mutants will form hyphae in air only on complex media and, when grown on minimal media, develop in the yeast phase in air. This class of mutants, referred to as *coy* (conditional yeast), were found to respond to methionine equally as well as casein hydrolysate or yeast extract, for development of hyphae. The *coy* mutants do not require methionine for growth, but only for completing the transition from yeast to hyphae. Of the several reactions involving methionine, it seems that its role in synthesis of polyamines via S-adenosylmethionine might prove to be the most fruitful point for investigating the *coy* mutant phenotype. It was subsequently found that putrescine and spermidine synthesis occurred at very low levels in the yeast phase of the wild-type organism but were considerably derepressed in the transition of yeasts to hyphae. Furthermore, it was found that, when the *coy* mutant was transferred from carbon dioxide to air, there was virtually no synthesis of spermidine in the absence of exogenous methionine. On the other hand, when methionine was added, synthesis of both putrescine and spermidine took place with the ensuing development of hyphae. During the analysis of mutant and wild-type extracts for polyamines, a third nitrogenous compound was discovered and shown to be cadaverine, the decarboxylation product of lysine. This compound, identified on thin-layer chromatograms following dansylation, appears transiently after the shift of wild-type cells from carbon dioxide to air. The rapid rise and fall in the concentration of cadaverine may be correlated with a similarly rapid rise and fall in the intracellular concentration of lysine following the shift of yeast cells into air (Orlowski and Sypherd, 1977a). Thus, the following overall pattern emerges. Upon a shift of yeast cells from carbon dioxide to air, the intracellular lysine pools increase 4–5 fold and then decrease to pre-shift levels. During that same interval, the decarboxylation product of lysine, namely cadavarine, also increases and then declines. Following the decline in the concentration of cadavarine, there is a large increase in putrescine and spermidine over the concentration found in carbon dioxide-grown yeasts. Although these data merely show a correlation between the morphogenetic transition and the appearance of three polyamines, it is possible that these compounds, known to have regulatory properties in mammalian cells, play a role in the re-arrangement of metabolic processes to accomplish the morphological change. We discuss one possible role for spermidine in Section X.

VI. Protein Synthesis and Control of Morphogenesis

Cellular morphogenesis involves the interaction of many types of molecules and probably the activation of new genetic information. Although there are alternative views about whether new gene products are involved in morphogenetic responses, similar activities such as embryogenesis and tissue differentiation are so profound as to leave little doubt that gene functions are turned off and on during the developmental process. Thus, one of the fundamental problems in studying morphogenesis is the mechanism for the selective retrieval of genetic information from the organism's genome. Of the myriad of events which accompany differentiation, such as morphological changes, subcellular changes, and changes in the activities of various molecules, it is necessary to identify those which are involved primarily in the massive switching from one form of metabolism to another. One approach to the study of dimorphism in fungi has been to examine in a rather detailed way the focus of that morphogenetic change, namely the cell wall. Another approach which seems more likely to reveal the fundamental mechanisms of *genetic* expression is to study synthesis of macromolecules. With the rather amazing success at describing the control of new enzyme synthesis in prokaryotes at the level of transcription, considerable effort has been spent on the analysis of new transcriptional events during morphogenesis and differentiation. These studies frequently have focused on the activity of the several RNA polymerases which can be recovered from a variety of eukaryotic cells. In general, it is found that multiple forms of this enzyme exist (Roeder and Rutter, 1969) in mammalian cells. Multiple forms of RNA polymerase have also been found in *Sacch. cerevisiae* (Adman *et al.*, 1972) in *M. rouxii* (Young and Whiteley, 1975) and in *Histoplasma* sp. (Boguslawski *et al.*, 1974). The RNA polymerases from eukaryotes invariably initiate transcription on either strand of DNA without regard to specific promoter sites, which results in an apparent nonspecificity. Moreover, with the possible exception of the enzyme from *Histoplasma* sp. (Boguslawski *et al.*, 1974), there is no clear evidence at present which implicates changes in the amounts of these multiple forms as a controlling feature of morphogenesis. In fact, the ratio of the multiple forms of RNA polymerase from *M. rouxii* did not change between the yeast and the mycelial forms (Young and Whiteley, 1975). At present, therefore, attempts to identify changes in RNA polymerases, *sigma*-like factors and other polymerase factors in the regulation of specific messenger-RNA synthesis have been inconclusive.

In attempts to discover the molecular level for expression of new genetic information, attention has also been focused on the translational system. Perhaps a precedent for such studies comes from the discovery that haem functions in controlling the rate of synthesis of rabbit globin (Bruns and London, 1965), and the finding that different species of tRNA transferred leucine into the different positions of the globin chain (Weisblum *et al.*, 1965). Accumulating evidence on the possible role that species of tRNA play in modulation of the rate of transcription, the effects of cytokinins and other hormones, and the various modifications of messenger-RNA such as polyadenylylation and "capping", make even more desirable a thorough understanding of the ways in which the regulation of translation might modify the distribution of gene products in the cell undergoing morphological change.

Investigators interested in the possibility that differential gene expression can take place by regulating the translational process are well versed in the catechism which seems to lay the experimental groundwork for such a notion. Perhaps the earliest indication that elements of the translational system contained qualitative information about the types of message to be translated came from the discovery by Lodish (1969) that ribosomes from *E. coli* were able to translate all three cistrons of the RNA from phage f2. Ribosomes from *Bacillus stearothermophilus*, however, were able to translate only the A gene effectively. Recently, this high degree of specificity has been shown to be a property of one of the ribosomal proteins of the 30 S subunit (Held et al., 1974; Goldberg and Steitz, 1974). In a similar fashion, the ribosomes from *Caulobacter crescentus* can translate Caulobacter phage CB5 RNA but are inactive on the RNA of *E. coli* phage MS2. Conversely, ribosomes from *E. coli* are active on the coli phage MS2 but inactive on CB5 RNA (Szer et al., 1975). Situations like this have led to exciting discoveries about the structure of messenger-RNA in the area surrounding the initiation codons. Moreover, these findings have given credence to the idea that elements of the translational system can be programmed to accept certain classes of messengers or, conversely, that messengers could be modified to be traslated more effectively by the cell's ribosomes.

The possible role of elements of the translational system in selective expression of genetic information has been pointed out in a series of studies performed in *M. racemosus*. In a study of the rates of protein synthesis during the yeast-to-mycelia shift of *M. racemosus,* Orlowski and Sypherd (1977a) showed that, shortly after exposure of cells to air, there was a rapid change in the specific rate of protein synthesis as measured by the incorporation of [^{14}C[L-leucine. The change in rate, even when corrected by the amount of cellular protein, was 3- to 5-fold higher and persisted during the period that germ tubes emerged from the yeast cells. The high specific rates of protein synthesis were followed by a decline until the rates returned to those characteristic of cells growing in carbon dioxide. From a variety of data it was concluded that incorporation of ^{14}C-leucine accurately reflected changing rates of protein biosynthesis and not fluctuations in the intracellular pool of leucine or in the preferential utilization of exogenously supplied amino acid.

The increase in the rate of protein synthesis during a morphogenetic change is not without precedence in other fungal species. The rate of protein synthesis during ascospore formation in *Saccharomyces cerevisiae* increased 6–8 hours following a transfer to sporulation medium (Hopper et al., 1974). Moreover, Lovett (1975) showed that there was a transient increase in the rate of leucine incorporation into protein in *Blastocladiella emersonii* during formation of zoosporangia. Transient increases in protein synthesis also occurred in *Achyla ambisexualis* during the steroid-induced development of antheridia (Timberlake, 1976).

Because of the rapid acceleration in the specific rate of protein synthesis during the transition of yeast to hyphae, Orlowski and Sypherd (1977b) examined the possibility that these changes represented the development of localized areas for high specific rates of protein synthesis. For example, Bartnicki-Garcia and Lippman (1969) demonstrated that new cell-wall material is made only at the hyphal tips of developing germ tubes of *M. rouxii*. By using whole-cell autoradiographic techniques, Orlowski and Sypherd (1977b) demonstrated that protein synthesis occurring during emergence of germ tubes from yeast cells was not restricted to the developing germ tube but occurred in all regions of the cell.

Because of the potential importance of this "burst" of protein synthesis during the transition of *M. racemosus* from yeast to hyphal morphology, Orlowski and Sypherd (1977c) pursued the nature of the acceleration of protein synthesis through an analysis of the distribution of ribosomes between polysomes and inactive subunits. Their results demonstrated that the burst in protein synthesis was accompanied by a concomitant change in both the percent active ribosomes and in the rate of polypeptide chain elongation. Table 4 shows these comparisons. It can be seen that, at the height of the burst (about 2–4 h), the rate of amino-acid addition into peptide linkage had increased from about 2.2 in yeast cells to 6.9 in cells with emerging germ tubes. Conversely, the number of ribosomes engaged with messengers, as represented by the polysome fraction, began to decline from about 82% in yeast to a low of 45% in developing hyphae. Although a decreasing number of active ribosomes and an increasing rate of amino-acid addition into peptide linkage would tend to show cancelling effects, the increase in elongation rate proceeds even faster than the decrease in the percentage of active ribosomes. Separate measurements were made on the time required for completion of a polypeptide chain (the transit time). These data were consistent with those obtained by measuring the rate of polypeptide chain elongation,

TABLE 4. Protein synthesis, rate of polypeptide chain elongation and percent active ribosomes during the dimorphic transition of *Mucor racemosus*

Time relative to shift (h)	Doublings/h	Rate of protein synthesis (c.p.m./min/µg protein)	Polypeptide chain elongation rate (amino acids/sec/ribosome)	Percent ribosomes in polysomes
−4	0.18	3.2	2.3	82
−2	0.18	3.2	2.2	82
0	—	3.1	—	87
1	0.48	8	5.3	81
2	0.48	10	5.3	80
4	0.48	9	6.9	62
6	0.48	7.5	8.2	51
8	0.48	3.8	8.3	45

Data taken from Orlowski and Sypherd (1977a, c).

and showed that the transit time had decreased (i.e. peptides were completed more rapidly) by a factor of three or more from yeast to hyphal growth.

From a study of the rates of protein synthesis alone, therefore, it can be seen that the transition of yeast to hyphal form involves a rather dramatic change in the translocational process to accommodate an acceleration of protein synthesis, brought about by a change in the rate of peptide elongation. It cannot be overemphasized that this burst of protein synthesis, which is caused by a change in the rate of peptide elongation, is not related to the simple fact that the cells grow faster in air. That is, it represents an absolute quantitative increase in the rate of formation of covalent bonds. But, it might reasonably be asked, how can a change in the rate of protein synthesis affect the quality of the proteins being made? Can a change in the rate of synthesis alone be responsible for changing the *ratio* of various gene products (proteins) and therefore affect the morphological state of the organism? Before answering these questions, it might be worthwhile to consider a few of the possibilities. First of all, we would like to distinguish between the gene products which are *developmentally controlled* from those gene products which *control development*. For example, in *Dictyostelium discoideum*, formation of UDP-galactose polysaccharide transferase precedes the appearance of the fruiting body (Loomis and Sussman, 1969). Because this enzyme is required for synthesis of cell-wall material which is part of the fruiting body, it might be regarded as an essential enzyme for morphogenesis. However, such a gene product, although required for morphogenesis to proceed, is probably not involved in any way in the complex array of biochemical signals which initiate the morphogenetic sequence. Such a protein is *developmentally controlled*. Other types of proteins exist, at least theoretically. Any protein that is responsible for interpretation of the environmental signal, for example, by raising the levels of cyclic AMP, polyamines or hormones, or by providing receptors for such molecules, would be one that *controls development*. The only reason for attempting to distinguish between these types of gene products is because their modes of regulation might be quite different. Proteins which are developmentally controlled might be regulated *en bloc* by either transcriptional or translational controls. On the other hand, proteins which control development can easily be imagined to respond by a process analogous to induction where control would be exerted at the transcriptional level in response to a specific stimulus. The second point to consider is that not all messenger-RNAs or all proteins are metabolically equivalent. Some RNAs are translated more rapidly or more slowly than others (Herskowitz and Signer, 1970; Atkins *et al.*, 1975) and some proteins have turnover rates quite different from the average. Lodish (1976) has pointed out that changing the overall rate of protein synthesis with a mRNA which is poorly translated results in a differential effect on the product of that mRNA. Conversely, increasing the overall rate of protein synthesis preferentially increases the rate of translation of the more rapidly translated mRNA. This would result in a change in the ratios of the gene products of these two theoretical mRNAs. In addition, if proteins are turning over at different rates, depending on their function, changing the rate of protein synthesis would also result in a change in the ratios of enzymes which undergo slow or rapid turnover, respectively.

Although we do not have quantitative data on how many gene products might change as a consequence of morphogenesis, or how dramatically such changes would be, it is theoretically possible that a change in the overall rate of peptide-bond formation could result in a change in the ratios of gene products even if no other mechanism were operative. Of course, if one adds to this mechanism the possibility of control at the level of transcription, where certain environmental signals would cause induction or derepression of some gene products, the array of regulatory programmes increases greatly. It is this array, and the manner in which such programmes interact, that present a large and exciting task for future experimentation.

We have been impressed by the rapid quantitative change in the rate of protein synthesis following a shift of *Mucor* sp. from carbon dioxide to air, owing, of course, to the central position that the translational system occupies in the expression of genetic information. Moreover, the correlation between the burst of protein synthesis and the morphological change is quite good. For example, when cells are transferred from carbon dioxide to 100% nitrogen, under conditions which lead to formation of mycelia, the burst of protein synthesis occurs to the same extent as if the cells had been transferred to oxygen for hyphal development. On the other hand, if cells were transferred from carbon dioxide to nitrogen under conditions which lead to continued yeast growth, there was no burst of protein synthesis (Orlowski and Sypherd, 1977a). In addition, morphology mutants respond in a predictable way. In the case of the *coy* mutant, which fails to form hyphae in oxygen in the absence of exogenous methionine, transfer of the organism from carbon dioxide to air in the absence of methionine is not accompanied by the burst of protein synthesis. However, if exogenous methionine is added, with the resultant formation of hyphae (and a concomitant formation of putrescine and spermidine), the burst in protein synthesis occurs (J. Peters, M. Orowski and P. S. Sypherd, unpublished observations). Under the several conditions that we have examined, therefore, the transition of cells from yeast to hyphal form is invariably accompanied by an acceleration in protein synthesis.

Although we have considerable information about the rates of protein synthesis in yeast and hyphal cells, and a change in the *activity* of the translational system during the morphogenetic transition, there is no clear understanding of the mechanism of the acceleration in the rate of peptide-bond formation, since detailed studies of the various components of the translational system have yet to be carried out. Along these lines, the structure of the ribosomes from *M. racemosus* has been studied to some extent. A. Larsen and P. S. Sypherd (unpublished observations) have developed methods for obtaining ribosomal subunits from both phases of the organism. Treating ribosomes with puromycin results in the quantitative separation of the ribosomes into 40S and 60S subunits in sucrose gradients containing 0.8 M NaCl.

Proteins were then extracted by 66% glacial acetic acid and subjected to two-dimensional polyacrylamide gel electrophoresis using the Kaltschmidt-Wittmann (1970) procedure. After staining, it was found that a total of 63 protein "spots" were observed, with 29 of the proteins in the 40S subunit and 34 proteins in the 60S subunit. When the proteins were prepared from cells grown in medium containing $^{32}PO_4$, two proteins of the 40S particle were radioactive. One of these (S6) was present in multiple forms, depending on the extent of phosphorylation (Fig. 4). It was further found that ungerminated spores lack phosphorylated forms of S6 but, as germination proceeded, the protein became highly phosphorylated. The extent of phosphorylation of protein S6 of the small subunit was also found to increase during a carbon dioxide-to-air shift. Moreover, the extent of phosphorylation became maximal at the peak of the burst in protein synthesis and then declined as the rate of protein declined. Although the significance of this correlation is not understood at the present time, it does demonstrate that rather subtle changes in a component of the translational system occur during morphogenesis. An analysis of initiation and elongation factors, and perhaps the distribution of various iso-accepting species of tRNA, is expected to reveal the mechanism for the change in the rate of protein synthesis which accompanies yeast-hyphal morphogenesis. In this regard, other studies should undoubtedly concentrate on the role polyamines might play in acceleration of protein synthesis. It is known from other systems that polyamines stimulate protein synthesis on a variety of messengers, and in at least one case (Atkins et al., 1975) addition of polyamines to an *in vitro* translation system increases the fidelity with which the translation proceeds. It has also been shown that polyamines play a role in chain initiation in the *in vitro* synthesis of globin peptides (Konecki et al., 1975). It has been documented that the structure of mRNA in the region of the initiating codon plays an important role in regulating the extent of translation (Lodish, 1971). One possible role of polyamines in a eukaryotic system could be to modify mRNA structure in the initiating region to alter the rate of translation of various mRNAs. Because of the correlation of high polyamine levels with morphogenesis on the one hand, and an acceleration of protein synthesis on the other, it should be worthwhile to pursue the effect of polyamines on the *in vitro* synthesis of proteins from *Mucor* species.

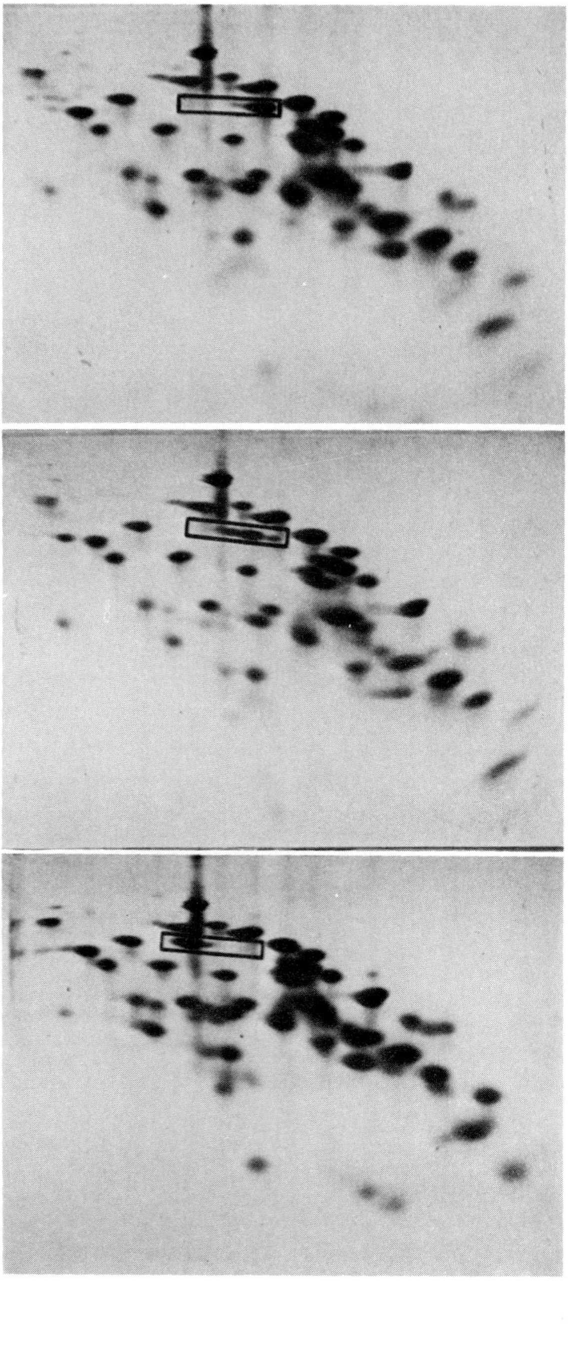

a

b

c

VII. Genetics of Mucor

In large measure, the success of molecular biology has been due to the judicious application of genetic techniques to the solution of biochemical questions. The melding of these two experimental tools is best exemplified in the amazingly detailed pictures of gene-product interaction which have emerged from the bacteriophage (especially T_4 and λ) and the *lac* operon. It may be axiomatic that genetic analyses must accompany a biochemical study in order to gain a full understanding of the molecular events surrounding a process.

The study of morphogenesis in *Mucor* species has the advantage that genetic studies are possible. Many species of *Mucor* are heterothallic and, when mated under suitable conditions, produce sexual spores, called zygospores (Blakeslee, 1904; Burgeff, 1915; Van den Ende, 1968; Plempel, 1963). Sex, in members of the Mucorales, has been surveyed in several reviews (Van den Ende and Stegwee, 1971; Bergman *et al.*, 1969). However, a major problem in the use of recombinational analyses is the period of dormancy which zygospores of *Mucor* species exhibit. This period may range from 27 days for *Rhizopus stolonifer* (Gauger, 1961) to as long as eight months for *Phycomyces blakesleeanus* (Hocking, 1967). Moreover, formation of zygospores in the various *Mucor* species requires exacting conditions which are often difficult to reproduce. The major requirements for genetic analyses in support of biochemical studies of macromolecular metabolism are the availability of mutants and a method for carrying out complementation studies. The latter are critical for answering questions about the possible number of genes controlling a certain process and whether the mutation which has altered a physiological process is dominant or recessive.

Since, to our knowledge, *M. rouxii* does not exhibit heterothallism in the laboratory, we have chosen to investigate dimorphism in *M.*

FIG. 4. Patterns of ribosomal proteins of *Mucor racemosus* following two-dimensional electrophoresis in polyacrylamide gels. (a) Proteins from spore ribosomes; (b) proteins from ribosomes of carbon dioxide-grown yeast cells; (c) proteins from seven-hour germlings grown in air. The spots outlined are S6 and its derivatives. Protein S6 is non-phosphorylated in spores and migrates at the extreme right of the outlined area (a). In carbon dioxide-grown yeasts, S6 is present in the non-phosphorylated state, and the mono- and diphosphorylated state, to give three "spots" (b). In young germlings (c) the protein is all in the triphosphorylated state. Unpublished results of A. Larsen and P. S. Sypherd.

racemosus which does contain + and − strains. A method for selecting both nutritional and morphological mutants of *M. racemosus* was developed by Peters and Sypherd (1977a). The procedure was based on an observation by Duane Mooney that spores of the organism were quite stable to freezing, while hyphal cells rapidly lost viability. Thus, by incorporating freeze-thaw cycles with conditions which prevented germination of mutant spores, Peters and Sypherd (1977a) were able to produce up to 1000-fold enrichment for auxotrophs. This procedure also works well for isolating yeast mutants (those which cannot form hyphae) since yeasts are nearly as resistant to freezing and thawing as are spores. Several strict and conditional morphology mutants have been isolated on our laboratory using this method (J. Peters, unpublished observations).

Formation of heterokaryons for complementation testing was accomplished by Ganthner and Borgia (1977) by adapting the procedure of protoplast fusion which was successful in *Phycomyces* spp. (Binding and Weber, 1974). This technique involves removal of the cell wall under hypertonic conditions with a mixture of chitinase and chitosanase and then fusing the protoplasts by centrifugation. Heterokaryons which showed the proper segregation patterns were isolated on suitable media.

VIII. Concluding Remarks

As promised, this review has not comprehensively covered the existing literature on dimorphism in *Mucor* species. Our goal of providing a review of more recent biochemical approaches to this interesting subject contained a hidden motive, namely to provoke investigators in related fields to see the utility of the Mucor morphogeneic system for examining questions on how selective genetic information-retrieval occurs during development in morphogenesis. In order to obtain a thorough biochemical understanding of this complex process, it will be necessary to pursue descriptions of the morphological transition at the metabolic and molecular level. It might also be necessary to get beyond the descriptive cataloguing of events and consider other features of the cell which might be involved in regulating the perception of environmental signals and the transference of that perception to the metabolic machinery itself. To our knowledge, no studies have been carried out on the possible role of changes in membrane structure and activity in the morphogenic system. The possible role of microtubules and microfilaments has not been examined, and the nature and activity of the purported volatile "hormone" in the triggering of morphogenesis has yet to be elucidated. From studies which have been carried out in the past few years, one general concept has become clear. Morphogenesis does not involve a *sequence* of events but rather a network of interrelated and interdependent metabolic processes. If we are fortunate, the complexity of this network will not be too great to be sorted out and understood in a coherent way. Moreover, if it is important to understand how genes are regulated in the process of development and differentiation, it seems likely to us that microbial systems offer the best hope of describing the events which take place during development and clearly identifying those which are essential and which control the developmental process. The emerging patterns of regulation in the Mucor development system and the rapidity with which new information is being gained, make us more than optimistic about the potential for a detailed understanding of how the complex metabolic activities interrelate to permit a cell to respond to its environment through differential expression of its genetic information.

IX. Acknowledgements

We wish to thank our colleagues who shared their ideas and data with us, without regard to claims of priority or possession. We also acknowledge the support of the Brown-Hazen Fund of the Research Corporation, and the California Heart and Lung Association for their financial support of the earlier phases of our work on Mucor. We also thank the Institute of general Medical Sciences, National Institutes of Health, for their support of work which is reported here for the first time.

REFERENCES

Adman, R., Schultz, L. and Hall, B. (1972). *Proceedings of the National Academy of Sciences of the United States of America* 69, 1702.
Araki, Y. and Oto, E. (1974). *Biochemical and Biophysical Research Communications* 56, 669.
Atkins, J., Lewis, J., Anderson, C. and Gesteland, R. (1975). *Journal of Biological Chemistry* 250, 5688.
Ballou, C. E. (1974). *Advances in Enzymology* 40, 239.

Bartnicki-Garcia, S. (1968a). *Annual Review of Microbiology* **22**, 87.
Bartnicki-Garcia, S. (1968b). *Journal of Bacteriology* **96**, 1586.
Bartnicki-Garcia, S. (1971). *Carbohydrate Research* **23**, 75.
Bartnicki-Garcia, S. (1973). *Symposium of the Society for General Microbiology* **23**, 245.
Bartnicki-Garcia, S. and Lippman, E. (1969). *Science, New York* **165**, 302.
Bartnicki-Garcia, S. and Lippman, E. (1972). *Journal of General Microbiology* **71**, 301.
Bartnicki-Garcia, S., Nelson, N. and Cota-Robles, E. (1968). *Archives of Microbiology* **63**, 242.
Bartnicki-Garcia, S. and Nickerson, W. J. (1961). *Journal of Bacteriology* **82**, 142.
Bartnicki-Garcia, S. and Nickerson, W. J. (1962a). *Journal of Bacteriology* **84**, 829.
Bartnicki-Garcia, S. and Nickerson, W. J. (1962b). *Journal of Bacteriology* **84**, 841.
Bartnicki-Garcia, S. and Nickerson. W. J (1962c). *Biochimica et Biophysica Acta* **64**, 548.
Bartnicki-Garcia, S. and Reyes, E. (1964). *Archives of Biochemistry and Biophysics* **108**, 125.
Bergman, K., Burke, P. V., Cerdá-Olmedo, E., David, C. N., Delbrück, M., Foster, K. W., Goodell, E. W., Heisenberg, M., Meissner, G., Zalokar, M., Dennison, D. S. and Shropshire, W. Jr. (1969). *Bacteriological Reviews* **33**, 99.
Binding, H. and Weber, H. J. (1974). *Molecular and General Genetics* **135**, 273.
Blakeslee, A. F. (1904). *Proceedings of the American Academy of Arts and Sciences* **40**, 205.
Bloxham, D. P. and Lardy, H. A. (1973). *The Enzymes* **8**, 247.
Boguslawski, G., Schlessinger, D., Medoff, G. and Kobayashi, G. (1974). *Journal of Bacteriology* **118**, 480.
Borgia, P. and Sypherd, P. S. (1977). *Journal of Bacteriology* **130**, 812.
Brody, S. (1972). *In* "Developmental Regulation, Aspects of Cell Differentiation", (S. Coward, ed.), p. 107. Academic Press, New York.
Bruns, G. and London, I. (1965). *Biochemical and Biophysical Research Communications* **18**, 236.
Burgeff, H. (1915). *Flora (Tena)* **108**, 353.
Clark-Walker, G. D. (1972). *Journal of Bacteriology* **109**, 399.
Clark-Walker, G. D. (1973). *Journal of Bacteriology* **116**, 972.
Elmer, G. W. and Nickerson, W. J. (1970). *Journal of Bacteriology* **101**, 592.
Flawia, M. M. and Torres, H. N. (1972). *Journal of Biological Chemistry* **247**, 6873.
Flores-Carreon, A., Ryes, E. and Ruiz-Herrera, J. (1969). *Journal of General Microbiology* **59**, 13–19.
Ganthner, F. and Borgia, P. (1977). *Journal of Bacteriology* **134**, 349.
Gauger, W. L. (1961). *American Journal of Botany* **48**, 427.
Goldberg, M. and Steitz, J. (1974). *Biochemistry, New York* **13**, 2123.
Haidle, C. W. and Storck, R. (1966). *Journal of Bacteriology* **92**, 1236.
Held, W., Gette, W. and Nomura, M. (1974). *Biochemistry, New York* **13**, 2115.
Herskowitz, I. and Signer, E. (1970). *Journal of Molecular Biology* **47**, 545–556.
Hocking, D. (1967). *Transactions of the British Mycology Society* **50**, 207.
Hopper, A., Magee, P., Welch, S., Friedman, M. and Hall, B. (1974). *Journal of Bacteriology* **119**, 619.
Inderlied, C. and Sypherd, P. S. (1978). *Journal of Bacteriology* **133**, 1282.
Kaltschmidt, E. and Wittmann, H. (1970). *Analytical Biochemistry* **36**, 401.
Konecki, D., Kramer, G., Pinphanichakarn, P. and Hardesty, B.(1975). *Archives of Biochemistry and Biophysics* **169**, 192–198.
Lara, S. L. and Bartnicki-Garcia, S. (1974). *Archives of Microbiology* **97**, 1.
Larsen, A. and Sypherd, P. S. (1974). *Journal of Bacteriology* **117**, 432.
Lodish, H. (1969). *Nature, London* **224**, 867.
Lodish, H. (1971). *Journal of Molecular Biology* **56**, 627.

Lodish, H. (1976). *Annual Review of Biochemistry* **45**, 39.
Loomis, W. and Sussman, M. (1969). *Journal of Molecular Biology* **22**, 401.
Lovett, J. (1975). *Bacteriological Reviews* **39**, 345.
Mooney, D. T. and Sypherd, P. S. (1976). *Journal of Bacteriology* **126**, 1266.
Moreno, S., Paveto, C. and Passeron, S. (1977). *Archives of Biochemistry and Biophysics* **180**, 225.
McMurrough, I. and Bartnicki-Garcia, S. (1971). *Journal of Biological Chemistry* **246**, 4008.
McMurrough, I. and Bartnicki-Garcia, S. (1973). *Archives of Biochemistry and Biophysics* **158**, 812.
McMurrough, I., Flores-Carreon, A., and Bartnicki-Garcia, S. (1971). *Journal of Biological Chemistry* **246**, 3999.
Orlowski, M. and Sypherd, P. S (1976). *Journal of Bacteriology* **125**, 1226.
Orlowski, M. and Sypherd, P. S (1977a). *Journal of Bacteriology* **132**, 209.
Orlowski, M. and Sypherd, P. S (1977b). *Journal of Bacteriology* **133**, 399.
Orlowski, M. and Sypherd, P. S (1977c). *Biochemistry, New York* **17**, 569.
Passeron, S. and Roselino, E.(1977). *Federation of European Biochemical Societies Letters* **18**, 9.
Passeron, S. and Terenzi, H. (1970). *Federation of European Biochemical Societies Letters* **6**, 213.
Paveto, C., Epstein, E. and Passeron, S. (1975). *Archives of Biochemistry and Biophysics* **169**, 449.
Paznokas, J. L. and Sypherd, P. S. (1975). *Journal of Bacteriology* **124**, 134.
Paznokas, J. L. and Sypherd, P. S. (1977). *Journal of Bacteriology* **130**, 661.
Peters, J. and Sypherd, P. S. (1977a). *Journal of General Microbiology* **105**, 77.
Peters, J. and Sypherd, P. S. (1978). *Journal of Bacteriology* in press.
Plempel, M. (1963). *Planta* **59**, 492.
Robinow, C. F. (1957). *Canadial Journal of Microbiology* **3**, 771.
Roeder, W. and Rutter, W. (1969). *Nature, London* **224**, 234.
Rogers, P. J., Clark-Walker, G. D. and Stewart, P. R. (1974). *Journal of Bacteriology* **119**, 282.
Rubin, C. S. and Rosen, O. M. (1975). *Annual Review of Biochemistry* **44**, 831.
Ruiz-Herrera, J. and Bartnicki-Garcia, S. (1974). *Science, New York* **186**, 357.
Ruiz-Herrera, J. and Bartnicki-Garcia, S. (1976). *Journal of General Microbiology* **97**, 241.
Ruiz-Herrera, J., Sing, V. O., van der Woude, N. and Bartnicki-Garcia, S. (1975). *Proceedings of the National Academy of Sciences of the United States of America* **97**, 241.
Seiffert, W. E. and Rudland, P. (1974). *Nature, London* **248**, 138.
Sorrentino, A. P., Zorzopulos, J. and Terenzi, H. F. (1977). *Archives of Biochemistry and Biophysics* **180**, 232.
Steiner, A. L., Parker, C. W. and Kipnis, D. M. (1972). *Journal of Biological Chemistry* **247**, 1106.
Storck, R. and Morrill, R. (1971). *Biochemical Genetics* **5**, 467.
Szer, W., Hermoso, J. and Leffler, S. (1975). *Proceedings of the National Academy of Sciences of the United States of America* **72**, 2325.
Terenzi, H. F. and Storck, R. (1969). *Journal of Bacteriology* **97**, 1248.
Terenzi, H. F., Roselino, E. and Passeron. S. (1971). *European Journal of Biochemistry* **18**, 342.
Timberlake, W. (1976). *Developmental Biology* **51**, 202.
Ulane, R. E. and Cabib, E. (1976). *Journal of Biological Chemistry* **251**, 3367.
Van den Ende, H. (1968). *Journal of Bacteriology* **96**, 1298.

Van den Ende, H. and Stegwee, D. (1971). *Botanical Review* **37**, 22.
Walsh, D. G. and Krebs, E. G. (1973). *In* "The Enzymes", (P. D. Boyer, ed.), vol. 3, pp. 555–581. Academic Press, New York.
Watson, J. (1975). *Journal of Experimental Medicine* **141**, 97.
Weisblum, B., Gonans, F., von Ehrenstein, G. and Benzer, S. (1965). *Proceedings of the National Academy of Sciences of the United States of America* **53**, 328.
Young, H. and Whiteley, H. (1975). *Journal of Biological Chemistry* **270**, 479.
Zorzopulos, J., Jobaggy, A. and Terenzi, H. (1973). *Journal of Bacteriology* **115**, 1198.

Surface Extension and the Cell Cycle in Prokaryotes

MICHAEL G. SARGENT

National Institute for Medical Research, Mill Hill, London NW7 1AA, England

I.	A Survey of the Problem	106
II.	The Age Distribution in Exponential-Phase Populations	109
	A. Idealized and Real Age Distribution	109
	B. Uses of the Idealized Age Distribution: The Unit Cell . . .	111
	C. The Origin of the Dispersion of Generation Times . . .	112
III.	The Size of Exponential Phase Rod-Shaped Bacteria	117
	A. Observations on Growing Bacteria	117
	B. Interpretation of Size Distributions	118
	C. Factors Determining Average Cell Size	120
	D. Experimental Tests	123
	E. Agents Which Can Synchronize Cell Division	126
IV.	Environmental Effects on Cellular Dimensions in the Steady State .	128
	A. Growth Rate and Medium Composition	128
	B. Temperature	130
	C. Plasmids	130
	D. Morphological Effects Produced by Specific Nutritional Factors .	131
V.	Inferences from Synchronous Cultures and Age-Classified Cells .	132
VI.	Topography of the Bacterial Surface	134
	A. Cocci	135
	B. Rod-Shaped Bacteria	138
	C. Caulobacter and Budding Bacteria	150
VII.	Inferences from Physiological Studies	151
	A. Inhibition of Protein Synthesis	151
	B. Inhibition of DNA Synthesis	151
	C. Inhibition of Peptidoglycan Synthesis	154
VIII.	Genetic Approaches to the Analysis of Surface Extension . . .	156
	A. Rod Mutants	157
	B. Mutations Affecting Cell Width	160
	C. Mutants Affecting Cell Size	160
	D. Spiral Growth Mutants	161

	E. Septation Mutants	161
	F. Minicell-Producing Mutants	162
	G. Mutants Producing Large Anucleate Cells	163
	H. Initiation Mutants	165
IX.	Concluding Remarks	166
X.	Acknowledgments	168
	References	168

I. A Survey of the Problem

In this essay I shall describe the approaches that have been made to understanding the control of bacterial surface area during the cell cycle. Although morphological changes in the surface are in many ways the most obvious manifestation of bacterial growth and division, their relationship with the cell cycle remains obscure. On the other hand a great deal is known about the chemistry of the surface components that maintain cell shape, but this information provides a predominantly static viewpoint that cannot be easily related to surface extension. In general, the information available concerning control of surface area is fragmentary and in many instances inconclusive. This review has been written in the hope that it may help to expose these problems and that it may be superseded in the future by a more substantial work.

Five important facts about the growth and form of rod-shaped bacteria have been known, and widely quoted, for many years. These are: (1) during growth in a steady state, length extension is continuous without change in width; (2) the timing of division is relatively casual with quite large variations in age of individual organisms at division; (3) the variation in size at division is significant, but less so than age at division; (4) division septa are formed almost exactly at the centre of a cell; and (5) average cell length and width vary with the growth conditions.

A great deal of work on bacterial division has been concerned with finding formal explanations for these phenomena. Such investigations have shown that the apparently casual timing of division is a consequence of variation in cell size at division (Koch and Schaechter, 1962) (see Section II). Other theoretical investigations suggest how the average cell length can be determined by the kinetics of surface growth (see Section III). Hypotheses of this kind make predictions of the time of insertion of surface elongation sites and of their mode of operation during the cell cycle. These can be tested by analysis of the variation in cell size under different conditions and by topographic studies of sites of surface growth (see Section VI).

The profound sense of position shown by bacteria in their choice of sites of septum formation is understood only in streptococci, where it is clear that the exact symmetry of septum formation originates in the symmetric surface extension process that precedes it (Higgins and Shockman, 1976). A plausible, but as yet unproven, case can be made for a similar scheme in rod-shaped bacteria, although in the budding bacteria a totally different mechanism must operate (see Section VI-C). It has been proposed that the process of surface extension has an important role in separating newly replicated chromosomes and segregating them into separate packages (Jacob et al., 1963, 1966). If this proves correct, it is a mechanism which imposes very considerable constraints on possible models of surface growth. These are considered in Section VI-B-4.

Although this review is not directly concerned with the control of septum formation, it is none the less pertinent to consider the relationship between surface extension and septum formation. One widely held view maintains that there is a "constitutive" length-extension process punctuated at intervals by septum formation "triggered" by the chromosomal clock. The septum is apparently distinguished from the peripheral wall by chemical differences (Mirelman et al., 1976, 1977) and the fact that there exist mutants blocked specifically in septum formation may indicate an independent control of septum formation (see p. 161, Section VIII-E). In Gram-positive organisms, septal wall components are considerably less susceptible to turnover (Frehel et al., 1971; Fan et al., 1972; Fan and Beckman, 1973; Archibald and Coapes, 1976), while in Gram-negative organisms the incipient septum, though not the completed pole, is exquisitely sensitive to autolysis (Reeve et al., 1972; Burdett and Murray, 1974a, b; and see p. 154).

Another view of this relationship is that the septum is formed as a result of modification of a growth zone, as is clearly the case in streptococci (see p. 134). It has been argued that putative growth zones extend the length of the organism while the internal hydrostatic pressure is high, and grow inwards if the hydrostatic pressure falls (Previc, 1970; Pritchard, 1974). Such a change in "turgor" would occur if the differential rate of envelope synthesis exceeds the differential rate of mass synthesis. Pritchard (1974) has presented experimental evidence that is consistent with this view and has interpreted the behaviour of a number of division mutants on the basis of this hypothesis (see p. 124).

Yet another interpretation of the relationship is suggested by mutants which produce anucleate cells (see p. 163). Anucleate cells of uniform size are formed repeatedly in these mutants without specific coupling to chromosome replication, in a manner which suggests that the organism has a developmental programme which determines the cell size at which septum formation occurs. In the wild-type organism, this developmental sequence may be integrated with the chromosome cycle such that it is started and completed at chromosome termination. A mutant of *Salmonella typhimurium* has been described in which initiation of this developmental sequence may be blocked (Shannon and Rowbury, 1975, and p. 165). It has also been claimed that the antibiotic mecillinam may act in a similar fashion (see p. 154). In certain cases septum formation starts before termination (Woldringh, 1976; Woldringh *et al.*, 1977; Terrana and Newton, 1975), indicating that initiation of septum formation may not be triggered by termination (see p. 151). In streptococci, surface extension and septum formation are both clearly integrated parts of a developmental sequence (see p. 135).

The control of surface extension can be analysed by studying the variation in cell size and shape under different conditions of steady-state growth (see p. 128), and to some extent by the effect of specific inhibitors of protein, DNA, and peptidoglycan synthesis (see p. 151). The large variety of mutants known to have abnormal morphology must all be considered relevant to the study of surface extension (see p. 156) as each class of mutant reveals some aspect of the biological influences determining morphology. Thus, the proportion of septal to peripheral wall can be changed in some mutants (see p. 157), whilst another mutant has been described in which a constraint on length extension leads to increased width (see p. 160). There are also mutants in which length extension and septum formation are no longer co-ordinated with chromosome termination (see pp. 162 and 163). Mutants defective in septum formation can be used to demonstrate that the correct placing of septa does not depend on their position relative to existing septa.

II. The Age Distribution in Exponential-Phase Populations

A. IDEALIZED AND REAL AGE DISTRIBUTION

A fundamental characteristic of exponential-phase cultures of bacteria is that cells of different age do not occur with the same frequency (Powell, 1956). Thus, in such a culture, if all bacteria divide with the *same* generation time τ and cell numbers (n_t) increase such that $n_t = n_0 2^{t/\tau}$, then it follows that the frequency of all age classes relative to new-born cells is $n_{(a)} = 2^{-a}$ (where a = cell age as a fraction of the

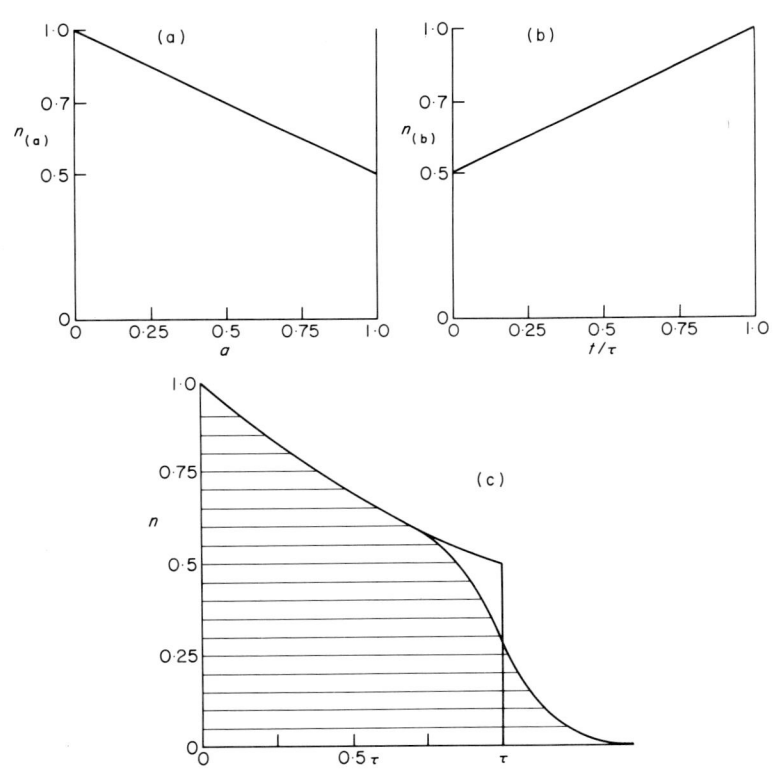

FIG. 1. Idealized and real-age distributions. (a). Frequency of all age classes of bacteria in an exponential-phase population plotted semilogarithmically relative to the number of new-born cells (i.e. $a = 0$), assuming all bacteria divide with the same generation time, i.e. $n_{(a)} = 2^{-a}$. (b). Growth curve that would be sustained if all bacteria in 1(a) divide in order of age (oldest first) i.e. $n_{(t)} = n_{(0)} 2^{t/\tau}$. (c). The idealized and real-age distribution plotted on an arithmetic scale. The real-age distribution is shown cross-hatched.

generation time; see Fig. 1). This relationship is usually known as the "idealized age distribution", as it is based on the assumption that all bacteria divide exactly one generation after the previous division. This is not strictly correct but is sufficiently close to the truth to be practically useful. A convention, used in most cell-division studies, is to regard the moment when daughter bacteria *separate* into independent entities as the termination of a cell cycle. Physiological separation of daughter cells by a completed septum may occur substantially earlier in the cell cycle, but this does not affect the use of the age-distribution theorem, providing the convention used is clearly appreciated.

Paradoxically, it has been understood since the earliest days of microbiology that all bacteria in a population do not have identical generation times (Frankland and Ward, 1895; Adolph and Bayne-Jones, 1932; Kelly and Rahn, 1932). Many early investigators noted a small proportion of bacteria in a growing culture that had very extended interdivision times, and whose daughters had abnormally short generation times. However, reliable estimates of the distribution of generation times under steady-state conditions (referred to hereafter as the τ distribution) were not made for some time (Powell, 1956, 1958; Powell and Errington, 1963; Errington *et al.*, 1965; Kubitschek, 1962, 1966; Schaechter *et al.*, 1962). An expression for the "real age distribution" which incorporated the τ distribution was derived by Powell (1956) and is illustrated in Fig. 1(c) together with the idealized age distribution. The extensive studies of Powell were made by phase contrast microscopy of bacteria growing on a cellophane membrane under which aerated growth medium was continuously circulated. Less sophisticated apparatus has been used by some authors, apparently without deleterious effects (Schaechter *et al.*, 1962). Large numbers of measurements of individual generation times were made on the descendants of single bacteria for up to six successive generations. Frequency distributions of the observed generation times were always unimodal. In most, but not all (Schaechter *et al.*, 1962) investigations, these distributions had a slight positive skew (as in $\lambda(L)$ Fig. 2). Values of the coefficients of variation of τ distributions taken from the literature are shown in Table 1.

The τ distribution has also been estimated in liquid culture by following the time-course of division of a population of bacteria of the same age. This can be done using new-born cells selected by the membrane-elution technique (Shehata and Marr, 1970; Harvey, 1972). These investigations have shown no appreciable skewness in the τ distribution. The possibility that the process of selecting synchronous cells eliminates the skew cannot be dismissed, but it seems more likely that the skewness found in micro-culture experiments is an artefact, possibly caused by micro-environments within the culture vessel that lead to slight increases in cell-separation time (Powell, 1958). However, in general, those using micro-culture have been satisfied that the bacteria were growing in a steady state, with average generation times identical to those in liquid culture and with no overt evidence for a change in average cell size.

B. USES OF THE IDEALIZED AGE DISTRIBUTION: THE UNIT CELL

The time in the cell cycle at which certain cytologically observable events occur, such as septation or nuclear division, can be determined from exponential-phase populations using the idealized age distribution. Thus, from an estimate of the average number (n) of nuclei per cell or cells separated by completed septa, the time elapsing (t) between the event (nuclear division or septum formation) and the moment when daughter bacteria separate into independent entities (i.e. cell separation) can be calculated from the relationship: $\log_2 n = t/\tau$ (see Paulton, 1970). This can be seen intuitively as follows: if at any point in an exponential growth curve there are x cells with, on average, n nuclei per cell, the time (t) is the time required for the cell number to increase to nx (e.g. if $n = 2$, $t = \tau$) and for each nucleus to segregate into a separate cell. The value of t/τ may be greater than unity, as is illustrated by the obvious example of organisms which grow in chains because a long time elapses between septum formation and the separation event (Paulton, 1970). If a culture grows on average in chains, of say eight cells, and each chain, once per generation separates into two, it would take three generations for each unit to segregate into its individual components. Similar arguments apply to all segregating entities such as chromosomes or possible surface markers (Eberle and Lark, 1967; Cooper and Helmstetter, 1968).

The use of the age-distribution theorem is also an essential element in constructing models of the processes determining average cell size (see p. 120). While it is clear that a chain of bacteria is composed of a number of units, it might be expected that, even within these units, there would be further units which are scheduled to divide at a later time. Donachie and Begg (1970) originally proposed that *E. coli* grown at any growth rate were composed of cell units whose completed length was 1.7 μm (see p. 147 for further discussion of evidence). New growth units were thought to start at chromosome initiation. A more precise definition of a "unit cell" is now possible on the basis of current understanding of the control of cell mass and length (see p. 120). Thus there is a unit of length, which when completed in *B. subtilis* and *E. coli* is 2.6–2.8 μm under all growth conditions, and which in both these organisms is started and completed at about the time of nuclear segregation (or chromosome termination) so that the average number of units per exponential-phase cells is equal to $2^{D/\tau}$ (Sargent, 1975a; Woldringh, 1976; Donachie *et al.*, 1976). Each bacterium also contains $2^{C+D/\tau}$ chromosome origins, which must also be regarded as "units" of growth as they ultimately segregate into separate units. There is thought to be a unit of mass associated with each origin (Donachie, 1968; Helmstetter *et al.*, 1968; Pritchard *et al.*, 1969; Bleecken, 1969) (see p. 120). Note that C is the chromosome replication time, and D the time elapsing between termination and the moment of cell separation.

C. THE ORIGIN OF THE DISPERSION OF GENERATION TIMES

Rahn (1932), and later Kendall (1948), proposed that the τ distribution arises because division is dependent on the completion of a fixed number of events (in random or fixed sequence respectively), whose timing in the cell cycle is subject to stochastic fluctuations. In a formal sense, both hypotheses can be modified to fit any data for the τ distribution, but are not based on the behaviour of any real entity in the bacterium. Both authors viewed the postulated sequence of events as a clock mechanism, which was reset after each division so that successive interdivision times were independent of each other. The inadequacy of this formal description of the origin of the τ distribution was demonstrated by the discovery that generation times were not independent. Substantial positive correlations between the generation times of sisters, and weaker (generally negative) correlations between mother and daughter generation times, have been shown by a number of authors (Table 1).

To try and bridge the gap between the stochastic view of cell division and the increasingly deterministic outlook of modern biology, Koch and Schaechter (1962) attempted to account for the τ distribution in

TABLE 1. Correlation coefficients of distribution of generation times and Mother–Daughter and Sister–Sister correlations

Organism	Technique	τ (min)	Coefficient of variation (standard deviation divided by the mean) of τ	Sister–Sister	Mother–Daughter	Reference
Escherichia coli	A	20	0.13	+0.4	−0.2	Powell (1964)
	A	30	0.19	+0.5	+0.12	
Escherichia coli B/r	A	24–32	0.14–0.2	+0.63 to 0.8	−0.18 to +0.54	Schaechter et al. (1962)
	A	27	ND	+0.54 to 0.72	0	Kubitschek (1962)
	B	41.5	0.17	ND	−0.4	Harvey (1972)
Escherichia coli strains	B	54–75	0.18–0.22	ND	ND	Shehata and Marr (1970)
Salmonella typhimurium	A	31–35	0.15–0.22	+0.56 to +0.68	−0.14 to −0.4	Schaechter et al. (1962)
	B	55	0.2	ND	ND	Shehata and Marr (1970)
Aerobacter cloacae	A	24	0.21	+0.6	−0.32	Powell (1956, 1958)
Streptococcus faecalis	A	26–30	0.3	+0.76 to 0.8	+0.14 to +0.18	Powell (1956, 1958)
Bacillus mycoides	A	28	0.5	ND	ND	Powell (1956)
Pseudomonas aeruginosa	A	31–38	0.14	0.58 to 0.78	−0.12 to +0.016	Powell (1956, 1958)

A indicates that the value was determined in microculture by microscopy, B that it was calculated from the time course of cell division in synchronous cultures, and ND that it was not determined.
All values shown were obtained in the temperature range 35–37°C.

terms of other features of the cell cycle. They noted that, whereas the coefficient of variation for the τ distribution was about 0.2, it was only about 0.1 for size at division. From this, they argued that the probability of division was more strongly related to cell size than to age. Using a theoretical model for the division process, the authors showed that the τ distribution could be a consequence of statistical fluctuations in a critical size (or some closely related property) that had to be reached for division to occur.

The negative correlations between the generation times of mothers and daughters, and the positive correlations between the generation times of sisters, have been regarded as tangible evidence of a self-regulating mechanism that compensates for fluctuations in the critical size (Koch and Schaechter, 1962). Thus, the life lengths of sisters may be positively correlated because they arise from a common mother, whose critical mass may have been smaller than average thus giving the daughters a handicap, or, larger than average, thus giving both daughters a head start.

Positive correlations between generation times of cousins and of second cousins were noted by Powell (1958), Powell and Errington (1963), and Errington *et al.* (1965) who regarded them as evidence that the influence of an ancestor was transmitted through at least three generations. Striking support for this was provided by Kubitschek (1962, 1966) and Hoffman and Frank (1965a, b) who showed an astonishing degree of synchrony of division for more than five generations during the development of microcolonies from a single bacterium.

A high degree of synchrony between generations of bacteria descended from the same ancestor is evident in colonial forms. In *Lampropedia hyalina*, a rigid two-dimensional array of cells can be maintained for up to 10 generations, although some distortions in the pattern are noted associated with short or long individual generation times (Kuhn and Starr, 1965). Mutants of species of *Micrococcus* and *Staphylococcus*, in which cell separation is inhibited, grow in ordered cubical arrays at least through six generations (Yamada *et al.*, 1977; Koyama *et al.*, 1977). In chains and filaments of rod-shaped bacteria, nuclear division is usually moderately synchronous (Lin *et al.*, 1971). These observations might suggest that the origin of the τ distribution lay in variation in the cell-separation stage of division. However, Schaechter *et al.* (1962) showed that the coefficient of variation between nuclear interdivision times was similar to that for the interdivision times of the cells, thus indicating that the separation phase does not contribute significantly, at least in this study, to the τ distribution.

An analysis of the relationship between cell size at birth and division and the generation time of *E. coli* by Kubitschek (1962, 1966), showed that if the mothers were abnormally small or big, strong negative correlations were found between mother and daughter generation times, but for the whole population the correlation was weak. The average growth rate (calculated as an exponential rate constant) was constant throughout the range of cell sizes. Generation times were most closely correlated to the generation time of an ancestor (the grandmother) and not the mother. Thus, deviations from the mean generation time in randomly selected bacteria were compensated for by a deviation of opposite sign within 2–3 generations, and furthermore, the variance in the sum of n generations rose linearly for 2–3 generations and then reached a plateau. In contrast, Harvey (1972) concluded that the only correlation between generation times of grandmother and granddaughter is due to their mutual correlation with the daughter cells. A negative correlation of -0.4 was obtained between the generation times of mother and daughter cells. The data were extracted from the time-course of division of a synchronous culture of *E. coli* B/r, obtained using the membrane-elution technique, growing with a slower generation time than that used by Kubitschek (1966) (41 min compared with 27 min).

Marr et al. (1969) suggested that the τ distribution originates from stochastic variations in the time of chromosome initiation and of the time between initiation and cell division $(C + D)$. Thus, if the coefficients of variation of interinitiation times and of $(C + D)$ were 0.072 and 0.075 respectively, these would generate a τ distribution with a coefficient of variation of 0.18. A satisfactory fit to the synchronous growth curve of *E. coli* B/r in glucose-minimal medium could be obtained using these measures of dispersion. They also briefly suggested that the relationship between grandmother and granddaughter, discussed above, could be explained by supposing that chromosome initiation is the agent which determines the compensatory adjustments in generation time. Thus, the bacteria used by Kubitschek (1966) had a mean generation time of 27 min, and the time elapsing between the most highly correlated generations was 54–81 min, covering the period between chromosome initiation and cell division (63 min; Helmstetter and Cooper, 1968). The hypothesis of Marr et al. (1969) does not clearly specify the primary source of the stochastic element in control of cell division. If indeed the initiation process compensates for stochastic fluctuations in generation times, then the source of this "stochastic element" may lie in the mass accumulation of the bacteria rather than in the chromosome cycle. However, they also tentatively suggested that the initiation process itself could be a rate-limiting step with first order kinetics that would generate the τ distribution. Similar views concerning animal cells have been put forward in more detail by Smith and Martin (1973).

There are other possible sources of stochastic variation which would lead to variation in size at division. Both Schaechter et al. (1962), and Errington et al. (1965) noted a variability in rate of length extension between cells (calculated as an exponential rate constant), which was uncorrelated with size at birth and which, in the hands of Errington et al. (1965), had a coefficient of variation of about 0.1. One source of the variance was a small, but statistically significant, difference in size and growth rate between sister cells, which they thought was related to the degree of rounding of the poles formed in previous divisions. The length at which nuclear segregation occurs also has a significant variance in B. subtilis and E. coli (Schaechter et al., 1962; Sargent, 1974; Woldringh, 1976).

Marr et al. (1969) calculated the τ distribution that follows from the theory of Koch and Schaechter (1962) as refined by Powell (1964), and also the time-course of division of a synchronous culture that follows from it. Instead of using the exponential growth curve assumed by Koch and Schaechter (1962), they used a growth curve which they calculated using the procedure of Collins and Richmond (1962; see p. 118). As developed by Marr et al. (1969), the model provides a very poor fit to the data and, in fact, does not predict synchronous growth. However, the conclusions may be suspect as it is not certain that the correct growth curve was used (see p. 118 for a full discussion). Powell (1964) argued that it was not a necessary assumption for the critical size to be attained at division, but that any size which every organism grew through could formally represent the critical size.

It should not be supposed that the attainment of a critical size must formally "trigger" division, but merely that there is a correlation between this size and a high probability of division. In Section III

(p. 117), it will be argued that the cell length attained per chromosomal terminus at nuclear segregation is in *B. subtilis* and *E. coli* a constant over a range of growth rates and may be regarded as the "adult size" of a growth unit; just as a phage or multicellular organism has a characteristic genetically determined "adult size". The existence of certain mutants (see p. 163) which in the absence of DNA synthesis, repeatedly produce anucleate cells of uniform size suggests that there is a developmental programme which determines the size at which septum formation occurs without the benefit of the chromosomal clock.

III. The Size of Exponential Phase Rod-Shaped Bacteria

A. OBSERVATIONS ON GROWING BACTERIA

Rod-shaped micro-organisms extend continuously in length and volume during the growth cycle with no perceptible change in cell width (Collins and Richmond, 1962; Schaechter *et al.*, 1962; Marr *et al.*, 1966; Errington *et al.*, 1965; Adler *et al.*, 1969), but it is not known with any certainty how the *rate* of length extension changes with time. This is a matter of some interest as it must reflect the formal mechanism governing surface extension. For example, if sites of length extension were being added at an exponentially increasing rate, the rate of extension would increase exponentially. Alternatively, if length extension occurred from a fixed number of sites at a fixed rate, then extension would be linear. Most authors have interpreted their observations in terms of exponential growth, probably because it accords well with the widely accepted view that macromolecular synthesis occurs at an exponentially increasing rate (p. 132). However, it is doubtful whether the two models discussed above (or a number of others) could be distinguished by direct observation, since the difference at the widest point is only 6 per cent (Mitchison, 1971). After septation, in many bacteria the pole may round-out and up to 0.5 μm may be added to the length of the cell during the next generation (Errington *et al.*, 1965; Burdett and Higgins, 1978). This will tend to complicate a simple growth pattern.

The average size and the distribution of sizes in an exponentially growing population of bacteria must obviously depend on the precise "growth law" (a term frequently used to dignify the pattern of growth;

Kubitschek, 1970) followed by individual bacteria with respect to age, and the relative number of bacteria in each cell age class. A number of attempts have been made to analyse the observed distributions and average size in terms of particular growth laws. The next two sections are devoted to these studies.

B. INTERPRETATION OF SIZE DISTRIBUTIONS

The range of cell lengths observed in an exponentially growing population of rod-shaped bacteria, owing to the variation in size at division, is always greater than the factor of two which might naively be expected of an organism that divides by binary fission. Theoretical size distributions of organisms that grow by binary fission have been calculated using the idealized age distribution and particular growth laws (Scherbaum and Rasch, 1957; McLean and Munson, 1961). Koch and Schaechter (1962), Koch (1966) and Kubitschek (1969) have made similar calculations using the estimated real-age distributions, and have argued that experimental error in determining the coefficient of variation at division would make it difficult to choose between alternative growth laws on the basis of size distributions.

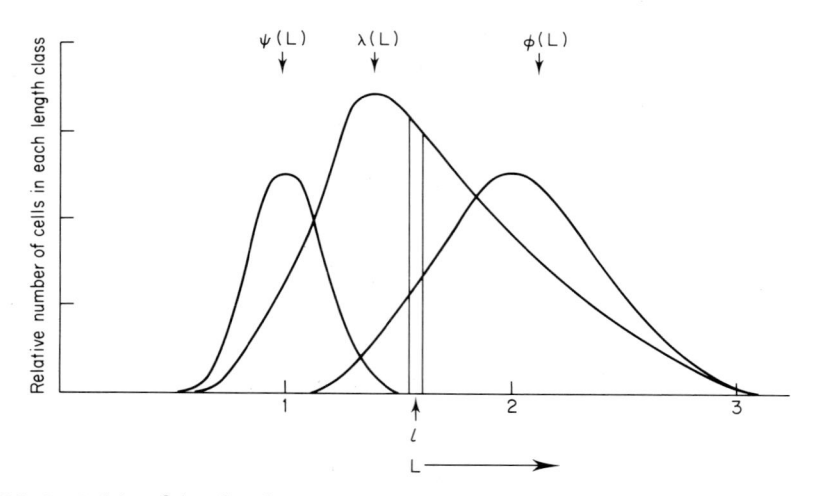

FIG. 2. Origin of size distributions. The distributions $\psi(L)$, $\lambda(L)$ and $\phi(L)$ represent the frequency distributions of cell lengths amongst new-born cells, extant cells and dividing cells, respectively (shown on an arbitrary scale). The number of cells in any size class of $\lambda(L)$ (i.e. l) depends on: (a) the rate at which cells are added from and lost to the adjacent size classes due to length extension; and (b) the number of cells added as new-born cells (given by $\psi(L)$) and the number of cells lost at division (given by $\phi(L)$).

A major advance was made however by Collins and Richmond (1962) when they proposed a method by which the "growth law" could be determined from observed length distributions. Their method also serves as a clear illustration of the origin of length distributions (see Fig. 2). They argued that the frequency of organisms in any length class is determined by three factors: (a) the increase in length of bacteria from the adjacent length classes; (b) the addition of new-born bacteria; and (c) the loss of bacteria due to division. A relatively simple argument then gives a formula for the rate of length increase at any cell length in terms of the three distributions shown in Fig. 2. Terms (b) and (c) were derived from the distribution of lengths at division which were obtained by observation of growing bacteria. The resultant plot of growth rate against cell length was then used to construct a time course of length extension. They concluded, on the basis of size distributions of *Bacillus cereus* involving almost 2000 measurements, that length extension occurred at a continuously increasing rate, but with a rather sharper acceleration in rate than would be expected on the basis of exponential growth and with an apparent decrease in rate immediately after cell division.

Koch (1966) and Harvey *et al.* (1967) have shown that the size distribution, predicted on the basis of linear or exponential growth models and realistic estimates of the τ distribution, would differ little in the mean or variance but substantially in the statistics skewness and kurtosis. Furthermore, very large samples would be needed to provide statistically significant conclusions. The advent of the Coulter electronic particle counter has made it possible to obtain a vastly increased sample which does not require "curve smoothing", although not without introducing fresh problems. Harvey *et al.* (1967) have carried out a more extensive investigation of volume increase of *E. coli* and *Azotobacter agilis*, applying the Collins–Richmond principle to the data obtained from a Coulter counter. The distribution of sizes of dividing and new-born bacteria were obtained from electron micrographs of septate cells and of new-born bacteria obtained by the membrane elution technique (Helmstetter and Cooper, 1968). The size distributions obtained differed markedly in skewness and kurtosis from those expected from an exponential or linear growth law. A plot of rate of volume increase against volume showed a continuous increase in rate of six-fold in the range of sizes between the average new-born bacterium and the average bacterium at division. Furthermore, the growth rate of the largest bacteria was lower than that of average cells at division. Both observations are in marked contrast to those made in the large number of studies of individual growing bacteria described above, and in particular to the findings of Hoffmann and Frank (1965a) who followed the growth of *E. coli* by time-lapse photography. A slight hesitation in growth rate at division was noted that lasted no more than one minute. The application of the Collins–Richmond principle by Harvey *et al.* (1967) therefore gives a picture of growth which is disconcertingly different from that suggested by direct observation.

The most likely source of this discrepancy is error in the observed particle-size distribution brough about by defects in the design of apertures and amplifiers (Grover *et al.*, 1969; Kubitschek, 1969; Thom *et al.*, 1969). The means of distributions can be calculated easily with standard latex beads, but the distribution of pulse heights is usually artefactually broader than microscopically sized standards. Harvey and Marr (1966) were able, using improved circuitry, to obtain distributions which apparently faithfully reflected the size distribution obtained by electron microscopy of a sample of 340 bacteria. However, Kubitschek (1969) has suggested that the path through the aperture used by Harvey *et al.* (1966, 1967) was too short for proper sizing. Kubitschek's (1969) size distributions were clearly less skewed than those of Harvey *et al.* (1967), although detailed statistics were not given.

Zusman *et al.* (1971) have conducted a similar analysis of the growth of *Myxococcus xanthus*, using measured cell lengths and an improvement in the numerical technique by which septate and non-septate bacteria were treated separately. They concluded that the rate of length extension per unit of cell length increased very slightly with length in non-septate cells, and was about 50 per cent greater in septate bacteria. Their data therefore strongly favour exponential growth without the ambiguities generated by the Coulter counter.

C. FACTORS DETERMINING AVERAGE CELL SIZE

1. *Cell Mass or Volume*

A quantitative relationship between average cell mass, or volume, and generation time was given by Donachie (1968) and by Helmstetter *et al.* (1968) who suggested that DNA replication was initiated when the bacterium reached a fixed size per chromosomal origin (Mi). It followed from this that the average mass per cell (M) in an exponential phase population would be $M = \text{Mi} \ln 2 . 2^{C+D/\tau}$. Some support for the constancy of Mi was obtained using values for ($C + D$) from *E. coli* B/r and average cell volumes measured with a Coulter counter (Grover *et al.*, 1977). However, Helmstetter (1974) and Woldringh *et al.* (1977) suggested that values for Mi may not be constant under all conditions.

2. Cell Length and Surface Area

The first explicit suggestion regarding the significance of the increased mass per unit length at high growth rates in enteric bacteria was that of Previc (1970). He proposed that growth zones are found at potential division sites in both rod and coccal forms, and that at termination of chromosome replication the activity of these zones had completed a unit cell length. At termination, surface extension at this site ceases while simultaneously two new growth zones are formed at the next potential division site again associated with chromosomal termini. Zaritsky and Pritchard (1973), and Pritchard (1974), clearly visualized a critical connection between replication of the chromosomal terminus and surface growth. They suggested that their own and other published data were consistent with models of surface extension in which sites of length extension appeared at termination but which could operate in two possible ways. These are: (i) at a constant rate proportional to the mass growth rate of the culture, such that $\bar{L} = K2^{D/\tau}$ where $\bar{L}$ = average length, τ is the generation time, and K is a constant, and that the circumference of the cell varied to accommodate the mass per unit length at the same density; or (ii) the rate of surface area increase was proportional to the output of an unregulated gene located in the middle of the chromosome. Before discussing the evidence for these statements and other experimental observations, it is desirable to discuss the theoretical possibilities in detail.

Zaritsky and Pritchard (1973) gave no formal proof of the postulated relationship between cell length and growth rate, but it can be derived by means of the idealized age distribution (Sargent, 1975a; Grover *et al.*, 1977). The means of distributions derived by the idealized age distribution are not significantly different to those derived using the real distribution (Koch, 1966). In deriving such expressions for the average length of bacteria in an exponential-phase population, the following variables should be considered: (i) The "growth law" followed, i.e. linear or exponential. In the former case, it is assumed that there are n sites of length extension which double in number at a specific time in the cell cycle. (ii) The growth rate per site, which may be a constant under all conditions, may vary in proportion with the mass growth rate of the population, or may vary in a more complex fashion (Pritchard, 1974). (iii) Whether the primary control of rate of length extension is over linear extension only, or over total rate of surface area increase. The former would require that the diameter of the bacterium was controlled by factors independently of length extension. The latter suggests that two variables would be interdependent. (iv) The cell age at which sites appear. In the case of exponential-growth models, this age factor is provided by defining a time in the cell cycle (say, x min before cell separation) at which the number of sites is constant. In all further discussions this is referred to as the "control point".

The relationships between length (or surface area) and growth rate that follow from these postulates are shown in Table 2. It is evident from these that the precise growth law followed is less important than

TABLE 2. Relationship between cellular dimensions and value for τ

Model	Formula
(1) Constant rate of length extension, with doubling in rate x min before cell separation. Rate of length extension proportional to mass growth rate.	$\bar{L} = K2^{x/\tau}$
(2) Exponential growth with cells reaching a constant length x min before cell separation	$\bar{L} = K2^{x/\tau}$
(3) Constant rate of length extension doubling in rate x min before cell separation. Rate of length extension constant.	$\bar{L} = K\tau 2^{x/\tau}$
(4) The rate of surface area increase is proportional to the mass growth rate and to the concentration of a gene located at a point on the chromosome a minutes from the origin	$\overline{SA} = K2^{\frac{C+a+D}{\tau}}$
(5) Growth sites appear at chromosome initiation and operate at a rate proportional to the velocity of chromosome replication and cease length extension at termination.	$\bar{L} = \dfrac{K\tau 2^{\frac{C+D}{\tau}}}{C}$

$\bar{L}$ = average length per cell.
τ = generation time.
$\overline{SA}$ = average surface area.
C = chromosome replication time.
D = time between termination and cell separation.
K = constant of proportionality.

other factors in the functional form of the expressions. For almost any set of postulates, the expression obtained is formally identical on a linear or exponential growth model. Other aspects of the models which can be tested are considered in the following section.

D. EXPERIMENTAL TESTS

1. *Variation in Size with Growth Rate*

Data showing the relationship between growth rate and cell length can be fitted to simple straight line plots to determine the best estimates of the parameters shown (Sargent, 1975a; Grover et al., 1977). However, even with fairly large numbers of data points, only moderate fits to any model are obtained (Zaritsky and Pritchard, 1973; Sargent, 1975a; Zaritsky, 1975; Donachie et al., 1976; Woldringh et al., 1977). The variations from a perfect fit are probably due to growth medium-specific variations in the value of x (defined on p. 122). In *E. coli*, the rather precise *linear* function of growth rate found by Donachie et al. (1976) could only arise if x decreased slightly with increasing growth rate. In view of the variability of x, it is difficult to eliminate models on the basis of goodness-of-fit. There are also differences in the relationship between length, width, and growth rate of a number of strains of *E. coli* B/r (Woldringh et al., 1977). In *B. subtilis*, variation in the time elapsing betwen nuclear division and cell separation is the source of variation in x (Sargent, 1975a). When this is considered, markedly better fits to certain equations are obtained. The data for *B. subtilis* and *E. coli* (Zaritsky and Pritchard, 1973; Sargent, 1975a; Donachie et al., 1976; Grover et al., 1977) favour models 1 and 2 rather than 3. On the basis of models 1 and 2, the x period in *B. subtilis* starts at nuclear division which probably co-incides with chromosome termination (Sargent, 1975a) and for *E. coli* too, the x period probably starts at or close to termination. A consequence of both models is that cell length at the start of the x period is a constant (Sargent, 1975a). In *B. subtilis* (Sargent, 1974, 1975a) and *E. coli* at fast growth rates (Woldringh, 1976) the average cell length at which nuclear divisions occur is approximately constant over a range of growth rates, although there is a range of cell lengths over which nuclear divisions occur.

There is also independent evidence that the rate of length extension per site is dependent on the growth rate as required by models 1 or 2. Thus during inhibition of DNA synthesis in *B. subtilis*, when length extension is linear, a nutritional shift-up causes an immediate increase in the rate of length extension (Sargent, 1975a). This is also true of *S. typhimurium* (Shannon and Rowbury, 1975; see p. 160) and probably of *E. coli* (Donachie *et al.*, 1976), although there are complications in the interpretation of this data (see p. 151).

2. *Effect of Reduced Velocity of DNA Replication*

Pritchard and his colleagues have devised a particularly useful system for investigating the relationship between dimension control and the chromosome cycle. By growing thymine-requiring strains of *E. coli* K12 at concentrations of thymine which decrease the velocity of chromosome replication without affecting the mass growth rate, the interval between chromosome initiation and cell separation was increased while the interval between termination and cell separation was decreased (Pritchard and Zaritsky, 1970; Zaritsky and Pritchard, 1973; Pritchard, 1974; Meacock and Pritchard, 1975). At low concentrations of thymine, when C may change from 41 to 74 min and D from 22 to 13 min, there is a 10 per cent decrease in cell length (Zaritsky and Pritchard, 1973; Meacock *et al.*, 1977). This indicates that the cell age of the "control points" for length extension (i.e. at which x starts) is almost unchanged by the lower velocity of DNA replication. The only point in the chromosome cycle where the time of replication is unchanged relative to cell separation is at, or close to, chromosome termination. Therefore, if length extension is dependent on the replication of a particular part of the chromosome, it is likely to be at termination. As these experiments are performed at constant growth rate, there is no ambiguity in this conclusion due to the "growth rate per site" as there is in a consideration of variation in length with growth rate.

As a result of the increased $(C + D)$ time, the cell mass increases, and the mass per unit length is accommodated by an increase in width (Zaritsky and Pritchard, 1973). Meacock *et al.* (1977) have formulated three alternative hypotheses (models 1, 4 and 5; Table 2) which could account for these and other observations (see p. 121). In order to test these hypotheses, very precise data for the dimensions of the peptidoglycan layer were obtained from measurements of isolated murein sacculi, but all three models appear to be consistent with the data. However, evidence from other sources tends to militate against models 4 and 5. Thus, in a mutant of *B. subtilis* that produces anucleate cells, the rate of length extension can be derepressed at the restrictive conditions in the absence of DNA synthesis (Sargent, 1975b). Although a relationship between the rate of length extension and gene dose at the permissive conditions is not precluded by this observation, it implies a degree of regulation which is the antithesis of the unregulated gene hypothesis (model 4). Model 5, which suggests that surface-extension sites might operate at a rate proportional to the velocity of chromosome replication, is undermined by the observation that, during inhibition of DNA synthesis, length extension continues at an almost unchanged rate which can be varied by the addition of nutrients that stimulate the mass growth rate (Sargent, 1975a; Shannon and Rowbury, 1975; Donachie *et al.*, 1976).

An unexpected, and as yet unexplained, feature of the thymine-limitation experiments is that the D time is shortened when the C time is extended. Previously, it had been supposed that the D period represented the time required to assemble a septum (Helmstetter and Cooper, 1968; Clark, 1968) and, more recently, it has been shown that morphologically identifiable stages in septum formation occur throughout the D period under normal growth conditions (Burdett and Murray, 1974a). There seems to be no evidence that early stages in septation occur prior to chromosome termination, as at low growth rates in *E. coli* and *Caulobacter* sp. (Woldringh, 1976; Woldringh *et al.*, 1977; Terrana and Newton, 1975). Thus, it seems likely that the rate of septum formation is accelerated at low concentrations of thymine, although the volume of septal material to be synthesized is increased. Pritchard (1974) and Meacock and Pritchard (1975) have suggested that this phenomenon is related to the decreased surface-to-volume ratio, and have argued that the *lon* mutant of *E. coli* (which is unable to form septa after ultraviolet irradiation) is defective in this self-regulatory mechanism. In this mutant, unlike the wild-type organism, a decreased DNA:mass ratio does not result in increased girth but to increased "hydrostatic pressure" which prevents septum formation. Wu and Pardee (1973) have noted that the *lon* mutant cannot "swell" as a result of osmotic shock.

Septum formation in *E. coli* may be accelerated, in some circumstances, where protein synthesis is slightly retarded. Also, in certain filamentous mutants, inhibition of protein synthesis may stimulate septum formation. (Zusman *et al.*, 1972; Jones and Donachie, 1973; Normark *et al.*, 1976; Ron *et al.*, 1977.) These effects may occur when the balance between wall synthesis and wall degradation is changed towards synthesis, thus favouring an acceleration in septum formation (see p. 151). In thymine-deficient cells there may be a slight decrease in protein synthesis which could change the balance between wall synthesis and wall degradation in favour of synthesis.

3. *Cell Density*

In each of the above test situations, the findings could equally well be interpreted by an exponential growth model as by a linear growth model. However, a cyclic oscillation in cell density would be expected on a linear growth model (Mitchison, 1971) but not on an exponential model. Thus, if volume increase is linear with doublings in rate at chromosome termination, and macromolecular mass accumulation is exponential, a point of maximum density would be reached at termination provided there are no fluctuations in pools of soluble material. Very few attempts to determine bacterial density have been recorded. Koch and Blumberg (1976) using D_2O gradients have claimed that the range of densities in an exponential phase culture of *E. coli* B/r is so narrow that it must be almost constant throughout the cell cycle. In contrast, Poole (1977) using a novel self-forming gradient system, has provided striking evidence for a fluctuation in cell density within the cell cycle. In *E. coli* grown with a generation time of 44 min, there is a peak in cell density at birth and a minimum in mid-cycle. If a doubling in rate of length extension occurred at chromosome termination, as suggested above, the maximum density would be expected at mid-cycle, in contrast to Poole's (1977) conclusion. There was no indication of whether the water content of the bacteria was affected by the gradient material. Kubitschek (1974) found no variation in the density of exponential-phase cells over a 30-fold range of growth rates.

E. AGENTS WHICH CAN SYNCHRONIZE CELL DIVISION

The earlier literature contains a number of examples of methods of inducing synchronized cell division. If indeed there is a relationship between cell size and the chromosome cycle, it is worth considering whether these methods of inducing cell division offer an opportunity to analyse the relationship between surface growth and the cell cycle. Mitchison (1971) has discussed in detail the rationale by which the action of synchronizing agents may be interpreted in formal terms. Synchronization occurs if the agent causes all of the bacteria in an exponential-phase population to behave as though they were at the same stage in the cell cycle. Such agents may prevent the completion of a particular stage or, alternatively, there may be a "setback" to an earlier time in the cycle. On restoration of normal growth conditions, the organisms proceed synchronously to the next stage in the cell cycle.

Two major block points have been recognized: (a) at the start of septation; and (b) at chromosome initiation. Under a number of conditions where division is inhibited in *E. coli*, the septum may be destroyed (presumably by autolysins), so that on restoration to normal growth conditions there is a synchronous burst of septum formation. This is clearly evident in filamentous mutants (Reeve and Clark, 1972) and in penicillin-induced filaments (Starka and Morava, 1970), and is likely to be the explanation of synchronized division induced by heat shock (Smith and Pardee, 1970), cold shock (McNair-Scott and Chu, 1958; Perry, 1959) and osmotic shock (Wu and Pardee, 1973). Synchronized cell divisions have also been observed at the permissive conditions after exposing chromosome-initiation mutants to the restrictive conditions to allow chromosome completion (Hirota *et al.*, 1968b; Abe and Tomizawa, 1971; Ohki, 1972; Laurent and Vannier, 1973). During both synchronization procedures mass accumulation and length extention continue, so that at division the cells are larger and more heterogeneous than normal (Perry, 1959). Similar observations have been made during induced synchrony in *B. subtilis* initiation mutants (M. G. Sargent, unpublished observation). A second synchronous division is seen in heat and cold shock experiments. This is probably caused by a secondary block on initiation resulting from decreased protein synthesis under the conditions of induction (Smith and Pardee, 1970; Perry, 1959). A similar double block on septation and initiation probably occurs during fluorophenylalanine-induced synchrony in *E. coli* (Brostrom and Brinkley, 1969).

A potentially more instructive example of synchronized division is that induced by amino-acid starvation in *E. coli* K12 and B (Matney and Suit, 1966; Edelmann and Echlin, 1974; Ron *et al.*, 1975, 1977). During inhibition of protein synthesis, rounds of chromosome replication complete and no further rounds initiate. On restoration of amino acids, the first division occurs at a substantially smaller cell volume than would be the case under exponential growth conditions.

During the amino-acid starvation period, the cell volume does not increase, but it must be assumed either that certain processes continue that are not involved in volume increase but are necessary for cell division, or that the time required for assembling the septum decreases. In support of the latter view, Jones and Donachie (1973) have shown that in *E. coli* K12 (under slightly different circumstances) septum assembly occurs in less than five minutes. It would not be surprising if, during inhibition of protein synthesis, when the balance between wall synthesis and autolysis favours synthesis (Rogers, 1967; Tomasz *et al.*, 1970), the time taken to synthesize the septum should shorten. In contrast to *E. coli* K12 and B, *E. coli* strain B/r requires a 40 min period of protein synthesis as well as the normal D period before cell division can occur (Pierrucci and Helmstetter, 1969), so that it is impossible for B/r either to divide synchronously after amino-acid starvation or to divide at a smaller than normal size. There are a number of unexplained published accounts of synchronization procedures which show synchronous division lasting two or more generations (Goodwin, 1969; Cutler and Evans, 1966; Paulton, 1971b).

IV. Environmental Effects on Cellular Dimensions in the Steady State

In the previous section, the variation in dimensions of bacteria under different growth conditions was interpreted in terms of simple relationships between the chromosome cycle and the control of mass and cell length. There are reports, described below, which show the effect of a greater range of environmental conditions on cell size and perhaps indicate the existence of more complex relationships. However, in general, these reports do not include detailed studies of cell dimensions and macromolecular composition, nor has any attempt been made to interpret the observed effects in terms of the controls discussed above.

A. GROWTH RATE AND MEDIUM COMPOSITION

The classic investigations of Schaechter *et al.* (1958) showed that the size and composition of *Salmonella typhimurium* could vary depending on the growth rate. Average cell mass, cell length, and other properties were approximately exponential functions of growth rate, but with differing slopes. The average mass increased more with increasing

SURFACE EXTENSION AND THE CELL CYCLE IN PROKARYOTES 129

growth rate than did length and the increased mass per unit length was accommodated as increased width. Clearly, the dimensions of the organism were controlled by separate factors. These data were obtained with batch cultures grown in many different media and in carbon- or nitrogen-limited chemostats. In later studies these conclusions have, in general, been supported (Herbert, 1961; Sud and Schaechter, 1964; Wright and Lockhart, 1965; Helmstetter et al., 1968; Helmstetter, 1974; Kubitschek, 1974), although further complications have emerged. The variation in cell size of *Aerobacter aerogenes* and *B. megaterium* at low growth rates has been more precisely studied using the chemostat (Herbert, 1961; Dean and Rogers, 1967). Only at the highest growth rates was an exponential relationship observed. At less than 0.5 generations per hour, the average mass per cell was larger than the value obtained if the exponential part of the curve was extrapolated to zero growth rate. Under these conditions, the $(C + D)$ period may be extended and the initiation mass increased (Helmstetter, 1974).

Initially, the relationship between size and growth rate at constant temperature was thought to be unaffected by the nature of the carbon source determining the growth rate, although it was known that, due to glycogen accumulation, nitrogen-limited enteric bacteria could have a greater mass per cell than carbon-limited cells at the same growth rate (Holme, 1957; Ecker and Schaechter, 1963; Wright and Lockhart, 1965). However, Dean and Rogers (1967) and Shehata and Marr (1971) have shown that the average size of *A. aerogenes* and *E. coli* is not uniquely determined by growth rate. Shehata and Marr (1971) grew *E. coli* in batch culture at very low densities in media in which specific nutrients were present in growth-limiting amounts, and measured cell volumes using a Coulter electronic particle counter. Under glucose limitation cell size was a function of the fourth power of the growth rate whereas, under phosphate and tryptophan limitation, the bacteria reached a maximum size at a growth rate of about 0.6 generations per hour. This was twice the size found under glucose limitation. Certain uracil-requiring auxotrophs of *E. coli* can be grown in a steady state on concentrations of orotic acid which limit the growth rate in batch cultures. The cell mass is a constant at all growth rates, and the chromosome replication time is proportional to generation time (Dennis and Herman, 1970). Deviations from an exponential relationship between average cell mass and growth rate have also been noted by Dennis and Bremer (1974).

The relationship between cell length and growth rate has been investigated in batch cultures of *B. subtilis thy⁻* (Sargent, 1975a) and *E. coli* (Donachie *et al.*, 1976; Grover *et al.*, 1977), as described on p. 121 and in *B. megaterium* (Sud and Schaechter, 1964). The effect of growing *thy⁻* strains of *E. coli* at thymine concentrations which limit DNA synthesis but not the mass growth rate has also been considered (see p. 124).

B. TEMPERATURE

Schaechter *et al.* (1958) showed that the average cell size of *S. typhimurium* varied very little in the temperature range 25–37°C, although the growth rate increased by a factor of 2.0. Similarly, in *E. coli* there is no significant difference in average cell volume between organisms grown at 15°C and 30°C, but above and below this range the bacteria becomes slightly smaller or larger respectively (Pierrucci, 1972; Shehata and Marr, 1975). In transitions between 22°C and 30°C, Shehata and Marr (1975) observed a transient change in cell size, which indicated that the processes contributing to cell growth and division did not have identical temperature coefficients and that feedback controls reversed the changes. Lags in their rates of growth have been noted during temperature transitions in *E. coli* (Ryu and Mateles, 1968) and *A. aerogenes* (Topiwala and Sinclair, 1971). In *B. subtilis*, there are similar oscillations in size during temperature changes from 30°C–45°C and also a lag in mass increase (Sargent, 1977b). A number of reports indicate that there is a tendency for several organisms to form filaments at the high and low extremes of their temperature ranges of growth (Hoffman and Frank, 1963; Terry *et al.*, 1966; Shaw, 1968; Bhatti *et al.*, 1976).

C. PLASMIDS

Escherichia coli carrying an R factor has a larger average size than the parent strain (Engberg *et al.*, 1975). The populations have a greater range of cell length and the cells are also wider. The long filaments contain the same number of nuclei per unit length as do the wild-type cells and tend to divide asymmetrically. Although most cells divide at the same length as the parent, a small proportion fail to form septa at this size and continue to grow; subsequent divisions usually occur at other sites.

D. MORPHOLOGICAL EFFECTS PRODUCED BY SPECIFIC NUTRITIONAL FACTORS

In members of certain genera of bacteria, quite distinct morphological forms can be obtained during steady-state growth, by deprivation or provision of specific nutritional factors.

Arthrobacter species can characteristically grow in a steady state either as Gram-positive cocci or as Gram-negative rods, depending on the composition of the growth medium (Krulwich *et al.*, 1967; Clark, 1972; Previc and Lowell, 1975). There appear to be marked chemical differences in wall structure and in metabolism between the two morphological forms. However, as these are associated with large differences in growth rate, it is not certain to which metabolic differences the morphological events are related. The importance of growth rate, rather than specific nutrients, in determining morphology was demonstrated by Luscombe and Gray (1974). Certain *Arthrobacter* strains were grown in chemostats under carbon-source limitation, using a medium which gave rod forms at maximum growth rate but coccal forms at low growth rate.

Although the rod and coccal forms are Gram-negative and Gram-positive, respectively, the wall and septal ultrastructure of both are typical of other Gram-positive organisms, although the rod forms have thinner walls than do the cocci (Ward and Claus, 1973). During a transition from the coccal to rod form, an intermediate stage can be found in which one half of a coccus extends in a cone perpendicular to the plane of the septum. The point of the cone appears to be the site of surface extension, as it has the characteristic thin wall of the rod form, i.e. apical growth (Ward and Claus, 1973). After several generations as a rod form, it still appears to grow unidirectionally (Kolenbrander and Hohman, 1977) as suggested for *E. coli* by Begg and Donachie (1977). Clark (1972) has claimed that filamentous forms are obtained if DNA synthesis is inhibited in the coccal form.

An enormous range of morphologies can be obtained in *Bifidobacterium bifidus* growing exponentially in different media, due to defects which render it dependent on exogenous sources of wall precursors (Sundman and Bjorksten, 1958; Glick *et al.*, 1960; Kojima *et al.*, 1970; Husain *et al.*, 1972). Thus N-acetylglucosamine is said to decrease the morphological aberration in one strain (Glick *et al.*, 1960). In another strain, highly branched forms are obtained during growth in a completely defined medium, which are suppressed by the presence of DL-alanine, DL-aspartate, L-glutamate and DL-serine but not by N-acetylglucosamine. Calcium ions are also required for normal rod form growth (Kojima *et al.*, 1970).

In the presence of vitamin K *Fusiformis nigrescens* grows characteristically as a coccus, but in the absence of the vitamin it grows as long filaments (Lev, 1968). No details of growth rate were given. As in *Arthrobacter* spp., it would be interesting to know how the patterns of growth in the two forms are related.

V. Inferences from Synchronous Cultures and Age-Classified Cells

There is overwhelming agreement, from experiments using synchronous cultures and autoradiography of single bacteria, that RNA and protein (most of the dry weight of the cell) accumulate exponentially during the cell cycle (Dennis, 1971; Ecker and Kokaisl, 1969). However, the kinetics of volume increase and surface extension have proved to be more difficult to study.

In synchronous cultures of *E. coli*, prepared by the membrane-elution technique (Clark, 1968; Ward and Glaser, 1971) or by velocity sedimentation (Kubitschek, 1968a), the cell volume and hence the cell length (as cell width remains constant) increase continuously throughout the cell cycle. Moreover, the rate of increase is claimed to be constant with a doubling in rate at the end of the cycle (Kubitschek, 1968a) or at chromosome initiation (Ward and Glaser, 1971). As with observations of individual cells (see p. 117). it is doubtful whether exponential and linear growth can be distinguished during one generation.

Measurement by pulse labelling of the rate of synthesis of a surface component is an appealing method of estimating the time-course of surface extension. Unfortunately, there is not sufficient reason to believe that any component of the surface can be regarded as an index of surface area. Total peptidoglycan synthesis, for example, cannot be exactly equated with surface extension, as it involves addition to existing glycan chains and initiation of new ones which probably make different contributions to surface extension (see p. 140). However, the rates of synthesis of a number of surface components have been measured in synchronous cultures. Hoffman *et al.* (1972) concluded that peptidoglycan synthesis in *E. coli* B/r measured as D-glutamate incorporation into synchronous cells obtained by the membrane elution technique, occurred at a constant rate with doublings in rate at about the same time as chromosomal termination.

Synthesis of proteins of the plasma membrane has been studied in synchronous cultures of *B. subtilis* obtained by filtration (Sargent, 1973, 1975c), and of *E. coli* obtained by the membrane-elution technique (Churchward and Holland, 1975; Hackenbeck and Messer, 1977). The rate of synthesis of the bulk of the membrane proteins of both organisms is apparently constant at the start of the cycle, increasing sharply at one time in the cycle, and then continuing at a constant rate which is approximately double that in the first period. In *B. subtilis* grown in two media in which nuclear division occurs either at the beginning of the cycle or in mid-cycle, the increase in rate is correlated with the time of nuclear division (Sargent, 1975c). Indirect evidence suggests that nuclear division occurs at chromosome termination in this organism (Sargent, 1975b, c). It has been suggested that in *E. coli* the increase in rate of membrane-protein synthesis occurs either in mid-chromosome cycle (Churchward and Holland, 1975), or at termination (Hackenbeck and Messer, 1977). Although a number of reports have suggested a constant rate of incorporation with doublings in rate at one time, this is not the only possible interpretation of the data. As the cultures are not perfectly synchronous, there may be a more complex pattern of rate changes which is disguised by the distribution of cell ages found at any point. Total lipid synthesis in both *B. subtilis* and *E. coli* appears to occur exponentially throughout the cell cycle (Sargent, 1973; Churchward and Holland, 1975; Hackenbeck and Messer, 1977).

Kubitschek (1968a, b, 1970, 1971) has argued that, although the macromolecular components of bacteria may increase exponentially, the dry weight and volume could increase linearly with discrete doublings in rate at one time in the cell cycle. A similar view was expressed previously by Mitchison and Cummins (1964), who showed that in *Schizosaccharomyces pombe* there was a linear increase in volume while macromolecular synthesis increased exponentially. There was a cyclic oscillation in pools of soluble components such that the total dry weight (macromolecules together with pools) increased linearly in parallel with the volume. Kubitschek (1970) has argued that the mass increase of an organism during any interval of time should equal the uptake of all nutrients used by the organism in that interval. This "equation" is clearly not strictly correct as it does not include a term for the efflux of growth products (such as carbon dioxide), exchange of pool substances or of cell water. None the less, using his velocity sedimentation method of size classification, Kubitschek (1968b) has claimed an apparently constant rate of uptake for a large number of metabolites, although the cyclic oscillation of pool sizes that is an essential feature of the Mitchison hypothesis was not demonstrated. Using synchronous cultures, obtained by velocity sedimentation, the rate of potassium uptake is apparently constant, with doublings in rate at one time in the cycle (Kubitschek *et al.*, 1971). If this is indeed so, the activity of permeases should increase discontinuously at one time in the cycle, in contrast to the continuous synthesis of plasma membrane noted above. Ohki (1972) and Shen and Boos (1973) have claimed that there are abrupt increases in activity of two permeases at cell division in mutants in which a synchronous division can be induced. However, Churchward and Holland (1975) were able to detect only one membrane protein whose synthesis was confined to a short time in the cell cycle.

VI. Topography of the Bacterial Surface

It will be clear from Section III (p. 117) that a knowledge of the number of growth sites in a bacterium, the rate of surface extension per site and the time at which the sites appear in the cell cycle would clarify our understanding of bacterial growth. In this section, methods of studying surface growth will be considered with a view to providing tests for the various theories. However, it is appropriate to consider these observations within a broader context. The almost exact placing of septa at the equator of rod-shaped bacteria indicates a profound, if simple, sense of position within an individual cell. This could be determined by the position relative to pre-existing septa or by proximity to the nucleus prior to septation. However, neither explanation is correct, as there are mutants which form potential division sites in the absence of fully formed septa (filamentous mutants, see p. 161) and mutants which form normal septa at sites without the presence of nucleus (anucleate cell-producing mutants, see p. 163). The absence of any fixed structure within bacteria, other than the envelope, suggests that the information determining the location of septa and chromosomes is likely to be provided by the pattern of surface growth.

A. COCCI

Cole and Hahn (1962) in their studies of *Streptococcus pyogenes* provided the first convincing direct demonstration of the pattern of surface growth in any micro-organism. By growing the organism in the presence of a fluorescent antibody (reactive principally against the group A-specific antigen, the M protein) and then transferring the bacteria to media containing non-fluorescent antibody, non-fluorescent areas appeared in the furrow and, to a smaller extent, at the circumference of the widest point in each bacterium (i.e. the equator). During subsequent growth, the fluorescent areas remained bright, but the non-fluorescent area in the furrow increased until septum formation. The minor (non-fluorescent) equatorial bands initially enlarged and became increasingly furrowed. The reciprocal experiment of observing the growth of unlabelled cells in fluorescent antibody confirmed that new growth occurred in the furrow. Similar conclusions were reached subsequently by Chung *et al.* (1964b) using the same technique with the same organism and by Wagner (1964) using *Diplococcus pneumoniae*.

Using electron microscopy, Shockman and his colleagues have confirmed and radically extended these conclusions. In negatively stained walls and thin sections of *Strep. faecalis*, a prominent band is observed traversing the equator at the site at which furrowing starts (Shockman and Martin, 1968; Higgins *et al.*, 1970). During subsequent growth, the band divides and surface extension occurs symmetrically between the two bands with only a very small change in surface area of the old poles. The newly inserted wall is laid down initially as an almost linear structure at an angle of about 60° to the equator (Fig. 3b) (Higgins and Shockman, 1970, 1976). In newly formed peripheral wall (P in Fig. 3b), the wall is thickest near to the equatorial band, and thinnest towards the centre of the furrow. This indicates that extension of the surface occurs from the centre of the groove and not from the equatorial band, thus corroborating the conclusion of Cole and Hahn (1962). A small amount of septal wall appears immediately after the new growth zone is formed, but thereafter does not increase significantly until the peripheral wall is almost complete. The peripheral wall becomes increasingly curved as it approaches its maximum length, and at this time the septum starts to close.

Walls formed during threonine starvation of *Strep. faecalis* are extremely thick. On restoration of threonine, new wall of normal thickness is laid down at a notch formed in the equatorial band while the previously formed thickened wall remains unchanged, thus demonstrating unequivocally that the equatorial band is present at the junction of old and new surface (Higgins *et al.*, 1971). In the next generation, a new growth site is formed at the junction of thick and thin surface. The evidence was strengthened further by the demonstration that newly synthesized peptidoglycan was preferentially released by autolysis (Shockman *et al.*, 1967) and that sites of autolysis could be correlated with the notch on the surface observed microscopically (Higgins *et al.*, 1970). Autoradiography has been used by Briles and Tomasz (1970) to demonstrate the conservation of older parts of the wall of *Diplococcus pneumoniae*, in confirmation of the conclusion drawn by Wagner (1964).

The growth of the surface relative to cell volume of *Strep. faecalis* has been determined very accurately by means of computer-assisted three-dimensional reconstructions of the volume occupied by the wall, made from the data provided by median sections (Higgins and Shockman, 1976). The increase in surface area during the cell cycle is approximately proportional to cell volume. It has not yet been possible to reconstruct the time-course of wall synthesis. Earlier studies (Mitchison, 1961), with light and interference microscopy, suggested that volume increase was sigmoidal and that cells reached a maximum density at furrow formation. This may indicate that mass increase is exponential throughout the cell cycle of *Streptococcus* but is not matched by volume increase in the later stages of septum formation.

The growth of cocci which divide in more than one plane has not been studied in detail. Autolysates of *Staphylococcus aureus* contain hemispherical shells which seemed to have originated from localized dissolution of the wall in an equatorial band around the coccus (Mitchell and Moyle, 1957). A thickened band was observed in some cells, crossing the equator at right angles. Possibly, as in streptococci, this marks the next site of surface growth. Recently published scanning electron micrographs of staphylococci and micrococci which are unable to separate, have revealed rigid cuboidal arrays of up to 64 cells, showing that, in six consecutive divisions, the succession and orientation of planes is rigidly maintained (Yamada *et al.*, 1977; Koyama *et al.*, 1977). The irregular planes of division, which were formerly thought to characterize division of staphylococci, are now

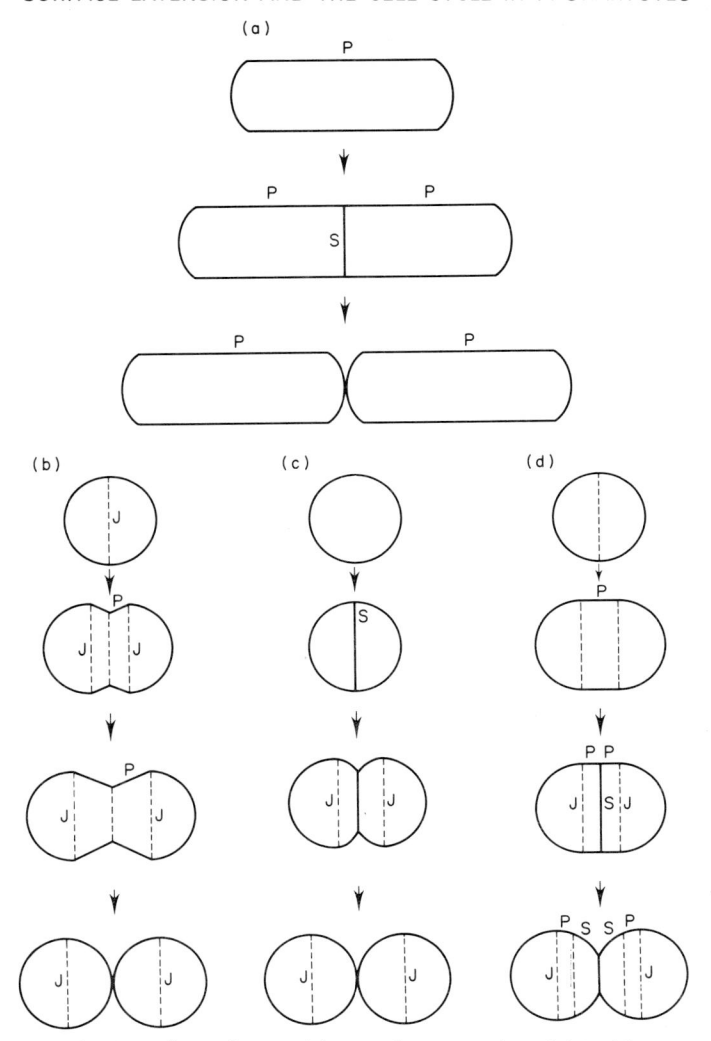

FIG. 3. Contribution of septal material to surface extension. (a) *Rod forms*. Peripheral wall (P) (85% of total) is synthesized throughout the cycle. The septum is morphologically distinct and synthesized during a restricted part of the cell cycle. (b). *Streptococci*. Peripheral wall (P) is laid down throughout the cell cycle. The next generation of division sites are formed at the junction of old and new wall (J). (c). *Staphylococci*. Septal wall (S) is laid down as an equatorial plate which then peels apart and "rounds out" to increase the volume of the cell, presumably by intercalation throughout the cell cycle. The next generation of septa are formed at the junction (J) of old and new "poles". (d). Theoretical situation in cocci or very short rods where septum formation could be preceded by extension of peripheral wall. (S, J and P as in Figs (b) and (c)).

seen to be due to changes in orientation of daughter cells during the separation stage (Tzagaloff and Novick, 1977).

In contrast to streptococci where extension of the peripheral wall can be clearly recognized prior to septation, there can be very little, if any, peripheral wall formed prior to septum formation in micrococci and staphylococci. The septum is laid down, rather as in bacilli, by centripetal growth to the centre of the cell, after which it peels apart as the two new poles round out to become hemispheres (Murray *et al.*, 1959; Suganama, 1962). The time taken from septation to complete rounding-out occupies one generation in *Micrococcus flavus* (Paulton, 1972). In this case, it seems that the completed pole must be formed by intercalation into the equatorial plate. No information is available as to whether septation is preceded by some circumferential extension as though these organisms were short rods (see Fig. 3). The degree to which septal material contributes to the surface area of the cell in bacilli, streptococci, and staphylococci is illustrated in Fig. 3.

B. ROD-SHAPED BACTERIA

The growth pattern of rod-shaped bacteria, unlike that of cocci, is not known with any certainty in spite of a voluminous literature. Some or all of the favourable circumstances found with streptococci (e.g. the lack of peptidoglycan turnover and the existence of wall bands) do not apply to rod-shaped bacteria. In the enteric bacteria, the antigens available for immunofluorescence studies are located in the outer membrane and are held by weak bonds which might not prevent lateral mobility. The very high rate of turnover of peptidoglycan found in some Gram-positive rods would also obscure any pattern of growth.

1. *Peptidoglycan Layer and Covalently Bound Accessory Polymers*

Fluorescent antibodies, autoradiography, and segregation of thickened wall have been used to detect areas of new growth in the peptidogylcan layer. When a diaminopimelic acid (DAP)-requiring auxotroph of *E. coli* is steady-state labelled with ^{3}H-DAP, autoradiography revealed that, during growth in unlabelled medium, the labelled material was transmitted to the progeny by a process which dilutes out the original label without conservation in units (Van Tubergen and Setlow, 1961; Lin *et al.*, 1971). As there is no measurable loss of peptidoglycan, the authors argued that the dilution must result from repeated subdivision of old wall by new wall material at a large number of sites (i.e. old wall is not conserved as a unit). In contrast, using high-resolution autoradiography to determine the location of ³H-DAP in isolated murein sacculi, Ryter et al. (1973) found that, after a pulse lasting one-eighth of a generation time, the sacculus became labelled in a fairly sharp equatorial band. A large amount of this represented septum formation during the pulse period. Thus Schwarz et al. (1975) subsequently reported that the equatorial zone was reduced by 50 per cent if similar autoradiographs of temperature-sensitive mutants defective in septum formation were prepared immediately after transfer to the restrictive temperature. It was not clear if there was any residual septum formation under these circumstances. After sustained growth at the restrictive temperature, there was little sign of localized incorporation of diaminopimelate. Clearly there was some incorporation into the peripheral wall, which could have represented incipient equatorial sites or intercalation of new material over the entire surface. The existence of length-extension sites should be demonstrable in principle during a chase period as, even if the labelled area is diluted by turnover synthesis, the area of new growth should be completely unlabelled. Furthermore, the appearance of the septum would not bias the observations. Ryter et al. (1973) reported that growth in the presence of non-radioactive diaminopimelate for one quarter of a generation, resulted in randomization of the equatorial zone of incorporation throughout the sacculus. However Braun and Wolff (1975) consider that the chase procedure was not effective in preventing incorporation. There may also be a problem relating to the compartmentalization of the pool of diaminopimelate (Leive and Davis, 1965). In *B. megaterium*, the distribution of ³H-DAP was random within the daughter cells after two generations of growth in non-radioactive medium with no indication of a zone of length extension (Mauck et al., 1972). In this organism there is a high rate of peptidoglycan turnover (Mauck et al., 1971) but, as suggested above, it should still be possible to detect growth zones in a chase experiment if indeed there are any. De Chastellier et al. (1975) have shown that ³H-DAP incorporated in two-minute pulses is found uniformly distributed along the cell, with a marked peak at the equator in cells containing an incipient septum (when examined in isolated walls and in thin sections) thus confirming that there are no specific zones of peptidoglycan synthesis in a high-turnover strain.

The *Bacillus subtilis* W23 phage SP50 receptor (teichoic acid) is produced only under phosphate-replete conditions and not during phosphate limitation. The localization of the receptor within the cell during derepression and repression of teichoic acid synthesis in chemostat culture has been studied by Archibald (1976) and Archibald and Coapes (1976). During the initial synthesis of teichoic acid, the phage receptors, visualized by specific attachment of the phage, were present on the inner surface of the wall throughout the length of the cell. When teichoic acid synthesis ceased, during phosphate limitation, the phage receptors were lost uniformly along the lateral walls, whereas those of the poles were conserved with very little dilution during subsequent growth.

Specific inhibition of protein synthesis causes the wall of most bacteria to thicken (see p. 151). The fate of such thickened walls during recovery from inhibition of protein synthesis has been studied in *B. subtilis* 168 (Frehel *et al.*, 1971). In general, the thickened wall very quickly becomes thinner by the loss of large pieces of wall from the outer surface. Some irregularity in the thinning process is observed, but not sufficient to suggest growth zones. Poles tend to remain thickened until a relatively late stage.

Studies of the growth of bacilli have been made using fluorescent antibodies (Chung *et al.*, 1964a; Hughes and Stokes, 1971). The conclusions of the former have been rejected on a number of grounds by Cole (1965). Hughes and Stokes (1971) used fluorescent antibody directed against peptidoglycan to show growth zones in a strain of *B. licheniformis* in which peptidoglycan turnover is very slow. Although not as clear as with the streptococcal system (Cole and Hahn, 1962), there were indications of conservation of fluorescent areas during growth in the absence of fluorescent antibody.

Although the bulk of the data for bacilli suggests that peptidoglycan synthesis occurs by intercalation of new material into old, the observations do not preclude the existence of specific surface-extension sites even when there is a high turnover rate. This can be seen from a brief consideration of peptidoglycan biosynthesis. Newly synthesized material is added to a rigid array of pre-existing peptidoglycan by two processes occurring simultaneously; (a) elongation of existing glycan strands by addition of disaccharides to their reducing termini which are probably held in the plasma membrane (Ward and Perkins, 1973); and (b) transpeptidation of **newly syn-**

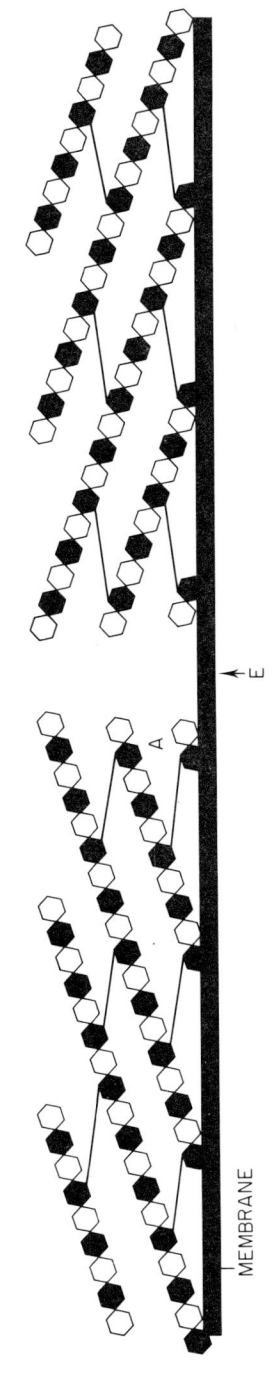

FIG. 4. Schematic representation in two dimensions of contribution of glycan chain initiation and elongation to surface extension. Black and white hexagons show N-acetylmuramic acid and N-acetylglucosamine residues, respectively. Bars between muramic acid residues represent cross-bridges (not intended to indicate actual measurements or position) E indicates the site of surface extension. Note addition of disaccharides by transpeptidation (A) extends the surface more by addition of the peptide than by addition of the disaccharide.

thesized glycan chains to pre-existing wall (Ward and Perkins, 1974). If a new chain is inserted into the wall by transpeptidation, the new reducing terminal will be displaced one peptide bridge length from an existing chain. The length of a fully extended peptide bridge in *B. subtilis* or *E. coli* is 4.68 nm, while one disaccharide unit is 1.03 nm (Keleman and Rogers, 1971). Therefore, if indeed the peptide bridge is in an extended configuration, the addition of one disaccharide unit by transpeptidation extends the area occupied by peptidoglycan by considerably more than does the extension of an existing chain (see Fig. 4). Thus, surface extensions may be more related to chain initiation than to chain extension, although it is not necessary to assume that surface extension always occurs as a result of chain initiation. Apart from the observation of Hughes and Stokes (1971) there have been no other studies of wall synthesis in low-turnover strains of bacilli. It is desirable that these observations be confirmed, as they imply that both chain elongation and initiation are confined to specific zones, presumably of length extension.

Ward (1973) has shown that the length of glycan chains in bacilli were severely underestimated in earlier work and are in fact 0.1–0.2 μm in *B. licheniformis* and 0.2–0.3 μm in *B. subtilis* (S. M. Fox and J. B. Ward, personal communication). Such long chains would need to be arranged in the plane of the wall, and their length also suggests that the peptidoglycan must be arranged in large blocks which are not subdivided by intercalation of new material. If all the chains were extending, but there were sites of cell-surface extension as suggested in Fig. 4, the sites could have some chemical identity as the meeting points of the non-reducing ends of glycan chains.

The existence of spiral-growth mutants of *B. subtilis* (Mendelson, 1976; Tilby, 1977) has been interpreted as indicative of a spiral arrangement of glycan chains along the longitudinal axis of the cell with annular growth sites at which they are extended. Indeed, it seems impossible to explain the necessary torque required for spiral growth without postulating annular length extension sites.

2. *The Outer Membrane of Gram-negative Bacteria*

On transferring *E. coli* and *S. typhimurium* grown in the presence of homologous fluorescent antibody to media containing non-fluorescent antibody, there was a general decline in fluorescence of the cell surface as though new material was being intercalated throughout the surface (May, 1963; Cole, 1964, 1965; Beachey and Cole, 1966). The polar caps, however, apparently retained their intense fluorescence for several generations. Contradictory results were reported by Chung et al. (1964b) who used a similar technique with *E. coli* and who claimed that there were growth zones just below the polar cap in small cells, with multiple growth zones in longer cells. Cole (1965) has criticized these observations on a number of grounds. In particular, the photographic technique used was thought to cause serious distortion of the image. Muhlradt et al. (1973, 1974) have subsequently demonstrated that lipopolysaccharide is synthesized at a large number of surface sites and is mobile within the plane of the membrane. Wagner (1967), using the fluorescent antibody technique with *Spirillum serpens*, concluded that new material was inserted initially at the poles and then at many sites which subdivided the fluorescent label.

The pattern of growth of *E. coli* has also been studied with inducible phage receptors found in the outer membrane. A strain of *E. coli* has been used which carried a nonsense mutation in the T6 receptor that made it unable to bind the phage, together with a temperature-sensitive nonsense suppressor which made phage binding temperature-conditional (Begg and Donachie, 1973, 1977). On transferring the bacteria from the permissive to the restrictive temperature, the phage receptors were retained in the equatorial region, while new surface without phage receptors was formed at the poles. Begg and Donachie (1977) argued that the unlabelled area represents true surface extension, as there is no change in the location of the existing phage receptors. The pattern of growth suggested is shown in Fig. 5. The experiments were performed in the presence of low concentrations of penicillin that inhibit cell division, but not growth, and appear not to affect the distribution of phage receptors. A feature of these observations which suggests that the interpretation of the experiments may be complex is that, in bacteria grown at the permissive temperature, the receptors are not evenly distributed but are more frequent by a factor of two at the centre of the cell compared with the poles.

Similar experiments have been performed using the phage λ receptor which is also the maltose permease and is inducible by maltose and cyclic AMP (Ryter et al., 1975). Immediately after

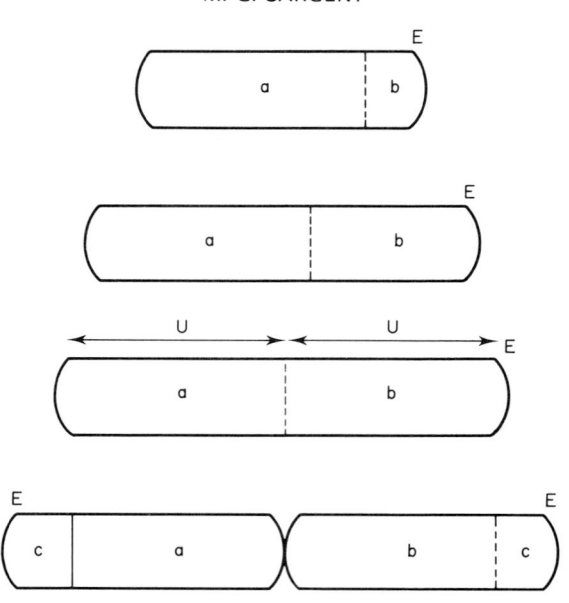

FIG. 5. Asymmetric growth model of Begg and Donachie. At low growth rates, extension occurs from one pole (E) until a "unit cell" is completed (U) after which a septum forms at the distal end of the unit. After septation is initiated, the old growth zone continues length extension while a new one starts at the other pole; a, b, and c represent growth units formed in three successive generations.

induction, the phage-binding sites were located at the equator. During further induction, the receptors spread over the surface by a process which is inhibited by chloramphenicol or azide. While the published photographs of the localized phage are spectacularly clear, there are a number of problems of interpretation. During de-induction, an unlabelled area does not appear in the equatorial region as might be expected if length extension continued from this site. However, there may be a pool of receptor precursors which continue to be incorporated into the outer membrane. Furthermore, there is considerable heterogeneity both in the amount of receptor per cell and the degree of localization within the cell. Like the T6 receptor, but unlike lipopolysaccharide, the λ receptor seems to be immobile within the membrane.

3. *Plasma Membrane—Phospholipids and Proteins*

Numerous techniques have been used to show that phospholipids are not conserved during growth in *E. coli* and *B. subtilis* (Lin *et al.*, 1971; Tsukayoshi *et al.*, 1971; Green and Schaechter, 1972; Mindich and Dales, 1972). However, it is widely believed that lipids are mobile within the plane of the membrane, so these studies do not clarify the growth process. There is very little information concerning the mobility of proteins within the membrane as they cannot, in general, be studied by direct techniques such as autoradiography. However, it would be unwise to assume that protein mobility is the same in bacteria as in eukaryotes. The ratio of protein to lipid in bacteria is generally twice that of eukaryote membranes, which may suggest that protein mobility could be less extensive as a result of protein-protein interactions. Furthermore, the bacterial membrane is held in close association with a rigid wall which may decrease the mobility of membrane components. The only unequivocal and direct demonstration of the pattern of growth of membrane protein is that of Green and Schaechter (1972). They showed that cytochromes of *E. coli* and *B. megaterium*, labelled with ^{3}H-α-aminolaevulinic acid, were transmitted to progeny during growth in a fashion which suggested that there were large numbers of sites within the membrane at which newly synthesized cytochromes were incorporated. Mindich and Dales (1972) have used transfer from deuterated to non-deuterated medium to detect segregation of conserved heavy membrane in *B. subtilis*. There was no segregation of heavy label in three generations, and the density of the membrane progressively decreased, as though membrane synthesis occurred by intercalation of new material into old. As the heavy label is not specific for protein (it will also label lipids), the report does not unequivocally demonstrate that proteins are not conserved during growth.

Kepes and Autissier (1972) have described an ingenious technique by which they claim to demonstrate conservation of a large number of inducible transport systems. They calculate that the number of growth zones required to achieve this is equal to the number of nuclei per cell. The apparent conservation of transport systems implies that there can be very little movement of the proteins within the plane of the membrane. The authors also claim that the glycerol residue of the phospholipid is segregated in a similar way to the permeases. This appears to be in such serious contradiction with the autoradiographic data (Lin *et al.*, 1971; Green and Schaechter, 1972) that conclusions from this method must be treated cautiously.

The mechanism of growth of the plasma membrane has also been studied using the minicell-producing mutant of *E. coli* (Wilson and Fox, 1971 Green and Schaechter, 1972). If there are conserved parts of the membrane, the isolated minicells which are derived from poles should be composed of these conserved units. Both authors have concluded that membrane proteins of the minicells must be a representative sample of the entire plasma membrane because, during pulse-chase experiments, both cell types have the same specific activity. A complication of interpretation in this type of experiment, which is not mentioned by the authors, is that half of the minicell must be laid down at the time of physiological separation (i.e. the septum) and this will bias the observed specific activities towards that of the nucleate cell. A similar approach has been employed by Sargent (1977a) using a temperature-sensitive initiation mutant of *B. subtilis* which produces large anucleate cells. No evidence for conservation of the bulk of membrane proteins could be found.

4. *Growth Relative to Internal Markers*

It has been claimed that growth must occur from just under the new pole formed at division in Gram-positive and Gram-negative rods (Bisset and Pease, 1957; Bisset and Hale, 1960). This was based on light-microscope observations of the segregation of flagella in strains temperature-sensitive for flagellum formation. However, the more careful analysis of Quadling (1958), using micro-manipulation, indicated that each flagellum could be segregated into separate descendants within a few generations, thus suggesting that growth occurs by intercalation between the flagella. More recently Ryter (1971), by electron microscopic studies with a mutant of *B. subtilis* temperature-sensitive for flagella production, reported that new growth occurs symmetrically around the nucleus.

Schwarz *et al.* (1969), using low concentrations of penicillin to cause autolytic damage in *E. coli*, found that potential sites of division were marked by splits in the murein sacculus. In whole cells, spheres (blebs) of outer membrane were formed which were attached to the surface by a narrow neck. At higher concentrations, extensive bulges were seen in the whole cell. The authors suggested the blebs marked the growth zones of the cell. Donachie and Begg (1970) also found that bleb formation was equatorial at high growth rates, but that at low growth rates the sites were present at the poles of newly divided cells. During further length extension, the blebs remained a constant distance (a "unit cell") from the distal pole, and the proximal part of the cell increased in length until it too was a unit cell, whereupon division occurred. Initially, it was supposed there was an annular growth zone at the penicillin-sensitive site which laid down new surface asymmetrically, they have subsequently suggested that the growth zone is, in fact, at the pole throughout the cell cycle (see p. 143 and Fig. 5; Begg and Donachie, 1973, 1977). Staugaard *et al.* (1976) have re-examined these observations by electron microscopy of bacteria grown under the same conditions and by a more extensive analysis of the distribution of blebs in cells of different size. They agreed that blebs were formed at the equator in large cells and at the poles of new born cells, but, unlike

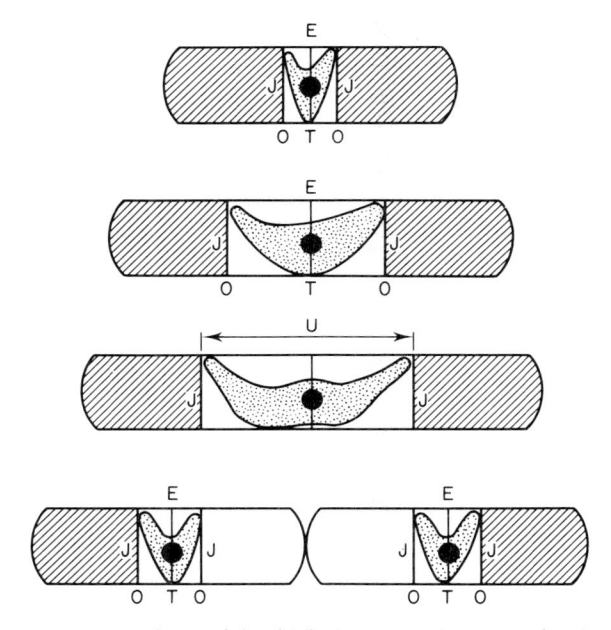

FIG. 6. Symmetric growth model which incorporates a mechanism of genome segregation; the replicon model. If chromosomes are separated and packaged into daughter cells solely by surface extension, all surface extension must occur between sites to which the daughter origins are attached (O). This could be achieved by specific zones of length extension (E) associated with the terminus (T). When the nucleus appears highly contracted, it appears to divide at termination and is seen either at the equator or at the next potential site of division (●). When in the dispersed state, it is seen extended between the origin attachment sites, as indicated by the stippled area (Ryter, 1967, 1968). At termination, the growth unit (U) is complete and new growth zones start at the junction of old and new wall started one generation earlier (J).

Donachie and Begg (1970), they concluded that blebs could be found at random in intermediate sized cells and not at a constant distance from one pole. They also argued that polar blebs were found in new born cells because the polar cap retained the sensitivity to penicillin, characteristic of the septum, for a short time after cell division. The experiments of Staugaard *et al.* (1976) were performed using synchronous cultures obtained by the membrane-elution method, and ampicillin instead of benzylpenicillin to potentiate "bleb" formation. No indication was given of whether this difference in technique is relevant to the controversy.

The only published proposal of a mechanism by which chromosomes are separated and packaged into daughter cells suggests that it is achieved by surface extension between membrane sites to which the chromosome is attached (Jacob *et al.*, 1963, 1966; Ryter, 1967, 1968). On the basis of a bidirectional model of chromosome replication, all surface growth must occur between chromosomal-attachment sites as shown in Fig. 6. The only visual indicator of the relationship between chromosome segregation and surface growth is the bacterial nucleus. This is a massive structure in comparison with any possible attachment site, and its size and appearance depend critically on the fixation conditions (Whitfield and Murray, 1956; Woldringh, 1973; Woldringh and Nanninga, 1976). This problem is illustrated by a comparison of the studies of Ryter (1967, 1968) and Mendelson (1968, 1969) on the location of nuclei in germinating spores of *B. subtilis*. Electron microscopic examination of serial sections revealed that all surface extension occurred within the area occupied by nuclear material (Ryter, 1967, 1968). In contrast nuclei, visualized by autoradiography of ^{3}H-labelled DNA, appeared very compact, and surface extension occurred symmetrically around each nucleus (Mendelson, 1968, 1969). The nuclei of suitably stained cells of *B. subtilis* from exponential phase cultures, when examined by light microscopy, were found to be arranged symmetrically either at the equator in mononucleate cells, or at 25 per cent from either pole in binucleate cells. This indicated that growth occurs symmetrically in each part of the cell, and also suggests that the daughter nuclei "jump" to the next potential division site at nuclear division (Sargent, 1974). In serial sections of *Strep. faecalis* examined by electron microscopy, the only points at which nuclei are found attached to the cell surface are the completing septum and newly initiated growth zones (Higgins and Shockman, 1971). This also suggests that the daughter nuclei "jump" from the almost completed division site to the next, in apparent contradiction to the Jacob *et al.* (1963, 1966) view of chromosome segregation (Sargent, 1974). However, this difficulty, and the discrepancy between Ryter's and Mendelson's observations, can be reconciled by supposing that the chromosome is attached to the cell surface at a future site of division (perhaps by the origin) while the terminus is attached at the equatorial division site. Then, under conditions where the nucleus is compact it will be contracted around the terminus while, under conditions where the nucleus is dispersed it is seen stretching to the origin. In *B. subtilis* there is clear evidence that the chromosome is attached most strongly by the terminus region (Yamaguchi and Yoshikawa, 1973) and it seems likely that, if the nucleus is contracted, it would be found at sites with which the chromosome interacts most strongly with the cell surface.

The time of nuclear division within the cell cycle can be estimated from the average number of nuclei per cell using the age-distribution theorem (see p. 111). When examined under conditions where the nucleus is contracted, nuclear division occurs at chromosome termination in *B. subtilis* (Sargent, 1975a, b) and in *E. coli* at high growth rates (Woldringh, 1976). If nuclear segregation is mediated by surface growth, as suggested by Jacob *et al.* (1963, 1966), there must be at least one growth zone per chromosomal terminus which extends the surface symmetrically as in Fig. 6. Calculations of the time at which putative growth zones are inserted in *B. subtilis* and *E. coli* (see p. 121) suggest that there is one growth zone per terminus. Furthermore, these calculations suggest that the site to which nuclei migrate is the junction of old and new surface (Sargent, 1975a). This has been clearly demonstrated in *Strep. faecalis* (Higgins and Shockman, 1971).

Support for the theory that genome segregation is achieved by surface extension between sites to which the chromosome is attached was originally based on four observations (Jacob *et al.*, 1963, 1966): (i) cosegregation of certain plasmids with the host chromosome during curing; (ii) association between the chromosome and the plasma membrane; (iii) non-random segregation of daughter chromosomes between progeny, (iv) indications of a conservative growth pattern. Since the original publications, powerful evidence for the association of the origin and terminus of the chromosome with the membrane has been reported for *B. subtilis* (Yamaguchi and Yoshikawa, 1973, 1977), and of the entire nucleoid with an envelope fraction in *E. coli* (Korch *et al*., 1976), but the cosegregation data can now be seen to be a result of specific associations between plasmids and the folded structure of the chromosome (Kline *et al*., 1976). There have also been fresh investigations of the randomness of chromosome segregation, but these remain contradictory (Eberle and Lark, 1967; Lin *et al*., 1971; Pierrucci and Zuchowski, 1973; Osley and Newton, 1974; Cooper and Weinberger, 1977).

Finally, the all important question of growth pattern has not been resolved in rod-shaped micro-organisms. It is clear however that growth from many sites located all over the bacterial surface would be incompatible with the theory propounded by Jacob *et al*. (Fig. 6).

C. CAULOBACTER AND BUDDING BACTERIA

Caulobacter spp. are of particular interest in studies of cell division as they show a primitive cellular differentiation and divide asymmetrically (Shapiro, 1976). The septum, which develops by a progressive pinching, is discernible well before chromosome termination as a constriction in the cell surface towards the non-stalked end of the cell (Terrana and Newton, 1975). The septum remains at the same relative position within the cell as it increases in length, indicating that the stalked part of the cell must have a slightly higher growth rate than the other part. When filaments are formed, during inhibition of cell division by low concentrations of penicillin, this asymmetry of growth persists, because on removing the inhibitor the septum still forms asymmetrically (Terrana and Newton, 1976). The non-stalked daughter cell has a longer interdivisional time than its "stalked" sister, during which time is also develops a stalk (Degnen and Newton, 1972). This stalk, which is a narrow tubular extension of the cell envelope, is generated by a growth zone at the base of the stalk (Schmidt and Stanier, 1966) which may contribute a small amount of surface to the stalked end of the cell, thus creating the differential in size and growth rate that distinguishes the daughters. Mutants have recently been described which are deficient in stalk formation but which still form characteristic asymmetric septa (Fukuda *et al*., 1977).

The budding bacteria (Hirsch, 1974), *Hyphomicrobium* sp. and *Rhodopseudomonas palustris* provide clear examples of a pattern of growth which is quite distinct from the streptococcal model. Daughter cells are formed by a budding process so that they contain only new surface, while the mother cell remains unchanged. The buds are formed at the end of a hyphum which in *Hyphomicrobium* sp. strain B-522 is long and slender (Moore and Hirsch, 1973a) and in *Rhodopseudomonas palustris* is short and fat (Westmacott and Primrose, 1976). Septa form at the junction of the hypha and bud. The sites of surface extension within the bud are not known. The nucleus divides in the mother cell, and one daughter nucleus moves through the hyphum, together with some parental RNA, into the developing bud (Moore and Hirsch, 1973b).

VII. Inferences from Physiological Studies

A. INHIBITION OF PROTEIN SYNTHESIS

The rate of peptidoglycan synthesis in members of many genera of bacteria is unaffected by complete inhibition of protein synthesis (Mandelstam and Rogers, 1958; Hancock and Park, 1958; Rogers and Mandelstam, 1962; Shockman, 1965; Hughes *et al.*, 1970) and, under these circumstances, a considerable thickening of the wall occurs (Hash and Davies, 1962; Shockman, 1965; Giesbrecht and Ruska, 1968; Hughes *et al.*, 1970; Frehel *et al.*, 1971). Such observations suggest that during inhibition of protein synthesis the bacterial surface area does not increase greatly and that peptidoglycan synthesis, normally involved in surface extension, is used in thickening the existing wall. Rather surprisingly, it is difficult to find quantitative evidence for the inhibition of surface-area increase. However, it seems likely in both streptococci and staphylococci that, under these conditions, any increase in surface area is small and is largely due to an increase in the volume occupied by the wall (Hash and Davis, 1962; Shockman, 1965; Giesbrecht and Ruska 1968; Higgins *et al.*, 1974). In *Strep. faecalis* the old hemispheres which would not have been thickened during exponential phase growth also became thickened (Higgins *et al.*, 1971). Inhibition of protein synthesis in *E. coli* (an organism in which wall synthesis continues after inhibition of protein synthesis, Rogers and Mandelstam, 1962) also inhibits volume increase completely (Ron *et al.*, 1977). Similar observations have been made with *B. subtilis* (M. G. Sargent, unpublished observations).

B. INHIBITION OF DNA SYNTHESIS

An important question in studying the control of surface growth is whether the rate of surface growth has a direct relationship with chromosome replication. It is well known that inhibition of DNA synthesis in *E. coli* prevents septum formation, and that filamentous forms result. A point in the cell cycle has been demonstrated in *E. coli* and *Caulobacter* sp. at which cell division becomes resistant to inhibitors of DNA synthesis, and which is co-incident with chromosome termination (Clark, 1968; Helmstetter and Pierucci, 1968; Degnen and Newton, 1972; Dix and Helmstetter, 1973). There is also a step immediately after chromosome completion that is required for septum formation, and which is sensitive to nalidixic acid and mitomycin C but not to thymine starvation of a thymine-requiring auxotroph (Dix and Helmstetter, 1973). In *E. coli* at low growth rates (Woldringh, 1976; Woldringh *et al.*, 1977) and in *Caulobacter* sp. (Terrana and Newton, 1975), incipient septa can be seen as constrictions before chromosome termination.

Donachie *et al.* (1971) have observed that a high proportion of anucleate cells are present after about two generations of thymine starvation of exponential-phase cultures of *thy⁻* stains of *B. subtilis*. They concluded that in this organism cell division is not coupled to chromosome replication. Anucleate cells are also produced when spores of *B. subtilis* are germinated during inhibition of DNA synthesis (Siccardi *et al.*, 1975; Van Iterson and Aten, 1976). Production of anucleate cells seems to be related to the growth medium used. In a basal medium lacking citrate but containing iron and manganese, which was specially formulated to minimize the time between nuclear division and cell separation, there is no significant anucleate cell production during thymine starvation (Sargent, 1975a). The conclusion that cell division is not coupled to chromosome replication in *B. subtilis* could not be strictly correct, as anucleate cells are virtually never seen in exponential-phase growth, thus indicating efficient coupling between chromosome termination and cell division. The differences in medium composition may well have created an interesting but pathological phenomenon.

While inhibition of DNA synthesis usually affects the initiation of septum formation but not its completion, the effect on growth of the peripheral wall is more complex. Long filaments are commonly observed in certain strains of enteric bacteria when DNA synthesis is inhibited, showing that cell elongation can continue in the absence of DNA synthesis. The question that must be answered, however, is whether DNA synthesis is a necessary condition to achieve an *increase in rate* of length extension. Inhibition of DNA synthesis in all prokaryotes may cause a large number of deleterious secondary effects, which may make it difficult to draw unequivocal conclusions (Stacey, 1976). The most important of these are induction of prophages and defective phages which lyse the cell, and degeneration of the DNA template for RNA synthesis. In addition, there are a large number of other consequences of inhibition of DNA synthesis such as changes in the size of nucleotide pools and induction of certain enzymes whose significance is unknown.

A number of authors have investigated the effect of thymine starvation on surface growth of rod-shaped bacteria. Kubitschek (1971) using *E. coli* 15T found that cell volume, measured using a Coulter electronic particle counter, increased linearly during thymine starvation of a synchronous culture. He suggested that inhibition of DNA synthesis prevented an acceleration of the rate of volume increase. Similar observations were made earlier (and published without comment) by Clark (1968) using *E. coli* B/r in which DNA synthesis was inhibited by nalidixic acid. In *E. coli*, synthesis of RNA and protein continues exponentially for about one and a half generations (Cohen and Barner, 1954) after which there is a marked decrease in rate of macromolecular synthesis.

In contrast, Donachie *et al.* (1976) using *E. coli* B/r thy^-, have repeated this experiment making length instead of volume measurements. They found that, although length extension is initially linear, there is a doubling in rate just before the time at which division *would* occur during normal growth. It is difficult to choose between these conflicting views, as there is some uncertainty about the induction of defective phages, and there is the possibility of incomplete inhibition of DNA synthesis by thymine starvation (Wilkins *et al.*, 1971; Hill and Fangman, 1973), and of osmotic shock during preparation of synchronous cultures (Wu and Pardee, 1973).

Sargent (1975a) performed a similar experiment using a strain of *B. subtilis* that does not lyse during thymine starvation and which is not leaky with respect to DNA synthesis (Sargent, 1977b). In this strain, length extension is linear for four hours at an average rate 20 per cent lower than the theoretical rate (Sargent, 1975a). During this period, cytoplasmic protein, RNA, peptidoglycan and teichoic acid increase exponentially for one generation, and then linearly (Sargent, 1975c,

1977b and unpublished observations). Inhibition of DNA synthesis therefore causes a specific constraint on length extension, but does not affect the increase in rate of synthesis of most cellular components. Unlike the synthesis of other macromolecules, the rate of membrane protein synthesis is constant during thymine starvation and may determine the rate of length extension (Sargent, 1975c).

In *Strep. faecalis*, inhibition of DNA synthesis by mitomycin C allows the completion of a considerable number of incipient septa, but seems to inhibit production of new sites at the equatorial wall band region (Shockman *et al.*, 1974). Bacteria at an early stage of peripheral wall formation continue to increase their peripheral wall far beyond the normal length at which the septum would form (Higgins and Shockman, 1971; Higgins *et al.*, 1974).

C. INHIBITION OF PEPTIDOGLYCAN SYNTHESIS

The availability of highly specific inhibitors of peptidoglycan synthesis offers an opportunity to investigate the role of wall synthesis in cell-surface extension. Gardner (1940) noted inhibition of septum formation and production of filaments in *E. coli*, *S. typhimurium* and *Clostridium perfringens*, and swelling of both streptococci and staphylococci with inhibition of cell separation at low penicillin concentrations, observations which have all been amply confirmed.

In Gram-negative bacteria, at least, a number of related antibiotics produce different morphological effects at their lowest inhibitory concentrations. Thus, it is claimed that cephalothin and cephaloridine do not cause filamentation but potentiate severe lysis (Perkins and Miller, 1973; Burdett and Murray, 1974b; Spratt and Pardee, 1975; Spratt, 1975). Mecillinam (FL1060), an antibiotic which inhibits *in vivo* peptidoglycan synthesis in *E. coli* up to 50 per cent but does not affect any known enzymic steps *in vitro* (Park and Burman, 1973; Matsuhashi *et al.*, 1974; Braun and Wolff, 1975), causes a slow conversion of exponential phase cells of *E. coli* to ovoid cell forms at low concentrations (Lund and Tybring, 1972; Greenwood and O'Grady, 1973; Tybring and Melchior, 1975; James *et al.*, 1975).

Spratt (1975, 1977), has divided the β-lactam antibiotics into three groups on the basis of the morphological effect they produce and their affinity for three of the penicillin-binding proteins found in the inner membrane of *E. coli*. Specific roles for these proteins in cell elongation (protein 1), shape maintenance (protein 2) and septation (protein 3) were suggested because specific binding to these proteins at the lowest inhibitory concentrations resulted in lysis, formation of round forms, or formation of filaments respectively. The characteristic equatorial bulges seen with ampicillin were thought to be a consequence of inhibition of both proteins 2 and 3. The relationship between binding of the antibiotics by these proteins and the morphological effect must be complex, as many of the consequences of inhibition of peptidoglycan synthesis result from unbalanced action of autolysins (Rogers, 1967; Tomasz et al., 1970). It could be argued that the effects observed are related to the degree of autolysis which, in turn, is probably related to the amount of inhibition of peptidoglycan synthesis. The role of autolysis is well illustrated by the demonstration that ampicillin inhibits division in synchronous cultures of E. coli at any point in the cell cycle almost until the point of cell separation, and even at such late times the nascent septum is destroyed (Burdett and Murray, 1974a,b). The effect of inhibition of protein 3 may not be entirely specific to the septum as there is thought to be a decrease in the amount of peptidoglycan in penicillin induced filaments (Starka and Morava, 1967; Schwarz et al., 1969; Burman et al., 1972). Nonetheless, length extension continues apparently at the same rate as in the control penicillin-untreated cells. Protein 1 may be more critical for the bulk of the cells in peptidoglycan synthesis, as total lysis results from inhibition of its synthesis.

The consequences of inhibition of protein 2 by mecillinam (FL1060) are quite distinct from inhibition of proteins 1 and 3. At the lowest inhibitory concentration, there is no turnover of peptidoglycan (Braun and Wolff, 1975), but within one generation the cells start to swell at the equator and become ovoid. James et al. (1975) suggested that division is inhibited by mecillinam up till 50–60 min before cell separation, but is insensitive thereafter. They suggest that the mecillinam-sensitive target appears in the cell cycle co-incident with chromosome initiation and that its appearance could represent the initiation of a specific developmental sequence leading to cell division. This potentially important discovery must be treated with caution, as there is no published evidence that the suggested "execution" point in the cell cycle is independent of drug concentration or of salt concentration in the medium. At the same mecillinam concentration, but in different media, continued division of the organism in its spherical cell form was noted by Goodel and Schwarz (1975), Tybring and Melchior (1975) and Greenwood and O'Grady (1973).

Relatively little is known about the effect of inhibition of wall synthesis on the morphology of Gram-positive or coccal bacteria. In *Streptococcus lactis*, penicillin inhibits septum formation and gives rise to irregular filaments (Dring and Hurst, 1969; Lorian, 1976), but in *Strep. faecalis* balloon-like swellings are found (Higgins and Shockman, 1971). In staphylococci, septum formation is relatively unaffected by penicillin treatment, but cell separation is inhibited. The cocci increase in volume, but the peripheral wall becomes thinner and seems to be more affected than the septum (Murray *et al.*, 1959; Suganama, 1962; Lorian, 1976; Tzagaloff and Novick, 1977). Lysis occurs at higher concentrations of penicillin, but not in an equatorial ring as during autolysis (Mitchell and Moyle, 1957). Distorted forms of *B. licheniformis* are obtained when grown in the presence of the minimal inhibitory concentration of penicillin, but growth continues without any specific effect on septum formation (Highton and Hobbs, 1971). In *B. megaterium*, Fitz-James and Hancock (1966) noted septal damage at penicillin concentrations that allowed continued increase in the cell volume. In the Gram-negative coccus *Neisseria gonnohorae*, benzylpenicillin induces evagination of the outer membrane at a potential septal site and inhibits cell division (as in *E. coli*) but the cell remains coccoid and increases in diameter (Westling-Haggstrom *et al.*, 1977).

VIII. Genetic Approaches to the Analysis of Surface Extension

The large variety of mutants known to have abnormal morphology must all be regarded as relevant to the study of surface extension in bacteria as they provide indications of specific controls operating over surface extension or division. In general, these can be interpreted at the biological level, but rarely at the level of specific enzymic processes. It will be evident that genes involved specifically in surface extensions have not been identified. The effect of inhibition of peptidoglycan synthesis by antibiotics (see p. 154) indicates that a conditional mutant affected in extension of the peptidoglycan sacculus (which must certainly be concerned in surface extension) is likely to lyse under the restrictive conditions. If it is possible to minimize these secondary effects by studying mutations in a genetic background deficient in autolysin, at intermediate temperatures or some other *condition of minimum insult*, it may be possible to establish their role in surface extension. The phenotypes which might be expected under these conditions are a matter of speculation, but if the cell had the capacity to adapt to the constraint it might adopt a form in which the surface to mass ratio was decreased (i.e. a spherical form). The observations of Zaritsky and Pritchard (1973) (see p. 124) and of Shannon and Rowbury (1975) (see p. 160) indicate that enteric bacteria can adapt to a potential increase in the mass:volume ratio by an increase in width.

An important corollary to this reasoning is that the phenotype of any morphological mutant could be one in which potential secondary effects of the primary lesion are minimized, either by incomplete inactivation of the gene product, or by bypass reactions that overcome the lesions by preventing autolysis or by providing an alternative "pathway". The importance of bypass reactions whereby the cell could avoid the lethal consequences of a particular lesion were recognized by Slater and Schaechter (1974) who drew attention to the investigations of Forsberg *et al.* (1973) with a phosphoglucomutase-deficient mutant of *B. licheniformis*. This mutant is deficient in its ability to glucosylate teichoic acid and to make teichuronic acid, and it has greatly diminished autolytic activity. It retains normal morphology except when grown under phosphate limitation, when it is also unable to make teichoic acid and then grows as small irregular coccal forms.

A. ROD MUTANTS

The mutants of rod-shaped bacteria, which are most acutely affected in the maintenance of correct shape, grow as osmotically stable irregular cocci at mass growth rates comparable to wild-type strains. The phenotype is generally described as Rod$^-$ and results from *rod* mutations. Mutants of this type have been isolated in *B. subtilis* (Rogers *et al.*, 1968; Boylen and Mendelson, 1969), *E. coli* (Adler *et al.*, 1968; Normark, 1969; Lazdunski and Shapiro, 1972; Henning *et al.*, 1972; Matsuzawa *et al.*, 1973; Matsuhashi *et al.*, 1974, Spratt, 1975), *Proteus* sp. (Martin, 1964; Hofschneider and Martin, 1968), *Klebsiella* sp. (Satta and Fontana, 1974a, b), and *Agrobacterium* sp. (Fugiwara and Fukui, 1972). Some grow as approximately spherical forms under all conditions, while others are temperature-dependent, or osmotically remedial, or pH-dependent, or are suppressed by specific chemicals. I shall not describe the phenotypes in detail but shall consider only the effect of the mutations on surface extension.

Amongst the *rod* mutants there is a range of morphologies from the almost perfectly spherical in which septa are formed more or less symmetrically to highly irregular structures in which the plane of division may apparently be at right angles to previous divisions (Adler et al., 1968; Forsberg et al., 1973). Studies of the growth requirements of some coccal forms have shown almost perfectly spherical forms can be obtained under certain conditions. The rod (lss 12) mutant of E. coli (Henning et al., 1972), for example, would lyse at the restrictive temperature unless grown in 2 mM mg^{2+}. The rod B mutant of B. subtilis also lyses at the lowest concentration of Mg^{2+} that supports growth of the wild-type, but at much higher concentrations of Mg^{2+} it grows well as either rods or cocci. Chloride, bromide, iodide, or nitrate are also required for growth in the rod or coccal form. By judicious choice of the concentrations of Mg^{2+} and anion, the cell shape obtained is dependent on the temperature at which the bacteria are grown and, furthermore, on changing temperature a complete change of morphology from rod to sphere, or vice versa, can be obtained in several generations. Under these conditions the irregularities of shape and septum location noted in earlier publications disappear, and almost perfect spherical forms are obtained (Rogers and Thurman, 1978). Apparently, the Rod$^-$ phenotype in lss 12 and rod B is expressed optimally when the "insult" caused by the mutational lesion is minimized by manipulating the environment. In such mutants, surface extension must be decreased relative to volume increase, but it is not known if volume increase can occur without concomitant mass increase. There are no recorded attempts to measure the density of Rod$^-$ forms. While it is possible that the mutants are specifically affected in surface extension, at least two other influences may have an important role in determining the morphology of these mutants. These are: (a) specific chemical changes in peptidoglycan and accessory polymers or in their rate of synthesis, which may alone be sufficient to alter the morphology of the cell wall directly (Rogers et al., 1974, Previc and Lowell, 1975); and (b) remodelling of the peptidoglycan as a result of an altered equilibrium between wall synthesis and the action of autolysins (Lazdunski and Shapiro, 1972). At present there is not sufficient information to conclude that the syndrome associated with one genetic lesion is of any one type and, furthermore, the three elements may overlap considerably.

The coccal form of most rod mutants retains certain characteristics of the growth pattern of Rod$^+$ cells. The septa are still found at the equator and the cells retain identifiable poles which are not subdivided by further growth. Some coccal forms may grow in chains, thus showing that the poles are conserved during growth (Reeve et al.,

1972). This can also be seen by direct observations of coccal forms growing as single cells. In these, irregular division planes probably arise because of unevenly distributed surface growth (Adler et al., 1968; Forsberg et al., 1973; Bloom et al., 1975; Goodel and Schwarz, 1975). When grown as a coccal form, a large part of the surface must be contributed by septum formation and modification of the septum after cell separation, whereas in the Rod⁺ form, the contribution of the poles to the surface area may be only 15 per cent (see Fig. 3). It is therefore pertinent to enquire whether septum formation in the Rod⁻ form is strictly comparable to the wild-type situation where septum formation is apparently distinguished from length extension by its dependence on chromosome termination (p. 151), its susceptibility to autolysis in Gram-negative bacteria (p. 154), and its apparently independent genetic control (p. 161). In Gram-positive bacteria the completed pole shows greatly decreased peptidoglycan turnover relative to peripheral wall (Frehel et al., 1971; Fan et al., 1972; Fan and Beckman, 1973; Archibald and Coapes, 1976) and the cross-linking of the peptidoglycan is possibly different from that in *E. coli* (Mirelman et al., 1976, 1977). This question has not been considered in any detail in the literature and, in particular, there has been no study of the effect of inhibition of DNA synthesis on the morphology of the Rod⁻ mutants.

Bloom et al. (1975) found long irregular filaments after treating the coccal form of the *E. coli envB* mutant with low concentrations of penicillin. However, the mutant grows as a mixture of short rods and cocci and has therefore not entirely lost its Rod⁺ character. Allison (1971) has constructed a strain of *E. coli* containing a *rod* mutation (Adler et al., 1968) together with the *lon* mutation (which causes filamentation after ultraviolet irradiation). When septation was inhibited by ultraviolet irradiation, growth continued in all dimensions until amorphous cells of almost one thousand times the volume of the parent were obtained. Clearly the double mutant was unable to extend sufficiently in the pole-to-pole axis to accommodate the excess mass. Such experiments should be performed with other combinations of *fts* (temperature sensitive filament forming mutants) and *rod* mutations.

In *rod* mutants of *B. subtilis*, the septum forms as a flat plate perpendicular to the pole-to-pole axis, as in the wild-type. This is subsequently cleaved from the periphery, allowing the daughter cells to separate and then to bulge out as in staphylococci (Fig. 3c; I. D. J. Burdett, personal communication). This differs slightly from the situation in streptococci (p. 135) where a clearly identifiable section of peripheral wall is laid down prior to septation. If septa are formed at the junction of old and new wall, it will be interesting to know whether this junction is found exactly at the equator or displaced to either pole as in Fig. 3d.

B. MUTATIONS AFFECTING CELL WIDTH

A temperature-sensitive mutant of *S. typhimurium* in which the cell width changes at the restrictive temperature has been described by Shannon *et al.* (1974), and by Shannon and Rowbury (1975). These authors consider it to be defective in establishing new putative growth zones. Cell division ceases after a small amount of residual division at the restrictive temperature, volume increases exponentially, while length extension continues at a constant rate. Under some conditions the cell width starts to increase after one generation, eventually becoming 50 per cent greater than that of wild-type cell. The authors argue that a doubling in rate of length extension which occurs late in the cell cycle is a prerequisite for cell division, and that those cells which divide at the restrictive temperature have already completed this preparation for division. The mutant continues to synthesize DNA for several hours at the restrictive temperature, but appears to be incapable of nuclear division. Mutants of *E. coli* defective in nuclear segregation have also been isolated (Hirota *et al.*, 1968a; Hirota and Ricard, 1972). The observations of Rowbury and Shannon (1975) are consistent with the hypothesis outlined above (p. 146 and Fig. 6), viz. that the nucleus is associated with a potential division site which is a growth zone that becomes committed to septum formation after nuclear division. Mutants of *B. subtilis* (Reeve, 1974) and of *E. coli* (Olden *et al.*, 1975) have been described which bulge irregularly and are corrected by D-alanine.

C. MUTANTS AFFECTING CELL SIZE

A variety of evidence described in Section III (p. 117) suggested that length, width and mass were the outcome of several interacting influences. Mutants with small changes in these parameters are frequently encountered. An increase in the chromosomal replication time would tend to increase cell width and mass but not length

(Zaritsky and Pritchard, 1973). Examples of this are the *rep* mutant of *E. coli* (Lane and Denhardt, 1974) and the *thy*⁻ mutants of *B. subtilis* (M. G. Sargent, unpublished observation). In contrast, an increase in the time between chromosome termination and cell separation would result in increased mass and length per cell, but no change in width. Mutants of this type are perhaps partially defective in septum formation (Weigand *et al.*, 1970). Other mutants having an abnormally high mass per cell at low growth rates may arise from protracted C and D periods (Kvetkas *et al.*, 1970; Newman and Bockrath, 1974).

D. SPIRAL GROWTH MUTANTS

Mendelson (1976) has recently described an unusual double mutant of *B. subtilis* that forms extended double helices when grown from germinating spores. He considers that glycan chains in the bacterial wall follow a spiral path along the length of the cylinder and are extended at annular surface elongation sites. During normal growth, the ends of the filament would rotate in opposite directions but, when the ends are trapped, as in this strain, the resulting torque generates a helical structure. Spiral forms of *B. subtilis* have been obtained by growing it in the presence of chlorpromazine (Tilby, 1977). Furthermore, a mutant has been isolated which is resistant to Triton-X 100 and grows as a spiral form.

Rhodospirillum rubrum normally grows with a helical form. Mutants of this organism have been isolated which grow as rods and have less cross-linked walls than the wild-type (Newton, 1967, 1968).

E. SEPTATION MUTANTS

Conditional septation mutants provide striking examples of how bacteria retain a pattern of where septa should be located, even while growing as a filament (Ricard and Hirota, 1973; Slater and Schaechter, 1974). At the restrictive temperature, division stops almost immediately, while DNA synthesis, length extension, and nuclear division continue at the preshift rate. While growing as a filament, potential septal sites form at the same rate as they would at the permissive temperature (Reeve *et al.*, 1970; Reeve and Clark, 1972), and on transfer to the permissive temperature, septa form synchronously at these sites. The cells formed are slightly smaller than cells grown at the permissive temperature (Reeve *et al.*, 1970). Potential sites of division are not used in order of age as they would be in normal growth (Allen et al., 1974; Reeve and Clark, 1972; Nagai and Tamura, 1972). As in the wild-type, daughter chromosomes can be partitioned by septation only when chromosome replication is complete (Reeve et al., 1970). The placing of potential septal sites is therefore as well controlled in a filament as in normal growth, and appears to be related to the position of nuclei (p. 148).

F. MINICELL-PRODUCING MUTANTS

Very small anucleate cells (termed minicells) are produced in large numbers by certain mutants of *E. coli* (*min*) (Adler et al., 1966; Hirota et al., 1968a; Hirota and Ricard, 1972) and *B. subtilis* (van Alstyne and Simon, 1970; Reeve et al., 1973). Characteristically, minicells are produced at the poles of nucleate cells or close to an equatorial septum. In exponential-phase cultures of an *E. coli min* mutant, the frequency of division is the same as in the wild-type, but a proportion of these divisions (one-third) are expended on minicell production (Adler and Hardigree, 1972; Teather et al., 1974). Where no equatorial septum forms, the cells continue to elongate exponentially and divide at a later time, thus giving the great range of sizes characteristic of such mutants. In the minicell-producing strains of *B. subtilis*, the frequency of division is considerably less than in the wild-type, per unit of cell length (Mendelson, 1975; Mendelson and Coyne, 1975).

As in septation mutants, the potential sites of division in minicell formers are laid down independently of existing septa. This has been shown in strains containing an *fts* (temperature-sensitive, filament-forming) and a *min* mutation (Khachatourians et al., 1973; Zusman and Krotski, 1974). At the restrictive temperature, all division is inhibited and filaments are produced. On restoration to the permissive temperature, normal division and production of minicells occurs in a synchronous burst, and minicells are produced at the same frequency and in the same pattern as during growth at the permissive temperature. The minicells are the same size as during exponential-phase growth. Suggesting that no significant length extension occurs between a potential minicell-septum and the cell pole at the restrictive temperature.

In strains of *B. subtilis* containing the *rod* A and *min* mutations, grown at the restrictive temperature, bulges appear at sites at which minicells form (Mendelson and Reeve, 1973; Mendelson et al., 1975).

Interestingly, the characteristic wall ultrastructure associated with the Rod A phenotype appears all over the cell surface and not just in the bulges. This may indicate that length extension sites are distinct from wall synthesis sites. Wall synthesis, but not length extension, continues in isolated minicells (Mertens and Reeve, 1977).

G. MUTANTS PRODUCING LARGE ANUCLEATE CELLS

A number of mutants of *E. coli*, *S. typhimurium*, and *B. subtilis* have been described, in which large anucleate cells are produced when a septum is formed at a characteristic site at one end of a cell (Hirota *et al.* 1968a, b; Hirota and Ricard, 1972; Inouye, 1969, 1971; Spratt and Rowbury, 1971b; Howe and Mount, 1975; Capaldo and Barbour, 1975; Sargent, 1975b, 1977b). These mutants are of special interest for three reasons: (a) the production of anucleate cells in the absence of DNA synthesis has been regarded as evidence that the gene products of the wild-type alleles act as repressors of septum formation (Inouye, 1969, 1971); (b) the choice of septal site at a rather precisely determined position in the cell surface has provided an intimation that the surface is topographically differentiated in the absence of nuclei; (c) the continued production of anucleate cells of uniform size in the absence of DNA synthesis suggests there is a developmental programme which, in the absence of the chromosomal clock, determines the size at which septum formation occurs.

Anucleate cell production in *E. coli* appears to require two mutations, viz. the *div* mutations (A, B and C) that have no demonstrable phenotype alone, and a mutation that affects DNA replication or segregation causing filaments to be formed with no nuclear material at the ends (i.e. *dna*, A, B and D or *par* A and B mutations which are defective in nuclear segregation but not DNA synthesis; Hirota *et al.*, 1968a, b; Hirota and Ricard, 1972). Most DNA-synthesis mutants of *E. coli* do not produce anucleate cells, but there are exceptions, such as strain Ts27 (Inouye, 1969) which is a *dna* B mutant (Wechsler and Gross, 1971). Anucleate cells are also produced during inhibition of DNA synthesis by *thy*⁻ strains of *E. coli* carrying the *rec*A or *lex* mutations, but do not require *div* mutations (Howe and Mount, 1975). A mutation in *S. typhimurium* similar to *dna* A of *E. coli*, causes anucleate cell production but there is no evidence for a *div* mutation in this organism (Spratt and Rowbury, 1970, 1971a, b). Two classes of *B. subtilis* chromosome-initiation mutants produce anucleate cells in growth media which would not allow the wild-type to produce anucleate cells (Gross et al., 1968; Sargent, 1975b, 1977b).

At the start of anucleate cell production, the nucleus is located approximately at the centre of the filaments and appears to block division at this site, while the next generation of potential division sites are mobilized for anucleate cell production (Hirota and Ricard, 1972; Shannon et al., 1972; Sargent, 1975b; Howe and Mount, 1975). In the dnaB mutants of B. subtilis, anucleate cells are formed initially at sites about 25 per cent from one end of the filament. However, their size appears to be more strongly controlled than the position of the septum in the filament, as at later stages the septa are formed closer to the end of the filament although the anucleate cells are the same size. In minimal media, the size of the anucleate cells is about 2.0 μm with a coefficient of variation of 20 per cent. Average exponential-phase cells are 2.4 μm at birth increasing to 4.8 μm at division (Sargent, 1975b and unpublished observations). Sargent (1975b) has argued that the appearance of a septum is preceded by a growth zone at which cell extension occurs. In the S. typhimurium mutant, the anucleate cells are larger than new-born cells but also have a co-efficient of variation of about 20 per cent, and the size is dependent on the medium in which the cells are grown at the permissive temperature (Shannon et al., 1972). Shannon et al. (1972) found that the long filaments contained several penicillin-sensitive sites, whereas the wild-type cell usually contains one equatorial site which is normally taken to indicate a potential site of division or possibly of growth (p. 146). This suggested that septum formation was slow relative to the formation of new growth sites which could account for the large size of anucleate cells of this mutant. The rate of anucleate cell formation in this mutant is low, and in other mutants (i.e. lex, Ts27) considerably less than the dna B mutant of B. subtilis which can produce an anucleate cell approximately once per generation time per nucleus (Sargent, 1975b, 1977b).

As was already suggested, the continued production of anucleate cells of uniform size in these mutants indicates that there is a mechanism which determines the size of cells without the use of the chromosomal clock. This could operate on the basis of time, but it seems more likely that division would be related to the completion of a unit of volume or length (p. 112). Thus the rate of anucleate cell production in the S. typhimurium mutant can be increased by a nutritional shift-up but there is no consequential change in size of the anucleate cells produced (Shannon and Rowbury, 1972). Similar observations have been made in B. *subtilis* (M. G. Sargent, unpublished observations).

In chromosome-initiation mutants that produce anucleate cells, the gene products are inactivated more or less immediately at the restrictive temperature, but anucleate cell production does not start for some time. However, when *dna*B mutants of *B. subtilis* are incubated at the permissive temperature for one generation time in the absence of DNA synthesis, there is a sparing effect on the length of time required at the restrictive temperature to allow anucleate cell production. This shows that anucleate cell production cannot occur until a period of normal growth has occurred, and may indicate that existing "growth units" must be completed before new ones can be started (Sargent, 1975b, 1977b).

Most mutants that produce anucleate cells characteristically continue increasing in mass during inhibition of DNA synthesis for a longer time than strains that do not produce anucleate cells (Inouye, 1969, 1971; Howe and Mount, 1975; Sargent, 1977b). Whereas, in wild-type strains, inhibition of DNA synthesis commonly causes extensive degenerative changes in the cell (Stacey, 1976), mutants that produce anucleate cells are able to overcome these deleterious effects and sometimes show greater resistance to thymine-less death (Bouvier and Sicard, 1975; Inouye, 1971). Methods of enrichment for *div, rec* A and initiation mutants have been described for *E. coli* and *B. subtilis*, using resistance to prophage induction by inhibition of DNA synthesis (Devoret and Blanco, 1970; Hirota and Ricard, 1972; Murakami *et al.*, 1976). Inactivation of the *dna*B and *dna*D gene products of *B. subtilis* stimulates RNA and protein synthesis, which presumably causes the increased rate of mass accumulation (Sargent, 1977b). A similar phenomenon may occur in other anucleate cell-producing strains. However, increased macromolecular synthesis caused by a nutritional shift-up will not stimulate anucleate cell formation in the wild-type during inhibition of DNA synthesis (Sargent, 1975a).

H. INITIATION MUTANTS

Temperature-sensitive chromosome-initiation mutants provide an important test case for any theory that attempts to explain the differences in control of mass growth and of surface extension in terms of a relationship with the chromosome cycle. At the restrictive temperature, the number of chromosomal origins does not change, whereas the number of chromosomal termini increases by a factor of $2^{C/\tau}$ (Sargent, 1975b). If the number of sites of length extension is dependent on the number of chromosomal termini, then there should be an increase in the rate of length extension associated with termination.

Using a *dna*B mutant of *B. subtilis*, Sargent (1975b) attempted to show this, by comparing the time course of length extension when the chromosome is allowed to complete, with a control in which DNA synthesis was inhibited. As a result of chromosome completion, the rate of length extension increases by a factor of two, which in this mutant is consistent with control of length extension at, or close to, termination. This interpretation is, however, complicated by two observations. Firstly, whereas during inhibition of DNA synthesis in a *dna*B$^+$ strain the rate of length extension was constant in a *dna*B$^-$ strain the rate of length extension gradually increased and was associated with anucleate cell formation. Secondly, inactivation of the *dna*B gene product stimulated RNA and protein synthesis (Sargent, 1977b). These observations suggest regulatory relationships of considerable complexity between initiation proteins, surface growth, the mass growth rate and the chromosome cycle, discussion of which should be deferred until a clearer picture emerges.

IX. Concluding Remarks

Although this review has been concerned only with identifying the physiological factors that control the surface area of bacteria, it has been written with the hope that ultimately these may be interpretable in biochemical terms. The target of these influences is the peptidoglycan layer which is undoubtedly the principal *shape maintaining* structure in most bacteria (Weidel and Pelzer, 1964). However, the synthesis in peptidoglycan does not, in itself, determine the morphology of the bacterial cell. This is aptly illustrated by the process of L-form reversion in which bacterial forms are obtained from naked protoplasts only after many hours of extensive wall synthesis on solid media (Ward, 1978). However, pre-existing peptidoglycan structures probably impose a characteristic form on newly synthesized peptidoglycan.

Surface extension is evidently closely co-ordinated with macromolecular synthesis in a number of ways and is not solely determined by peptidoglycan synthesis. Thus: (1) inhibition of protein synthesis, which does not prevent peptidoglycan synthesis, inhibits surface area increase. Furthermore, the rate of surface-area increase is dependent on the mass growth rate of the organism and is not constant irrespective of growth rate, in contrast to the polymerization of many macromolecules: (2) The width of enteric bacteria (and therefore the width of the peptidoglycan sacculus) responds to an increase in mass per unit of cell length by an increase in width, as though a constant density was maintained. The "sensing device" of the homeostatic mechanism remains unknown: (3) There exists a relationship between surface growth and the chromosome cycle, such that a growth unit is completed at chromosome termination, simultaneously with the initiation of two new growth units.

Surface extension would be easier to understand if it was known with certainty if the peptidoglycan layer was extended in area at many or few sites. This a question to which satisfactory answers have been obtained only for streptococci where there is clear evidence for zonal growth. At the risk of appearing perverse, a number of authors have argued that there may be specific sites of surface extension in rod-shaped organisms which would provide a mechanism of genome segregation and the topographical differentiation of the cell surface necessary to determine the location of septa. Although the bulk of the data concerning growth of the peptidoglycan layer favours intercalation, there is no critical evidence against a zone of length extension, while there is some suggestive evidence for growth zones (see p. 138 and Hughes and Stokes, 1970; Ryter, 1971). Moreover, an analysis of the mechanism of peptidoglycan biosynthesis suggests the existence of zones of length extension (p. 141).

A greater difficulty is encountered in deciding where the putative growth zones are located. The evidence from streptococci suggests that septa are modified growth zones, so that in rod-shaped bacteria an equatorial growth zone might be expected, and indeed there is some evidence for this (Ryter, 1971; Schwarz et al., 1975). In contradiction, however, Begg and Donachie (1977) have claimed that surface extension occurs from the poles of *E. coli* (p. 143).

The use of the Collins–Richmond principle (p. 118) and, to a lesser extent, the analysis of cell density in the chromosome cycle (p. 126) have considerable potential in elucidating the kinetics of bacterial growth. As yet the evidence does not unequivocally support any model, although, on balance, the data suggest that length extension in rod-shaped organisms is exponential. It has frequently been supposed that, if length extension occurs from a small number of growth zones, it is likely to occur at a linear rate with doublings in rate when the number of growth zones double (Kubitschek, 1970; Previc, 1970; Zaritsky and Pritchard, 1973; Sargent, 1975a; Grover et al., 1977). None the less, it seems just as plausible to imagine such growth zones operating at an exponentially increasing rate during normal growth. A constant density would be anticipated throughout the cell cycle on the basis of an exponential growth model, whereas a linear growth model predicts cyclic changes in density (see pp. 126 and 132). The quantitative relationship between average length and growth rate for linear growth is formally identical to that for an exponential growth model predicated on the cell attaining a constant length at chromosome termination (Table 2).

X. Acknowledgments

I gratefully acknowledge the advice of my colleagues P. Piggot, H. J. Rogers and J. B. Ward who made significant improvements to this article.

REFERENCES

Abe, M. and Tomizawa, J. (1971). *Genetics* 69, 1.
Adler, H. I., Fisher, W. D., Cohen, A. and Hardigree, A. A. (1966). *Proceedings of the National Academy of Sciences of the United States of America* 57, 321.
Adler, H. I., Terry, C. E. and Hardigree, A. A. (1968). *Journal of Bacteriology* 95, 139.
Adler, H. I., Fisher, W. D. and Hardigree, A. A. (1969) *Transactions of the New York Academy of Sciences* 31, 1059.
Adler, H. I. and Hardigree, A. A. (1972). *In* "Biology and Radiobiology of Minicells: in Biology and Radiobiology of Anucleate Systems" (S. Bonotto et al., eds) pp. 51–66. Academic Press, New York.
Adolph, E. F. and Bayne-Jones, S. (1932). *Journal of Cellular and Comparative Physiology* 1, 409.
Allen, J. S., Filip, C. C., Gustaphson, R. A., Allen, R. G. and Walker, J. R. (1974). *Journal of Bacteriology* 117, 978.
Allison, D. P. (1971). *Journal of Bacteriology* 108, 1390.
Archibald, A. R. (1976). *Journal of Bacteriology* 127, 956.
Archibald, A. R. and Coapes, H. E. (1976). *Journal of Bacteriology* 125, 1195.
Beachey, E. H. and Cole, R. M. (1966). *Journal of Bacteriology* 92, 1245.
Begg, K. J. and Donachie, W. D. (1973). *Nature, New Biology, London* 245, 38.

Begg, K. J. and Donachie, W. D. (1977). *Journal of Bacteriology* **129**, 1524.
Bhatti, A. R., DeVoe, I. W. and Ingram, J. M. (1976). *Journal of Bacteriology* **126**, 400.
Bisset, K. A. and Hale, C. M. F. (1960). *Journal of General Microbiology* **22**, 536.
Bisset, K. A. and Pease, P. (1957). *Journal of General Microbiology* **16**, 382.
Bleecken, S. (1969). *Journal of Theoretical Biology* **25**, 137.
Bloom, G. D., Gumpert, J., Normark, S., Schumann, E., Taubeneck, U. and Westling, B. (1975). *Zeitschrift fur Allgemeine Mikrobiologie* **14**, 465.
Bouvier, F. and Sicard, N. (1975). *Journal of Bacteriology* **124**, 1198.
Boylen, R. J. and Mendelson, N. H. (1969). *Journal of Bacteriology* **100**, 1316.
Braun, V. and Wolff, H. (1975). *Journal of Bacteriology* **123**, 888.
Briles, E. B. and Tomasz, A. (1970). *Journal of Cell Biology* **47**, 786.
Brostrom, M. A. and Binkley, S. B. (1969). *Journal of Bacteriology* **98**, 1271.
Burdett, I. D. J. and Higgins, M. L. (1978). *Journal of Bacteriology* **133**, 959.
Burdett, I. D. J. and Murray, R. G. E. (1974a). *Journal of Bacteriology* **119**, 303.
Burdett, I. D. J. and Murray, R. G. E. (1974b). *Journal of Bacteriology* **119**, 1039.
Burman, L. G., Nordstrom, K. and Bloom, G. D. (1972). *Journal of Bacteriology* **112**, 1364.
Capaldo, F. N. and Barlour, S. D. (1975). *Journal of Molecular Biology* **91**, 53.
de Chastellier, C., Hellio, R. and Ryter, A. (1975). *Journal of Bacteriology* **123**, 1184.
Chung, K. L., Hawirko, R. Z. and Isaac, P. K. (1964a). *Canadian Journal of Microbiology* **10**, 43.
Chung, K. L., Hawirko, R. Z. and Isaac, P. K. (1964b). *Canadian Journal of Microbiology* **10**, 473.
Churchward, G. G. and Holland, I. B. (1975). *Journal of Molecular Biology* **105**, 245.
Clark, D. J. (1968). *Journal of Bacteriology* **96**, 1214.
Clark, J. B. (1972). *Chemical Rubber Company Critical Reviews in Microbiology* **1**, 521.
Cohen, S. S. and Barner, H. D. (1954). *Proceedings of the National Academy of Sciences of the United States of America* **40**, 885.
Cole, R. M. (1964). *Science, New York* **143**, 820.
Cole, R. M. (1965). *Bacteriological Reviews* **29**, 326.
Cole, R. M. and Hahn, J. J. (1962). *Science, New York* **135**, 722.
Collins, J. F. and Richmond, M. H. (1962). *Journal of General Microbiology* **28**, 15.
Cooper, S. and Helmstetter, C. E. (1968). *Journal of Molecular Biology* **31**, 519.
Cooper, S. and Weinberger, M. (1977). *Journal of Bacteriology* **130**, 118.
Cutler, R. G. and Evans, J. E. (1966). *Journal of Bacteriology* **91**, 469.
Dean, A. C. R. and Rogers, P. L. (1967). *Biochemica et Biophysica Acta* **148**, 267.
Degnen, S. T. and Newton, A. (1972). *Journal of Molecular Biology* **64**, 671.
Dennis, P. P. (1971). *Nature, New Biology, London* **232**, 43.
Dennis, P. P and Bremer, H. (1974). *Journal of Bacteriology* **119**, 270.
Dennis, P. P. and Herman, R. K. (1970). *Journal of Bacteriology* **102**, 118.
Devoret, R. and Blanco, M. (1970). *Molecular and General Genetics* **107**, 272.
Dix, D. E. and Helmstetter, C. E. (1973). *Journal of Bacteriology* **115**, 786.
Donachie, W. D. (1968). *Nature, London* **219**, 1077.
Donachie, W. D. and Begg, K. J. (1970). *Nature, London* **227**, 1220.
Donachie, W. D., Martin, D. T. M. and Begg, K. J. (1971). *Nature, New Biology, London* **231**, 274.
Donachie, W. D., Begg, K. J. and Vicente, M. (1976). *Nature, London* **264**, 328.
Dring, G. J. and Hurst, A. (1969). *Journal of General Microbiology* **55**, 185.
Eberle, H. and Lark, K. G. (1967). *Proceedings of the National Academy of Sciences of the United States of America* **57**, 95.
Ecker, R. E. and Kokaisl, G. (1969). *Journal of Bacteriology* **98**, 1219.

Ecker, R. E. and Schaechter, M. (1963). *Annals of the New York Academy of Sciences* **102**, 549.
Edelmann, P. L. and Echlin, G. (1974). *Journal of Bacteriology* **120**, 657.
Engberg, B., Hjalmarsson, K. and Nordstrom. K. (1975). *Journal of Bacteriology* **124**, 633.
Errington, F. P., Powell, E. O. and Thompson, N. (1965). *Journal of General Microbiology* **39**, 109.
Fan, D. P. and Beckman, B. E. (1973). *Journal of Bacteriology* **114**, 790.
Fan, D. P., Pelvit, M. C. and Cunningham, W. P. (1972). *Journal of Bacteriology* **105**, 1266.
Fitz-James, P. and Hancock, R. (1965). *Journal of Cell Biology* **26**, 657.
Forsberg, C. W., Wyrick, P. B., Ward, J. B. and Rogers, H. J. (1973). *Journal of Bacteriology* **113**, 969.
Frankland, P. F. and Ward, W. H. (1895). *Proceedings of the Royal Society* **58**, 265.
Frehel, C., Beaufils, A.-M. and Ryter, A. (1971). *Annales de l'Institut Pasteur, Paris* **121**, 139.
Fugiwara, T. and Fukui, S. (1972). *Journal of Bacteriology* **110**, 743.
Fukuda, A., Iba, H. and Okada, Y. (1977). *Journal of Bacteriology* **131**, 280.
Gardner, A. D. (1940). *Nature, London* **146**, 837.
Giesbrecht, P. and Ruska, H. (1968). *Klinische Wochenscrift* **46**, 575.
Glick, M. C., Sall, T., Zilliken, F. and Mudd, S. (1960). *Biochimica et Biophysica Acta* **37**, 361.
Goodel, E. W. and Schwarz, U. (1975). *Journal of General Microbiology* **86**, 201.
Goodwin, B. C. (1969). *European Journal of Biochemistry* **10**, 511.
Green, E. W. and Schaechter, M. (1972). *Proceedings of the National Academy of Sciences of the United States of America* **69**, 2312.
Greenwood, D. and O'Grady, F. (1973). *Journal of Clinical Pathology* **26**, 1.
Gross, J. D., Karamata, D. and Hempstead, P. G. (1968). *Cold Spring Harbor Symposium of Quantitative Biology* **33**, 307.
Grover, N. B., Naaman, S., Ben-Sasson, S. and Doljanski, F. and Nadav, E. (1969). *Biophysics Journal* **9**, 1398.
Grover, N. R., Woldringh, C. L., Zaritsky, A. and Rosenberger, R. F. (1977). *Journal of Theoretical Biology* **67**, 181.
Hackenbeck, R. and Messer, W. (1977). *Journal of Bacteriology* **129**, 1234.
Hancock, R. and Park, J. T. (1958). *Nature, London* **181**, 1050.
Harvey, R. J. (1972). *Journal of General Microbiology* **70**, 109.
Harvey, R. J. and Marr, A. G. (1966). *Journal of Bacteriology* **92**, 805.
Harvey, R. J., Marr, A. G. and Painter, P. R. (1967). *Journal of Bacteriology* **93**, 605.
Hash, J. H. and Davies, M. C. (1962). *Science, New York* **138**, 829.
Helmstetter, C. E. (1974). *Journal of Molecular Biology* **84**, 21.
Helmstetter, C. E. and Cooper, S. (1968). *Journal of Molecular Biology* **31**, 507.
Helmstetter, C. E. and Pierucci, O. (1968). *Journal of Bacteriology* **95**, 1627.
Helmstetter, C. E., Cooper, S. Pierrucci, O. and Revelas, E. (1968). *Cold Spring Harbor Symposium on Quantitative Biology* **33**, 809.
Henning, U., Rehn, K., Braun, V., Hahn, B. and Schwarz, U. (1972). *European Journal of Biochemistry* **26**, 570.
Herbert, D. (1961). *Symposium of the Society for General Microbiology* **11**, 39.
Higgins, M. L. and Shockman, G. D. (1970). *Journal of Bacteriology* **101**, 643.
Higgins, M. L. and Shockman, G. D. (1971). *Chemical Rubber Company Critical Reviews in Microbiology* **1**, 29.
Higgins, M. L. and Shockman, G. D. (1976). *Journal of Bacteriology* **127**, 1346.

Higgins, M. L., Pooley, H. M. and Shockman, G. D. (1970). *Journal of Bacteriology* **103**, 504.

Higgins, M. L., Pooley, H. M. and Shockman, G. D. (1971). *Journal of Bacteriology* **105**, 1175.

Higgins, M. L., Daneo-Moore, L., Boothby, D. and Shockman, G. D. (1974). *Journal of Bacteriology* **118**, 681.

Highton, P. J. and Hobbs, D. G. (1971). *Journal of Bacteriology* **106**, 646.

Hill, W. E. and Fangman, W. L. (1973). *Journal of Bacteriology* **116**, 1329.

Hirota, Y. and Ricard, M. (1972). *In* "Production of DNA less Bacteria. Biology and Radiobiology of Anucleate Systems" (S. Bonotto *et al.*, eds.), pp. 29–50. Academic Press, New York.

Hirota, Y., Ryter, A. and Jacob, F. (1968a). *Cold Spring Harbor Symposium of Quantitative Biology* **33**, 677.

Hirota, Y., Jacob, F., Ryter, A. and Butlin, G. (1968b). *Journal of Molecular Biology* **35**, 175.

Hirsch, P. (1974). *Annual Review of Microbiology* **28**, 391.

Hoffman, H. and Frank, M. E. (1963). *Journal of Bacteriology* **85**, 1221.

Hoffman, H. and Frank, M. E. (1965a). *Journal of Bacteriology* **89**, 212.

Hoffman, H. and Frank, M. E. (1965b). *Journal of Bacteriology* **89**, 513.

Hoffman, B., Messer, W. and Schwarz, U. (1972). *Journal of Supramolecular Structure* **1**, 29.

Hofschneider, P. H. and Martin, H. H. (1968). *Journal of General Microbiology* **51**, 23.

Holme, T. (1957). *Acta Chemica Scandinavica* **11**, 763.

Howe, W. E. and Mount, D. W. (1975). *Journal of Bacteriology* **124**, 1113.

Hughes, R. C. and Stokes, E. (1971). *Journal of Bacteriology* **106**, 694.

Hughes, R. C., Tanner, P. J. and Stokes, E. (1970). *Biochemical Journal* **120**, 159.

Husain, I., Poupard, J. A. and Norris, R. F. (1972). *Journal of Bacteriology* **111**, 841.

Inouye, M. (1969). *Journal of Bacteriology* **99**, 842.

Inouye, M. (1971). *Journal of Bacteriology* **106**, 539.

Jacob, F., Brenner, S. and Cuzin, F. (1963). *Cold Spring Harbor Symposium of Quantitative Biology* **28**, 329.

Jacob, F., Ryter, A. and Cuzin, F. (1966). *Proceedings of the Royal Society, Series B,* **164**, 267.

James, R., Haga, J. Y. and Pardee, A. B. (1975). *Journal of Bacteriology* **122**, 1283.

Jones, N. C. and Donachie, W. D. (1973). *Nature, New Biology, London* **243**, 100.

Keleman, M. V. and Rogers, H. J. (1971). *Proceedings of the National Academy of Sciences of the United States of America* **68**, 992.

Kelly, C. and Rahn, O. (1932). *Journal of Bacteriology* **23**, 147.

Kendall, D. G. (1948). *Biometrika* **35**, 316.

Kepes, A. and Autissier, F. (1972). *Biochimica et Biophysica Acta* **265**, 443.

Khachatourians, G. G., Clark, D. J. and Hardigree, A. A. (1973). *Journal of Bacteriology* **116**, 226.

Kline, B. C., Miller, J. R., Cress, D. E., Wlodarzyk, M., Manis, J. J. and Otten, M. R. (1976). *Journal of Bacteriology* **127**, 881.

Koch, A. L. (1966). *Journal of General Microbiology* **45**, 409.

Koch, A. L. and Blumberg, G. (1976). *Biophysics Journal* **16**, 389.

Koch, A. L. and Schaechter, M. (1962). *Journal of General Microbiology* **29**, 435.

Kojima, M., Suda, S., Hotta, S., Hamadi, K. and Suganama, A. (1970). *Journal of Bacteriology* **104**, 1010.

Kolenbrander, P. E. and Hohman, R. J. (1977). *Journal of Bacteriology* **130**, 1345.

Korch, C., Ovrebo, S. and Kleppe, K. (1976). *Journal of Bacteriology* **127**, 904.
Koyama, T., Yamada, M. and Matsuhashi, M. (1977). *Journal of Bacteriology* **129**, 1518.
Krulwich, T. A., Ensign, J. C., Tipper, D. J. and Strominger, J. L. (1967). *Journal of Bacteriology* **94**, 734.
Kubitschek, H. E. (1962). *Experimental Cell Research* **26**, 439.
Kubitschek, H. E. (1966). *Experimental Cell Research* **43**, 30.
Kubitschek, H. E. (1968a). *Biophysics Journal* **8**, 792.
Kubitschek, H. E. (1968b). *Biophysics Journal* **8**, 1401.
Kubitschek, H. E. (1969). *Biophysics Journal* **9**, 792.
Kubitschek, H. E. (1970). *Journal of Theoretical Biology* **28**, 15.
Kubitschek, H. E. (1971). *Journal of Bacteriology* **105**, 472.
Kubitschek, H. E. (1974). *Biophysics Journal* **14**, 119.
Kubitschek, H. E., Freedman, M. L. and Silver, S. (1971). *Biophysics Journal* **11**, 787.
Kuhn, D. A. and Starr, M. P. (1965). *Archiv für Mikrobiologie* **52**, 360.
Kvetkas, M. J., Krisch, R. E. and Zelle, M. R. (1970). *Journal of Bacteriology* **103**, 393.
Lane, H. E. D. and Denhardt, D. T. (1974). *Journal of Bacteriology* **120**, 805.
Laurent, S. J. and Vannier, F. S. (1973). *Journal of Bacteriology* **114**, 474.
Lazdunski, C. and Shapiro, B. M. (1972). *Journal of Bacteriology* **111**, 499.
Leive, L. and Davis, B. D. (1965). *Journal pf Biological Chemistry* **240**, 4370.
Lev, M. (1968). *Journal of Bacteriology* **95**, 2317.
Lin, E. C. C., Hirota, Y. and Jacob, F. (1971). *Journal of Bacteriology* **108**, 375.
Lorian, V. (1976). *Antimicrobial Agents and Chemotherapy* **7**, 864.
Lund, F. and Tybring, L. (1972). *Nature, New Biology, London* **236**, 135.
Luscombe, B. M. and Gray, T. R. G. (1974). *Journal of General Microbiology* **82**, 213.
Mandelstam, J. and Rogers, H. J. (1958). *Nature, London* **181**, 956.
Marr, A. G., Harvey, R. J. and Tentini, W. C. (1966). *Journal of Bacteriology* **91**, 2389.
Marr, A. G., Painter, P. R. and Nilson, E. H. (1969). *Symposium of the Society for General Microbiology* **19**, 237.
Martin, H. H. (1964). *Journal of General Microbiology* **36**, 441.
Matney, T. S. and Suit, J. C. (1966). *Journal of Bacteriology* **92**, 960.
Matsuhashi, S., Kamiryo, T., Blumberg, P. M., Linnett, P., Willoughby, E. and Strominger, J. L. (1974). *Journal of Bacteriology* **117**, 578.
Matsuzawa, H., Hayakawa, K., Sato, T. and Imahori, K. (1973). *Journal of Bacteriology* **115**, 436.
Mauck, J., Chan, L. and Glaser, L. (1971). *Journal of Biological Chemistry* **246**, 1820.
Mauck, J., Chan, L., Glaser, L. and Williamson, J. (1972). *Journal of Bacteriology* **109**, 373.
May, J. W. (1963). *Experimental Cell Research* **31**, 217.
McLean, F. I. and Munson, R. J. (1961). *Journal of General Microbiology* **25**, 17.
McNair-Scott, D. B. and Chu, E. (1958). *Experimental Cell Research* **14**, 166.
Meacock, P. A. and Pritchard, R. H. (1975). *Journal of Bacteriology* **122**, 931.
Meacock, P. A., Pritchard, R.H. and Roberts, E. M. (1978). *Journal of Bacteriology* **133**, 320.
Mendelson, N. H. (1968). *Cold Spring Harbor Symposium of Quantitative Biology* **33**, 313.
Mendelson, N. H. (1969). *Biochimica et Biophysica Acta* **190**, 132.
Mendelson, N. H. ((1975). *Journal of Bacteriology* **121**, 1166.
Mendelson, N. H. (1976). *Proceedings of the National Academy of Sciences of the United States of America* **73**, 1740.
Mendelson, N. H. and Coyne, S. I. (1975b). *Journal of Bacteriology* **121**, 1200.
Mendelson, N. H. and Reeve, J. N. (1973). *Nature, New Biology, London* **243**, 62.
Mendelson, N. H., Keener, S. and Cole, R. M. (1975c). *Microbios* **13**, 175.
Mertens, G. and Reeve, J. N. (1977). *Journal of Bacteriology* **129**, 1198.

Mindich, L. and Dales, S. (1972). *Journal of Cell Biology* **55**, 32.
Mirelman, D., Gan, Y. Y. and Schwarz, U. (1976). *Biochemistry, New York* **15**, 1781.
Mirelman, D., Gan, Y. Y. and Schwarz, U. (1977). *Journal of Bacteriology* **129**, 1593.
Mitchell, P. and Moyle, J. (1957). *Journal of General Microbiology* **16**, 184.
Mitchison, J. M. (1961). *Experimental Cell Research* **22**, 208.
Mitchison, J. M. (1971). "The Biology of the Cell Cycle", 201 pp. Cambridge University Press.
Mitchison, J. M. and Cummins, J. E. (1964). *Experimental Cell Research* **35**, 394.
Moore, R. L. and Hirsch, P. (1973a). *Journal of Bacteriology* **116**, 418.
Moore, R. L. and Hirsch, P. (1973b). *Journal of Bacteriology* **116**, 1447.
Muhlradt, P. F., Menzel, J., Golecki, J. R. and Speth, V. (1973). *European Journal of Biochemistry* **35**, 471.
Mulhradt, P. F., Menzel, J., Golecki, J. R. and Speth, V. (1974). *European Journal of Biochemistry* **43**, 533.
Murakami, S., Murakami, S. and Yoshikawa, H. (1976). *Nature, London* **259**, 215.
Murray, R. G. E., Francombe, W. H. and Mayall, B. H. (1959). *Canadian Journal of Microbiology* **5**, 641.
Nagai, K. and Tamura, G. (1972). *Journal of Bacteriology* **112**, 959.
Newman, C. N. and Bockrath, R. C. (1974). *Journal of General Microbiology* **85**, 203.
Newton, J. W. (1967). *Biochimica et Biophysica Acta* **141**, 633.
Newton, J. W. (1968). *Biochimica et Biophysica Acta* **165**, 534.
Normark, S. (1969). *Journal of Bacteriology* **98**, 1274.
Normark, S., Norlander, L., Grundstrom, T., Bloom, G. D., Boquet, P. and Frelat, G. (1976). *Journal of Bacteriology* **128**, 401.
Ohki, M. (1972). *Journal of Molecular Biology* **68**, 249.
Olden, K., Ito, S. and Hastings Wilson, T. (1975). *Journal of Bacteriology* **122**, 1310.
Osley, M. A. and Newton, A. (1974). *Journal of Molecular Biology* **90**, 359.
Park, J. T. and Burman, L. (1973). *Biochemical and Biophysical Research Communications* **51**, 863.
Paulton, R. J. L. (1970). *Journal of Bacteriology* **104**, 762.
Paulton, R. J. L. (1971b). *Canadian Journal of Microbiology* **17**, 119.
Paulton, R. J. L. (1972). *Canadian Journal of Microbiology* **18**, 1721.
Perkins, R. L. and Miller, R. A. (1973). *Canadian Journal of Microbiology* **19**, 251.
Perry, R. P. (1959). *Experimental Cell Research* **17**, 414.
Pierrucci, O. (1972). *Journal of Bacteriology* **109**, 848.
Pierrucci, O. and Helmstetter, C. E. (1969). *Federation Proceedings, Federation of American Societies for Experimental Biology* **28**, 1755.
Pierrucci, O. and Zuchowski, C. (1973). *Journal of Molecular Biology* **80**, 477.
Poole, R. K. (1977). *Journal of General Microbiology* **98**, 177.
Powell, E. O. (1956). *Journal of General Microbiology* **15**, 492.
Powell, E. O. (1958). *Journal of General Microbiology* **18**, 382.
Powell, E. O. (1964). *Journal of General Microbiology* **37**, 231.
Powell, E. O. and Errington, F. P. (1963). *Journal of General Microbiology* **31**, 315.
Previc, E. P. (1970). *Journal of Theoretical Biology* **27**, 471.
Previc, E. P. and Lowell, N. (1975). *Biochimica et Biophysica Acta* **411**, 377.
Pritchard, R. H. (1974). *Philosophical Transactions of the Royal Society, London* **267**, 303.
Pritchard, R. H. and Zaritsky, A. (1970). *Nature, London* **226**, 126.
Pritchard, R. H., Barth, P. T. and Collins, J. (1969). *Symposium of the Society for General Microbiology* **19**, 263.
Quadling, C. (1958). *Journal of General Microbiology* **18**, 227.

Rahn, O. (1932). *Journal of General Microbiology* **15**, 257.
Reeve, J. N. (1974). *Journal of Bacteriology* **119**, 560.
Reeve, J. N. and Clark, D. J. (1972). *Journal of Bacteriology* **110**, 117.
Reeve, J. N., Groves, D. J. and Clark, D. J. (1970). *Journal of Bacteriology* **104**, 1052.
Reeve, J. N., Mendelson, N. H. and Cole, R. M. (1972). *Molecular and General Genetics* **119**, 11.
Reeve, J. N., Mendelson, N. H., Coyne, S. I., Hallock, L. L. and Cole, R. M. (1973). *Journal of Bacteriology* **114**, 860.
Ricard, M. and Hirota, Y. (1973). *Journal of Bacteriology* **116**, 314.
Rogers, H. J. (1967). *Nature, London* **213**, 31.
Rogers, H. J. and Mandelstam, J. (1962). *Biochemical Journal* **84**, 299.
Rogers, H. J. and Thurman, P. F. (1977). *Journal of Bacteriology* **133**, 298.
Rogers, H. J., McConnell, M. and Burdett, I. D. J. (1968). *Nature, London* **219**, 285.
Rogers, H. J., Thurman, P. F., Taylor, C. and Reeve, J. N. (1974). *Journal of General Microbiology* **85**, 335.
Ron, E. Z., Rozenhak, S., Grossman, N. (1975). *Journal of Bacteriology* **123**, 374.
Ron, E. Z., Grossman, N. and Helmstetter, C. E. (1977). *Journal of Bacteriology* **129**, 569.
Ryter, A. (1967). *Folia Microbiologica, Praha* **12**, 283.
Ryter, A. (1968). *Bacteriological Reviews* **32**, 39.
Ryter, A. (1971). *Annales de l'Institut Pasteur, Paris* **121**, 271.
Ryter, A., Hirota, Y. and Schwarz, U. (1973). *Journal of Molecular Biology* **78**, 185.
Ryter, A., Shuman, H. and Schwarz, M. (1975). *Journal of Bacteriology* **122**, 295.
Ryu, D. Y. and Mateles, R. I. (1968). *Biotechnology and Bioengineering* **10**, 385.
Sargent, M. G. (1973). *Journal of Bacteriology* **116**, 397.
Sargent, M. G. (1974). *Nature, London* **250**, 252.
Sargent, M. G. (1975a). *Journal of Bacteriology* **123**, 7.
Sargent, M. G. (1975b). *Journal of Bacteriology* **123**, 1218.
Sargent, M. G. (1975c). *Biochimica et Biophysica Acta* **406**, 564.
Sargent, M. G. (1977a). *Proceedings of the Society for General Microbiology* **4**, 81.
Sargent, M. G. (1977b). *Molecular and General Genetics* **155**, 329.
Satta, G. and Fontana, R. (1974a). *Journal of General Microbiology* **80**, 51.
Satta, G. and Fontana, R. (1974b). *Journal of General Microbiology* **80**, 65.
Schaechter, M., Maaløe, O. and Kjeldgaard, N. O. (1958). *Journal of General Microbiology* **19**, 592.
Schaechter, M., Williamson, J. P., Hood, J. R. and Koch, A. L. (1962). *Journal of General Microbiology* **29**, 421.
Scherbaum, O. and Rasch, G. (1957). *Acta Pathologica et Microbiologica Scandinavica* **41**, 161.
Schmidt, J. M. and Stanier, R. Y. (1966). *Journal of Cell Biology* **28**, 423.
Schwarz, U., Asmus, A. and Frank, H. (1969). *Journal of Molecular Biology* **41**, 419.
Schwarz, U., Ryter, A., Rambach, A., Hellio, R. and Hirota, Y. (1975). *Journal of Molecular Biology* **98**, 749.
Shannon, K. P. and Rowbury, R. J. (1972). *Molecular and General Genetics* **115**, 122.
Shannon, K. P. and Rowbury, R. J. (1975). *Zeitschrift für Allgemeine Mikrobiologie* **15**, 447.
Shannon, K. P., Spratt, B. G. and Rowbury, R. J. (1972). *Molecular and General Genetics* **118**, 185.
Shannon, K. P., Armitage, J. and Rowbury, R. J. (1974). *Annales de Microbologie* **125B**, 233.
Shapiro, L. (1976). *Annual Review of Microbiology* **30**, 377.
Shaw, M. K. (1968). *Journal of Bacteriology* **95**, 221.

Shehata, T. E. and Marr, A. G. (1970). *Journal of Bacteriology* **103**, 789.
Shehata, T. E. and Marr, A. G. (1971). *Journal of Bacteriology* **107**, 210.
Shehata, T. E. and Marr, A. G. (1975). *Journal of Bacteriology* **124**, 857.
Shen, B. H., P. and Boos, W. (1973). *Proceedings of the National Academy of Sciences of the United States of America* **70**, 1481.
Shockman, G. D. (1965). *Bacteriological Reviews* **29**, 345.
Shockman, G. D. and Martin, J. T. (1968). *Journal of Bacteriology* **96**, 1803.
Shockman, G. D., Pooley, H. M. and Thompson, J. S. (1967). *Journal of Bacteriology* **94** 1525.
Shockman, G. D., Daneo-Moore, L. and Higgins, M. L. (1974). *Annals of the New York Academy of Sciences* **235**, 161.
Siccardi, A. G., Galizzi, A., Mazza, G., Clivio, A. and Albertini, A. M. (1975). *Journal of Bacteriology* **121**, 13.
Slater, M. and Schaechter, M. (1974). *Bacteriological Reviews* **38**, 199.
Smith, J. A. and Martin, L. (1973). *Proceedings of the National Academy of Sciences of the United States of America* **70**, 1263.
Smith, H. S. and Pardee, A. B. (1970). *Journal of Bacteriology* **101**, 901.
Spratt, B. G. (1975). *Proceedings of the National Academy of Sciences of the United States of America* **72**, 2999.
Spratt, B. G. (1977). *Journal of Bacteriology* **131**, 293.
Spratt, B. G. and Pardee, A. B. (1975). *Nature, London* **254**, 516.
Spratt, B. G. and Rowbury, R. J. (1970). *Journal of General Microbiology* **64**, 127.
Spratt, B. G. and Rowbury, R. J. (1971a). *Journal of General Microbiology* **65**, 305.
Spratt, B. G. and Rowbury, R. J. (1971b). *Molecular and General Genetics* **114**, 35.
Stacey, K. (1976). *Symposium of the Society for General Microbiology* **26**, 365.
Starka, J. and Morava, J. (1967). *Folia Microbiologica, Praha* **12**, 240.
Starka, J. and Morava, J. (1970). *Journal of General Microbiology* **60**, 251.
Staugaard, P., Berg, F. M., Woldringh, C. L. and Nanninga, N. (1976). *Journal of Bacteriology* **127**, 1376.
Sud, I. J. and Schaechter, M. (1964). *Journal of Bacteriology* **88**, 1612.
Suganama, A. (1962). *Journal of Infectious Diseases* **111**, 8.
Sundman, V. and Bjorksten, K. (1958). *Journal of General Microbiology* **19**, 491.
Teather, R. M., Collins, J. F. and Donachie, W. D. (1974). *Journal of Bacteriology* **118**, 407.
Terrana, B. and Newton, A. (1975). *Developmental Biology* **44**, 380.
Terrana, B. and Newton, A. (1976). *Journal of Bacteriology* **128**, 456.
Terry, D. R., Gaffar, A. and Sagers, R. D. (1966). *Journal of Bacteriology* **91**, 1625.
Thom, R., Hampe, A. and Saverbrey, G. (1969). *Zeitschrift fur die Gesamte Experimentelle Medizin* **151**, 331.
Tilby, M. J. (1977). *Nature, London* **266**, 450.
Tomasz, A., Albino, A. and Zanati, E. (1970). *Nature, London* **227**, 138.
Topiwala, H. H. and Sinclair, C. G. (1971). *Biotechnology and Bioengineering* **13**, 795.
Tsukayoshi, N., Fielding, P. and Fox, C. F. (1971). *Biochemical and Biophysical Research Communications* **44**, 497.
Tybring, L. and Melchior, N. H. (1975). *Antimicrobial Agents and Chemotherapy* **8**, 271.
Tzagaloff, H. and Novick, R. (1977). *Journal of Bacteriology* **129**, 343.
Van Alstyne, D. and Simon, M. I. (1970). *Journal of Bacteriology* **108**, 1366.
Van Iterson, W. and Aten, J. A. (1976). *Journal of Bacteriology* **124**, 384.
Van Tubergen, R. P. and Setlow, R. B. (1961). *Biophysics Journal* **1**, 589.
Wagner, M. (1964). *Zantralblatt fur Bakteriologie* **195**, 87.

Wagner, M. (1967). *Zantralblatt fur Bakteriologie* **203**, 378.
Ward, J. B. (1973). *Biochemical Journal* **133**, 395.
Ward, J. B. (1978). *Symposium of the Society for General Microbiology* **28**, 249.
Ward, C. M. and Claus, G. W. (1973). *Journal of Bacteriology* **114**, 378.
Ward, C. B. and Glaser, D. A. (1971). *Proceedings of the National Academy of Sciences of the United States of America* **68**, 1061.
Ward, J. B. and Perkins, H. R. (1973). *Biochemical Journal* **135**, 721.
Ward, J. B. and Perkins, H. R. (1974). *Biochemical Journal* **139**, 781.
Wechsler, J. A. and Gross, J. D. (1971). *Molecular and General Genetics* **113**, 273.
Weidel, W. and Pelzer, H. (1964). *Advances in Enzymology* **26**, 193.
Weigand, R. A., Shively, J. M. and Greenawalt, J. W. (1970). *Journal of Bacteriology* **102**, 240.
Westling-Haggstrom, B., Elmros, T., Normark, S. and Winblad, B. (1977). *Journal of Bacteriology* **129**, 333.
Westmacott, D. and Primrose, S. B. (1976). *Journal of General Microbiology* **94**, 117.
Whitfield, J. F. and Murray, R. G. E. (1956). *Canadian Journal of Microbiology* **2**, 245.
Wilkins, A., Gallant, J. and Harada, B. (1971). *Journal of Bacteriology* **108**, 1424.
Wilson, G. and Fox, C. F. (1971). *Biochemical and Biophysical Research Communications* **44**, 503.
Woldringh, C. L. (1973). *Cytobiology* **8**, 97.
Woldringh, C. L. (1976). *Journal of Bacteriology* **125**, 248.
Woldringh, C. L. and Nanninga, N. (1976). *Journal of Bacteriology* **127**, 1455.
Woldringh, C. L., de Jong, M. A., van den Berg, W. and Koppes, L. (1977). *Journal of Bacteriology* **131**, 270.
Wright, D. N. and Lockhart, W. R. (1965). *Journal of Bacteriology* **89**, 1026.
Wu, P. C. and Pardee, A. B. (1973). *Journal of Bacteriology* **114**, 603.
Yamada, M., Koyama, T. and Matsuhashi, M. (1977). *Journal of Bacteriology* **129**, 1513.
Yamaguchi, K. and Yoshikawa, H. (1973). *Nature, New Biology, London* **244**, 204.
Yamaguchi, K. and Yoshikawa, H. (1977). *Journal of Molecular Biology* **110**, 219.
Zaritsky, A. (1975). *Journal of Theoretical Biology* **54**, 243.
Zaritsky, A. and Pritchard, R. H. (1973). *Journal of Bacteriology* **114**, 824.
Zusman, D. R. and Krotski, D. M. (1974). *Journal of Bacteriology* **120**, 1427.
Zusman, D. R., Gottlieb, P. and Rosenberg, E. (1971). *Journal of Bacteriology* **105**, 811.
Zusman, D. R., Inouye, M. and Pardee, A. B. (1972). *Journal of Molecular Biology* **69**, 119.

For Note added in Proof see p. 312.

Physiology and Biochemistry of Bacterial Phospholipid Metabolism

W. R. FINNERTY

Department of Microbiology, University of Georgia, Athens, Georgia 30602, U.S.A.

I. Introduction	177
II. Structural, Functional and Biological Characteristics of Microbial Phospholipids	178
A. General Considerations of Phospholipid Structure and Composition	178
B. Asymmetry of Phospholipids in Membranes	181
C. The Role of Phospholipids in Enzyme Activity	183
D. Phospholipids and Microbial Viruses	187
E. Regulation of Phospholipid Composition	188
III. Biosynthesis of Microbial Phospholipids	194
A. Phosphatidic Acid Biosynthesis	194
B. Cytidine Diphosphate Diglyceride Biosynthesis	198
C. Phosphatidylglycerophosphate and Phosphatidylglycerol Biosynthesis	199
D. *O*-Amino-Acyl Phosphatidylglycerol Biosynthesis	200
E. Fatty-Acyl Phosphatidylglycerol Biosynthesis	201
F. Diphosphatidylglycerol (Cardiolipin) Biosynthesis	203
G. Phosphatidylserine and Phosphatidylethanolamine Biosynthesis	204
H. Methylated Derivatives of Phosphatidylethanolamine	207
I. Ornithine-Containing Lipids	208
IV. Catabolism of Microbial Phospholipids	210
A. Phospholipase A_1	213
B. Phospholipase A_2	219
C. Lysophospholipases	219
D. Phospholipase C	220
E. Phospholipase D	221
F. Regulation	222
V. Conclusion	225
References	226

I. Introduction

The role of lipids in diverse biological systems has received increasing attention throughout the last decade. This resurgence of interest in lipid research stems, in part, from advances in methodological, analytical and preparative techniques allowing for greater sensitivity, precision and accuracy in lipid analyses. The application of sophisticated instrumentation, such as gas-liquid chromatography, mass spectrometry, nuclear magnetic resonance spectrometry and liquid chromatography, has enabled the lipid chemist more effectively to probe the structural complexities of lipids.

The study of lipids necessarily originated in their purification and structure elucidation, an effort vigorously pursued even today. With the accumulation of a substantial body of structural knowledge, attention has turned to those functional interrelationships relating the role of lipids to cellular physiology. This effort has culminated today in studies of membranes and membrane-associated activities.

The phospholipid composition of diverse numbers of microorganisms has been subjected to intensive study over the past several years. These investigations have demonstrated that phospholipids are dynamic biochemical entities exhibiting important functional roles in cellular physiology. The interaction that exists between proteins and lipids comprises the organelle termed the membrane. This complex structure has assumed somewhat of a paradoxical role in such diverse biological phenomena as sporulation, initiation of DNA synthesis, electron transport, active transport, phospholipid biosynthesis, phospholipid catabolism and ribosomal RNA synthesis, to name a few. Such a diversity of biological attributes serves to emphasize the dynamic characteristics of the biological membrane. The realization that significant portions of the biosynthetic and metabolic functions of the living cell are performed by membrane-associated proteins has complemented the fact that cellular lipids are predominantly localized in the membrane. As our body of knowledge concerning lipids has accumulated, it becomes increasingly possible to relate and identify structures and biosynthetic processes participating in the various aspects of cellular metabolism.

II. Structural, Functional and Biological Characteristics of Microbial Phospholipids

A. GENERAL CONSIDERATIONS OF PHOSPHOLIPID STRUCTURE AND COMPOSITION

The major phospholipids characterized from a variety of microorganisms are shown in Table 1. Comparative analyses of microbial

TABLE 1. Structures of microbial phospholipids

Name	Structure

Phosphatidylglycerol	H$_2$COOCR
	|
	RCOOCH O
	| ||
	CH$_2$–O–P–O–CH$_2$–CH–CH$_2$
	        | | |
	       OH OH OH

Diphosphatidylglycerol (cardiolipin)	H$_2$COOCR RCOOCH$_2$ RCOOCH O O HCOOCR CH$_2$–O–P–O–CH$_2$–CH–CH$_2$–O–P–O–CH$_2$ OH OH OH
Phosphatidylethanolamine	H$_2$COOCR RCOOCH O CH$_2$–O–P–O–CH$_2$–CH$_2$–NH$_2$ OH
Phosphatidylserine	H$_2$COOCR RCOOCH O CH$_2$–O–P–O–CH$_2$–CH–COOH OH NH$_2$
Phosphatidylcholine	H$_2$COOCR CH$_3$ RCOOCH O CH$_2$–O–P–O–CH$_2$–CH$_2$–N–CH$_3$ OH CH$_3$
Phosphatidyl-N,N'-dimethylethanolamine	H$_2$COOCR CH$_3$ RCOOCH O CH$_2$–O–P–O–CH$_2$–CH$_2$–N–CH$_3$ OH
Phosphatidyl-N-monomethylethanolamine	H$_2$COOCR CH$_3$ RCOOCH O CH$_2$–O–P–O–CH$_2$–CH$_2$–NH OH

TABLE 1 (*continued*)

Name	Structure
bis-Phosphatidic acid	H$_2$COOCR RCOOCH$_2$ RCOOCH O HC–OOCR CH$_2$–O–P–O–CH$_2$ OH
Semilyso-*bis*-phosphatidic acid (fatty-acyl phosphatidylglycerol)	H$_2$COOCR RCOOCH$_2$ RCOOCH O HCOH CH$_2$–O–P–O–CH$_2$ OH
Lyso-*bis*-phosphatidic acid	H$_2$COOCR RCOOCH$_2$ HOCH O HCOH CH$_2$–O–P–O–CH$_2$ OH

phospholipids have shown that Gram-negative bacteria are characterized by phosphatidylethanolamine and its monomethyl- and dimethyl derivatives (Goldfine, 1972). The Enterobacteriaceae have a phospholipid composition consisting of phosphatidylethanolamine, phosphatidylglycerol and cardiolipin. Gram-positive nonsporogenous bacteria are characterized by phospholipids consisting of phosphatidylglycerol, cardiolipin and O-amino-acyl esters of phosphatidylglycerol. Phosphatidylethanolamine and its methylated derivatives are rarely observed in this group of bacteria. The sporogenous Bacillaceae characteristically contain phosphatidylethanolamine, but normally at lower concentrations than is found in Gram-negative bacteria.

The lipids of bacteria are predominately comprised of phospholipids which are localized almost exclusively in the cell membranes. Biosynthesis of phospholipids is catalysed by integral membrane proteins (McCaman and Finnerty, 1968; Carter, 1968; Chang and Kennedy, 1967a, b, c; Patterson and Lennarz, 1971; DeSiervo and Salton, 1971; Short and White, 1972; White *et al.*, 1971; Bell *et al.*, 1971).

The distribution of the membrane-associated proteins catalysing biosynthesis of bacterial phospholipids remains less well resolved. The random distribution of phospholipid biosynthetic enzymes throughout the membrane offers as feasible an alternative as multiple sites of synthesis localized in specific regions of the membrane. The latter alternative offers the possibility of specific membrane structures responsible for phospholipid biosynthesis with subsequent rapid diffusion of the phospholipid into contiguous membranes. Evidence supporting the dispersive growth of membranes tends to agree, however, with the presence of multiple sites of phospholipid biosynthesis throughout the membrane (Mindich and Dades, 1972; Green and Schaechter, 1972).

B. ASYMMETRY OF PHOSPHOLIPIDS IN MEMBRANES

The biological membrane is composed basically of phospholipids organized into a lipid bilayer with interdispersed proteins. As such, this complex structure presents one monolayer surface to the cytoplasm and the opposite monolayer surface to the external environment. Recently, the important feature of asymmetry in the lipid bilayer has been recognized and described (Bretscher, 1972a). Lipids which occupy one side of the bilayer are different from the lipids on the opposing side. Bilayer asymmetry now appears to be a general property of biological membranes, although the asymmetry does vary from one membrane species to another (Bretscher, 1972b). Recognition of the asymmetric distribution of phospholipids in the membrane dictates the need for cellular mechanisms for generating this topological organization. This aspect of membrane organization remains unresolved. Methods available for the study of asymmetric phospholipid distribution in membranes include chemical labelling of phospholipids, phospholipid exchange proteins, phospholipase digestion of membranes, immunochemical techniques and nuclear magnetic resonance spectroscopy (Bergelson and Barsukov, 1977).

Chemical labelling of membrane-localized phospholipids can be accomplished with such membrane penetrating probes as 1-fluoro-2,4-dinitrobenzene and acetic anhydride which react with aminophospholipids. Non-penetrating probes are sulphanilic acid diazonium chloride (Schafer et al., 1974) which reacts with aminophospholipids as well as phosphatidylglycerol, fluorescamine

(Hawkes *et al.*, 1976), cyanogen bromide-activated dextran (Kamio and Nikaido, 1976) and the lactoperoxidase-catalysed iodination reaction (Mersel *et al.*, 1976).

Phospholipid exchange proteins which effect the translocation of specific phospholipids between two membrane structures have been described in eukaryotic cells (Helmkamp *et al.*, 1974; Wirtz, 1974). It appears that the translocation reaction involves only one-half of the lipid bilayer. When phospholipid translocation proteins are reacted with donor membranes, only the phospholipids of the outer surface of the donor membrane are exchanged. A few of the donor membrane systems studied include liposomes and mitochondria (Johnson and Zilversmit, 1975), erythrocytes (Bloj and Zilversmit, 1976) and bacterial protoplasts (Barsukov *et al.*, 1976). Lipid-transfer proteins exhibit a number of positive advantages over other techniques employed for assessing membrane asymmetry. They are non-permeating so that only the outer surfaces of the respective donor and acceptor membranes interact; no lytic activity has been observed towards vesicular membranes and only minor perturbations of membrane structure result, allowing for better interpretation of results.

Phospholipases are considered non-penetrating proteins which hydrolyse those phospholipids on the outer surface of either liposomes or membrane vesicles. By appropriate manipulation of membranes, it is possible to prepare inside-out vesicles as contrasted to right-side out, enabling selective action of phospholipases at the opposing sides of the bilayer. Interpretation of phospholipase-digestion data is often difficult due to the variety of actions of phospholipases against liposomes and biomembranes. The diversity of cofactor requirements, substrate specificities and physical conditions of reaction often result in structure perturbations of membranes making interpretation difficult.

A number of phospholipids are immunogenic in that antibodies are formed to the polar head group of cardiolipin, phosphatidylinositol, phosphatidylglycerol and sphingomyelin (Teitelbaum *et al.*, 1973; Inoue and Nojima, 1967; Guarneri *et al.*, 1971; Schiefer *et al.*, 1975). Immunological information is only qualitative, and is limited by the small number of immunologically active phospholipids.

Lipid asymmetry in membranes of bacteria has been documented in several micro-organisms. Chemical labelling of the amino-acyl esters of phosphatidylglycerol of *Acholeplasma* sp. has indicated their localization on the inner surface of the cytoplasmic membrane (Marinetti and

Love, 1976). Membranes of *Mycoplasma capricolum* were shown to be asymmetric with respect to cholesterol, the greater portion being on the outer surface of the membrane (Bittman and Rottem, 1976). The cytoplasmic membrane of *M. lysodeikticus* exhibits asymmetric distribution of its phospholipids, with cardiolipin evenly divided between the inner and outer surface of the membrane. Phosphatidylglycerol was localized predominantly on the outer surface and phosphatidylinositol on the inner surface (Barsukov *et al.*, 1976). Rothman and Kennedy (1977) have studied the asymmetric distribution of phospholipids in the membrane of *B. megaterium* using chemical probes. The inner leaflet of the bilayer contained twice as much phosphatidylethanolamine as the outer leaflet. It was inferred that the other major phospholipid, phosphatidylglycerol, was localized predominantly on the outer surface of the cytoplasmic membrane.

The implications attached to asymmetric distribution of phospholipids in biomembranes are considerable. The possibility that asymmetric lipid distribution regulates differential membrane fluidity in each half of the bilayer poses interesting questions. Does the outer half of the lipid bilayer which is exposed to the external environment require a more rigid state for proper functional attributes than the inner half which is exposed to the cytoplasm? Does a higher proportion of charged lipids on one side of the bilayer facilitate interactions with extrinsic membrane proteins and represent a necessary micro-environment for the activity of membrane-bound enzymes? The proposition has been advanced that asymmetrically distributed phospholipids can operate as bilayer couples (Sheetz and Singer, 1974). The microheterogeneity of membrane-lipid structure remains to be resolved, but promises to yield many new concepts and insights into the structure–function relationships of biological membranes.

C. THE ROLE OF PHOSPHOLIPIDS IN ENZYME ACTIVITY

An increasing number of enzymes exhibit a stringent requirement for lipids either as cofactors or activators of enzyme activity. Many of these enzymes have been further observed to exhibit different chemical and physical characteristics in the soluble form as compared to the particulate or membrane-bound form. Enzymes previously considered refractory to solubilization have further become useful experimental models once their lipid requirements were recognized and fulfilled. The realization that lipids exhibit functional relevance to enzyme activity and that specific protein-lipid interactions alter a variety of physical properties of a protein led to the concept of allotropy (Racker, 1967). Some of the physical changes that occur which reflect allotopy are inhibitor sensitivity, cold lability, pH optimum, Michaelis–Menten constants, oxidation-reduction potential, reconstitution and substrate specificity. Consequently, caution must be exercised in the interpretation and application of *in vitro* experiments particularly where radical modifications of the native system occurs.

Criteria for demonstrating lipid dependency have been proposed, requiring that removal of the lipid from the protein results in loss of enzyme activity. Restoration of enzyme activity is obtained on reassociation of the lipid to the protein (Fleischer *et al.*, 1962).

Studies of the role of lipids in the activity of membrane-bound enzymes require that the proteins be highly purified. Unfortunately, only a few membrane-bound enzymes or complexes have been isolated and their lipid requirements determined. Most of the enzymes that have been studied are of bacterial origin or are from mitochondria or microsomes of higher organisms. Some of the better studied enzymes of bacterial origin are listed in Table 2.

The murein lipoprotein of *Escherichia coli* exists in the outer membrane in two forms, namely a free form and one covalently linked to the peptidoglycan (Bosch and Braun, 1973; Braun and Sieglin, 1970; Inouye *et al.*, 1972; Lee and Inouye, 1974). The estimated number of lipoprotein molecules is 7.5×10^5 and it appears to be the most abundant protein in the cell (Inouye, 1974). The lipoprotein consists of 58 amino-acid residues and its cell-free biosynthesis and assembly have been studied (Hirashima *et al.*, 1974; Inouye *et al.*, 1972). The function of this well studied protein remains, however, totally unresolved.

The C_{55}-isoprenoid alcohol phosphokinase represents one of the few examples of an intrinsic membrane protein purified to homogeneity (Sandermann and Strominger, 1971, 1972). This enzyme catalyses the ATP-dependent phosphorylation of C_{55}-isoprenoid alcohols, and is insoluble in water but soluble in organic solvents. Enzyme activity is absolutely dependent on lipid activators. A number of amphipathic lipids function as effective activators although no particular chemical structure or charge appears optimal (Gennis and Strominger, 1976).

TABLE 2. Lipid-requiring enzymes of microbial origin

Protein	Microorganisms	Reference
Murein lipoprotein	*Escherichia coli*	Hantke and Braun (1973); Inouye (1974)
Bacteriorhodopsin	*Halobacterium halobium*	Racker and Hinkle (1974); Knowles *et al.* (1975); Racker and Stoeckenius (1974)
C_{55}-Isoprenoid alcohol phosphokinase	*Staphylococcus aureus*	Gennis and Strominger (1976)
Phosphoenolpyruvate phosphotransferase	*Escherichia coli*	Kundig and Roseman (1971); Kundig (1974); Milner and Kaback (1970)
Galactosyl transferase and glucosyl transferase	*Salmonella typhimurium*	Endo and Rothfield (1969); Hinckley *et al.* (1972); Rothfield and Romeo (1971)
Pyruvate oxidase	*Escherichia coli*	Cunningham and Hager (1975)
Malate-vitamin K reductase	*Mycobacterium phlei*	Imai and Brodie (1973)
Adenosine triphosphatase	*Escherichia coli* *Micrococcus lysodeikticus* *Acholeplasma laidlawii*	Peter and Ahlers (1975) Salton (1974) Bevers *et al.* (1977)
Penicillinase	*Bacillus licheniformis*	Aiyappa and Lampen (1976); Yamamoto and Lampen (1977)
ω-Hydroxylase	*Pseudomonas putida*	Ruettinger *et al.* (1974)

The most effective activators relative to egg phosphatidylcholine were lysophosphatidylethanolamine, bovine lysophosphatidylserine and bovine phosphatidylcholine.

The phosphotransferase system of *E. coli* has been obtained in soluble form, and the specific involvement of phosphatidylglycerol demonstrated in the vectorial transport of sugars. This system represents the first demonstration of the involvement of a specific phospholipid in a membrane-localized complex effecting transport and serves to emphasize that a variety of phospholipids are involved in the activation of various membrane reactions with an inordinate amount of specificity.

Glucosyl- and galactosyltransferases are involved in lipopolysaccharide biosynthesis and require phospholipids for activity. One role for phospholipid in this series of complex reactions is that of substrate activation (Rothfield and Romeo, 1971). The phospholipids react physically with the lipopolysaccharide, thereby forming an activated intermediate. Phosphatidylethanolamine was shown to be the most effective phospholipid in reconstitution experiments using both monolayers and aqueous lipid dispersions.

The pyruvate-oxidase system of *E. coli* consists of the flavoprotein, pyruvate oxidase, and a membrane-associated terminal electron-transport system. A series of studies have demonstrated that phospholipids modulate the activity of purified pyruvate oxidase (Cunningham and Hager, 1971a, b). Phospholipids were also shown to affect the structural integrity of the electron-transport portion of the membrane. Neutral lipid constituents were recently shown to re-activate pyruvate oxidation. Ubiquinone-6 exhibited a specificity in re-activation of lipid-depleted pyruvate oxidase electron-transport system (Cunningham and Hager, 1975).

A malate-vitamin K reductase purified from *Mycobacterium phlei* requires FAD and phospholipid for optimal activity (Imai and Brodie, 1973). The monomeric molecular weight was assessed at 51,000–53,000 and the aggregated form at approximately 164,000. The activity of the enzyme was found to be a function of the degree of aggregation, a phenomenon in which phospholipids appeared to play a role. The aggregation is further affected by salt concentration in that high concentrations of salt favoured the monomeric form of the enzyme and low concentrations the aggregated form.

The adenosine triphosphatase (ATPase) of micro-organisms has been extensively studied and observed to be influenced by the presence of phospholipids (Salton, 1974). The membrane-localized ATPase of *Strep. faecalis* was demonstrated as reacting with phospholipid bilayers which effected changes in enzyme properties (Redwood *et al.*, 1973). Munoz *et al.* (1969) described a soluble ATPase from *Micrococcus lysodeikticus* as phospholipid free. A recent study of the ATPase of *E. coli* has demonstrated a phospholipid requirement (Peter and Ahlers, 1975). Phosphatidylethanolamine was shown to represent the main phospholipid component of the solubilized enzyme.

A requirement for phosphatidylglycerol has been demonstrated for the Mg^{2+}-dependent ATPase of *Acholeplasma laidlawii*. Treatment of *A. laidlawii* membranes with phospholipase A_2 or phospholipase C results in the 90 per cent hydrolysis of phosphatidylglycerol without a corresponding loss of activity in the ATPase. However, if the remaining 10 per cent phosphatidylglycerol is hydrolysed, Mg^{2+}-dependent ATPase activity is appreciably decreased. The inactivated

Mg^{2+}-dependent ATPase can be re-activated by adding phosphatidylglycerol, phosphatidic acid or phosphatidylserine, but not phosphatidylcholine or phosphatidylethanolamine (Bevers et al., 1977). This ATPase bears close similarities to the $Na^+ + K^+$-dependent ATPase of eukaryotic cells with respect to dependence on acidic phospholipids and the intrinsic nature of enzyme in the membrane.

The membrane-bound penicillinase of B. licheniformis is a phospholipoprotein which differs from the extracellular penicillinase (Yamamoto and Lampen, 1975). The membrane-bound form contains 25 extra amino-acid residues with phosphatidylserine at the amino-terminus. The hydrophobic portion of the membrane-bound enzyme has recently been assessed in the context of the amino-acid sequence (Yamamoto and Lampen, 1977). It has been further established that several proteins are present in the membranes of B. licheniformis which contain covalently-linked phosphatidylserine (Aiyappa and Lampen, 1976). Penicillinase, therefore, becomes the first example of a group of membrane-localized proteins, the phospholipoproteins, which may contribute significantly to our knowledge of the mechanism of protein secretion in micro-organisms.

A phospholipid requirement for the ω-hydroxylase of Ps. putida was demonstrated by Ruettinger et al. (1974). The ω-hydroxylase is a yellow protein of molecular weight 42,000 containing one atom of iron per peptide chain. Lipid-free hydroxylase preparations exhibited decreased activity unless supplemented with phospholipid derived from the ω-hydroxylase or with dilauroylglyceryl-3-phosphorylcholine.

D. PHOSPHOLIPIDS AND MICROBIAL VIRUSES

That phospholipids may play a role in viral infectivity has received experimental support in recent years. The discovery of specific bacteriophage-associated phospholipids serves to emphasize the importance of further studies on the role of lipid metabolism in the infectious process. For example, the lipid-containing bacteriophage PM2 contains phosphatidylglycerol (67%) and phosphatidylethanolamine (27%) as contrasted to its host, Pseudomonas BAL-31, which contains phosphatidylglycerol (23%) and phosphatidylethanolamine (75%; Braunstein and Franklin, 1971). As a consequence of bacteriophage infection, the cellular levels of phosphatidylglycerol increase, with a concomitant rise in the levels of phosphatidic acid.

As to the origin of the viral phospholipids, several possibilities present themselves for consideration. (1) Viral phospholipids result from selective uptake of pre-existing cellular phospholipids. (2) Viral phospholipids are synthesized *de novo* following infection. (3) Viral phospholipids result from a modification of pre-existing cellular phospholipids. The results of dual-isotope, pulse-labelling experiments suggest that the cellular phospholipid polar groups are cleaved and replaced by glycerol during bacteriophage infection (Snipes *et al.*, 1974). Tsukagoshi and Franklin (1974) have demonstrated that one-third of the viral phospholipids are obtained from pre-existing cellular phospholipids and two-thirds are synthesized during viral replication. Synthesis of bacteriophage PM2 has been accomplished in *in vitro* reconstitution studies (Schäfer and Franklin, 1975) and the effect of modifying the fatty-acyl residues in the viral phospholipids has been assessed with respect to membrane composition and structure (Tsukagoshi *et al.*, 1975; Scandella *et al.*, 1974).

A bacteriophage has recently been isolated which infects Gram-negative bacteria, including *E. coli*, and carries a drug-resistance plasmid (Bradley and Rutherford, 1975). This virus, PR4, produced in *E. coli* contains phospholipid (Sands and Cadden, 1975). Virus PR4 has a lipid-rich region that contains a large amount of phosphatidylglycerol and lower amounts of phosphatidylethanolamine and phosphatidylserine (Sands, 1976). Synthesis of the viral phospholipids does not alter the composition of the host cellular phospholipids as a result of infection. More than one-half of the viral phospholipids are synthesized in the host following infection with no effect on host phospholipid composition. This contrasts with the alteration of phospholipids effected by PM2 infection.

E. REGULATION OF PHOSPHOLIPID COMPOSITION

The physical and metabolic characteristics of phospholipids localized in biological membranes have received a significant amount of attention throughout the past decade. Many of these studies have attempted to gain insight into the regulatory aspects of membrane biogenesis as related to phospholipid composition and their associated physical parameters. An earlier review assessed the status of phospholipid metabolism in *E. coli* (Cronan and Vagelos, 1972). A review has appeared which serves to complement the previous summary by presenting a critical evaluation of the *E. coli* literature concerning phase-transition characteristics of lipids, experimental systems from which these properties were derived, and the study of these properties in *E. coli* membranes (Cronan and Gelman, 1975).

The physical properties of phospholipids in the *E. coli* membrane appear to be controlled by manipulation of either endogenous or supplemental fatty acids. This phenomenon serves to effect a relatively constant fluidity state in the membrane in response to physical changes such as temperature. Initial studies indicated that the specificity of the acyltransferase was responsible for the fatty-acyl composition of cellular phospholipids. It has now been demonstrated that the ratio of saturated fatty acids to unsaturated fatty acids represents a regulatory mechanism for maintaining proper membrane lipid fluidity relationships (Cronan, 1974). Lipid-phase characteristics, therefore, appear to be regulated at both fatty acid and phospholipid biosynthetic levels (Cronan *et al.*, 1975). Alteration of the polar head groups of phospholipids has yet to be studied in the context of regulatory parameters.

The variety of chemical, physical and cultural factors influencing the phospholipid composition of micro-organisms is extensively documented in the literature. In spite of this vast body of knowledge, much remains to be learned as to the mechanism of regulation of phospholipid metabolism. The effect of temperature on phospholipid metabolism probably represents the best studied physical variable (Fulco, 1974; Finnerty and Makula, 1975). Glycerol auxotrophs of *E. coli* defective in β-oxidation accumulate free fatty acids (Cronan *et al.*, 1975). The ratio of saturated fatty acid to unsaturated fatty acid, accumulating as free fatty acid, was dependent upon the incubation temperature (Cronan, 1975). Incorporation of exogenously supplied fatty acids into the membrane phospholipids showed the same temperature dependency relationships. These data indicate that fatty-acid synthesis represents one site involved in the temperature-mediated regulation of phospholipid fatty acid composition, a second site being the regulation of phosphatidic acid biosynthesis.

The phospholipids of *Thermus aquaticus* increased two-fold when the growth temperature was increased from 50°C to 75°C (Ray *et al.*, 1971). The proportions of individual phospholipids remained constant throughout the temperature transition. De Siervo (1969) demonstrated that the phospholipids of *E. coli* B decreased throughout the growth cycle when grown at 37°C. However, when growth was at 27°C, phospholipids increased throughout the growth cycle.

Culture age influences the quantitative composition of a bacterium's phospholipids. Qualitative and quantitative differences were documented in *M. lysodeikticus* in the exponential and stationary phases of batch culture (DeSiervo and Salton, 1973). Cardiolipin increased to a maximum concentration in exponential growth and decreased to a minimum concentration in the stationary phase. Phosphatidylglycerol was low in early exponential growth and increased to a maximum concentration in the stationary phase. A complex combination of physical and chemical factors appears operative in controlling the lipid composition of micro-organisms. Unfortunately, clear and concise information regarding these mechanisms has yet to be obtained.

The effect of cellular metabolites on phospholipid synthesis has been documented. Glycero-3-phosphate acylation in *E. coli* is inhibited 25 per cent by CTP and 60 per cent by ATP, while other nucleoside triphosphates exhibited no effect (Kito and Pizer, 1969). Glaser *et al.* (1973) provided evidence that phospholipid biosynthesis was coupled to DNA, RNA and protein biosynthesis. When temperature-sensitive glycero-3-phosphate acyltransferase mutants of *E. coli* were shifted to the restrictive temperature, all macromolecular and phospholipid biosynthesis ceased immediately. Stringent cells stopped synthesis of protein, RNA and phospholipid following starvation for a required amino acid. Relaxed cells, however, stopped growth and protein synthesis, but continued to synthesize RNA at normal rates and phospholipids at approximately one-half the normal rate observed in amino acid-supplemented cultures (Golden and Powell, 1972). Phosphatidylethanolamine exhibited a rapid turnover in amino acid-deprived cells suggesting that its turnover is regulated somehow by the lesion in the RNA control gene.

The molecular mechanism of the decreased rate of phospholipid biosynthesis has been considered as involving inhibition of fatty acid and phospholipid biosynthetic enzymes by guanosine 3′,5′-bis(diphosphate) (ppGpp). Cashel and Gallant (1969) showed that ppGpp accumulated during amino-acid starvation of rel^+ strains, but not rel^- strains of *E. coli* (Cashel, 1975). The apparent involvement of ppGpp in *rel* A gene control of phospholipid biosynthesis was demonstrated in both *in vivo* and *in vitro* experiments. An inverse relationship between the rate of phospholipid biosynthesis and the presence of ppGpp was demonstrated during the onset and release of stringency (Merlie and Pizer, 1973). Further *in vitro* studies showed that ppGpp inhibits a number of phospholipid biosynthetic enzymes (Polakis et al., 1973; Lueking and Goldfine, 1975; Ray and Cronan, 1975). Evidence indicating that inhibition of phospholipid biosynthesis is not due solely to inhibition of fatty acid biosynthesis has been presented by Nunn and Cronan (1974, 1976a, b). A recent report by these authors showed a quantitative relationship between intracellular levels of ppGpp and the rate of phospholipid biosynthesis (Nunn and Cronan, 1976a).

Crowfoot et al. (1972a, b) studied an unsaturated fatty-acid auxotroph of E. coli which exhibited stringent control of RNA synthesis but a relaxed response for phospholipid biosynthesis upon amino-acid starvation. The phospholipids of this mutant turned over continuously, with the turnover being stimulated by amino-acid starvation.

A variation in light intensities has been shown to induce changes in the phospholipid content of specific physiological groups of microorganisms. *Chromatium* strain D, grown under varying light intensities, exhibited 40 per cent more phospholipid at 7500 ft-candles than at 100 ft-candles (Steiner et al., 1970). Cells grown under low light intensity contained 3.5 times more bacteriochlorophyll and significantly more internal membranes than cells grown under high light intensity. Thus, the amount of bacteriochlorophyll in the cells did not correlate with the concentration of phospholipids. In contrast, *Rhodopseudomonas spheroides* undergoes inductive changes accompanied by increased phospholipid synthesis during the transition from aerobic growth to semi-anaerobic growth in the light. Gorchein et al. (1968a) observed that the specific activity of individual phospholipids in the chromatophore was identical with that of induced cells. It was concluded that this transition to anaerobic growth in the light resulted in new phospholipid which became evenly distributed throughout the membranes of the cell. Formation of photosynthetic vesicles complete with accessory photosynthetic pigments parallels the increase in new phospholipid synthesis.

An example of chemically induced inhibition of phospholipid biosynthesis is phenethanol. This compound has been shown to inhibit phospholipid biosynthesis in E. coli (Nunn and Tropp, 1972; Nunn, 1975). Phenethanol alters the phospholipid composition by decreasing the rate of phosphatidylethanolamine and phosphatidylglycerol biosynthesis as well as inhibiting *de novo* synthesis of saturated and unsaturated fatty acids (Nunn, 1975). Phenethanol appears to cause inhibition of phospholipid synthesis at the level of the acyltransferase. The activities of subsequent phospholipid biosynthetic enzymes were not affected by phenethanol, indicating the inhibitory action to be at the level of glycero-3-phosphate acyltransferase (Nunn et al., 1977).

Studies on induction of electron-transport chains in facultative micro-organisms has provided additional insight into the integration of phospholipids in specific membrane-bound activities. *Staphylococcus aureus* exhibits no detectable electron-transport chain when grown anaerobically. Introduction of oxygen results, however, in an increased rate of growth and sequential induction of a membrane-bound electron-transport chain consisting of flavoproteins, cytochrome $b-b_1$, cytochrome a and cytochrome oxidase o (Taber and Morrison, 1964; Frerman and White, 1967b). During the shift from anaerobic to aerobic metabolism, cytochrome b_1 and cytochrome o appeared first, followed four hours later by cytochrome a. A two-fold increase in phospholipids occurred during this metabolic transition (Frerman and White, 1967a). The increase in phospholipid was not uniform, in that phosphatidylglycerol increased two-fold, cardiolipin increased 1.6-fold and lysylphosphatidylglycerol remained constant. The molar ratio between phosphatidylglycerol and lysyl-phosphatidylglycerol varied from 4.1:1 to 8.9:1 throughout the aerobic transition period indicating functional electron-transport membranes with varying molar proportions of specific phospholipids (Frerman and White, 1967b).

In addition to the two-fold increase in phospholipids, a corresponding two-fold increase in phospholipid fatty acids with disproportionate turnover kinetics in individual phospholipids was documented during induction of the electron-transport chain (Frerman and White, 1968). These studies provided evidence for functional mosaicism in membranes differing in lipid composition, a subject of greater importance today in terms of the asymmetric distribution of phospholipids in membranes and their functional relevance to biological membranes.

A further analysis of phospholipid metabolism in relation to electron-transfer chains has appeared in studies with *Haemophilus parainfluenzae*. The electron-transport chain of this organism consists of dehydrogenases, cytochromes, cytochrome oxidases, quinone and phospholipids (White, 1962). A 50-fold increase in cytochrome content occurs following the transition from high to low oxygen tension

(White, 1963). Concomitant with increased cytochrome synthesis was a significant decrease in the turnover of phosphatidylethanolamine and cardiolipin and an increase in the turnover of phosphatidylglycerol. An overall net 5–10 per cent increase in total phospholipid was noted during the transition period (White and Tucker, 1969a). These changes in phospholipids occur with no alteration in growth rate in contrast to *Staph. aureus* which exhibited a faster rate of growth following the introduction of oxygen.

An example of carbon-induced modification of the cellular phospholipid content has been described by Makula and Finnerty (1970). Growth of *Acinetobacter* species HO1-N on a number of long-chain *n*-alkanes, including hexadecane, resulted in a two-fold increase in the total cellular phospholipid content. The phospholipid composition was qualitatively identical in both hydrocarbon- and non-hydrocarbon-grown cells, but quantitatively doubled in alkane-grown cells. The major phospholipid was phosphatidylethanolamine (61.5%) which did not exhibit a significant rate of turnover (Makula and Finnerty, 1971). Similar results for turnover of phosphatidylethanolamine in Gram-negative bacteria have been reported for *Salmonella typhimurium* (Ames, 1968) and *E. coli* (Kanfer and Kennedy, 1963). In contrast *H. parainfluenzae* exhibited significantly greater turnover kinetics of the ethanolamine head-group (White and Tucker, 1969b). Turnover of phosphatidylglycerol in an *Acinetobacter* sp. was less than 10 per cent per hour (Makula and Finnerty, 1971) as contrasted to a 50 per cent turnover in *S. typhimurium* (Ames, 1968) and *E. coli* (Kanfer and Kennedy, 1963). White and Tucker (1969b) described a disproportionate turnover of the non-acylated glycerol of phosphatidylglycerol in *H. parainfluenzae* as did Makula and Finnerty (1971) in *Acinetobacter* sp.

The phospholipid fatty acids of *Acinetobacter* sp. grown at the expense of hexadecane were derived from direct oxidation of the alkane substrate. These fatty acids, palmitic acid and palmitoleic acid, exhibited rapid turnover kinetics in growing cells (Makula and Finnerty, 1971). In addition to a two-fold increase in cellular phospholipid content in alkane-grown cells, there was also induction of intracytoplasmic membrane systems (Kennedy *et al.*, 1974; Kennedy and Finnerty, 1974). The membranes of alkane-grown *Acinetobacter* sp. have been isolated and characterized, and serve, in part, to explain the two-fold increase in cellular phospholipid content (Scott *et al.*, 1976).

Scott and Finnerty (1976) have characterized a membrane-limited intracellular inclusion induced only by growth on alkanes. These studies have documented a carbon-induced synthesis of new intracellular membrane systems concomitant with an increase in phospholipids (Finnerty, 1977).

III. Biosynthesis of Microbial Phospholipids

Definitive work on the pathways of phospholipid biosynthesis in micro-organisms has developed relatively rapidly. The metabolic routes shown in Fig. 1 represent the current pathways for biosynthesis of the common bacterial phospholipids in Gram-positive and Gram-negative micro-organisms. Relatively few of the designated enzymes have been purified to significant degrees of homogeneity, and less is understood about regulation of these enzymes.

A. PHOSPHATIDIC ACID BIOSYNTHESIS

The initial reaction in phospholipid biosynthesis is sequential acylation of the water-soluble sn-glycero-3-phosphate to yield phosphatidic acid. This phospholipid occupies a key position as the *de novo* precursor for all subsequent phospholipid biosynthetic reactions in well studied microbial systems. The acylation reaction proceeds through a two-stage sequence as represented by the following equations:

1. glycero-3-phosphate + fatty-acyl-CoA →

    lysophosphatidic acid + CoA

2. lysophosphatidic acid + fatty-acyl-CoA → phosphatidic acid + CoA

Fatty acid thio-esters of CoA or acyl-carrier protein are capable of functioning *in vitro* as fatty-acyl donors (Ailhaud and Vagelos, 1966; van den Bosch and Vagelos, 1970). The enzyme catalysing the first reaction is glycero-3-phosphate acyltransferase, and has been studied in crude membrane preparations (Kito and Pizer, 1969; Okuyama and Wakil, 1973; van den Bosch and Vagelos, 1970). The second enzyme is a lysophosphatidic acid acyltransferase which, in co-ordination with the previous acyltransferase, is responsible for the asymmetric distribution of fatty acids in the phospholipids of *E. coli*. Saturated fatty acids occur predominantly in the 1-position of phosphatidic acid and

sn-glycero-3-phosphate + 2 fatty acyl-S-CoA or 2 fatty-acyl-S-acyl carrier protein

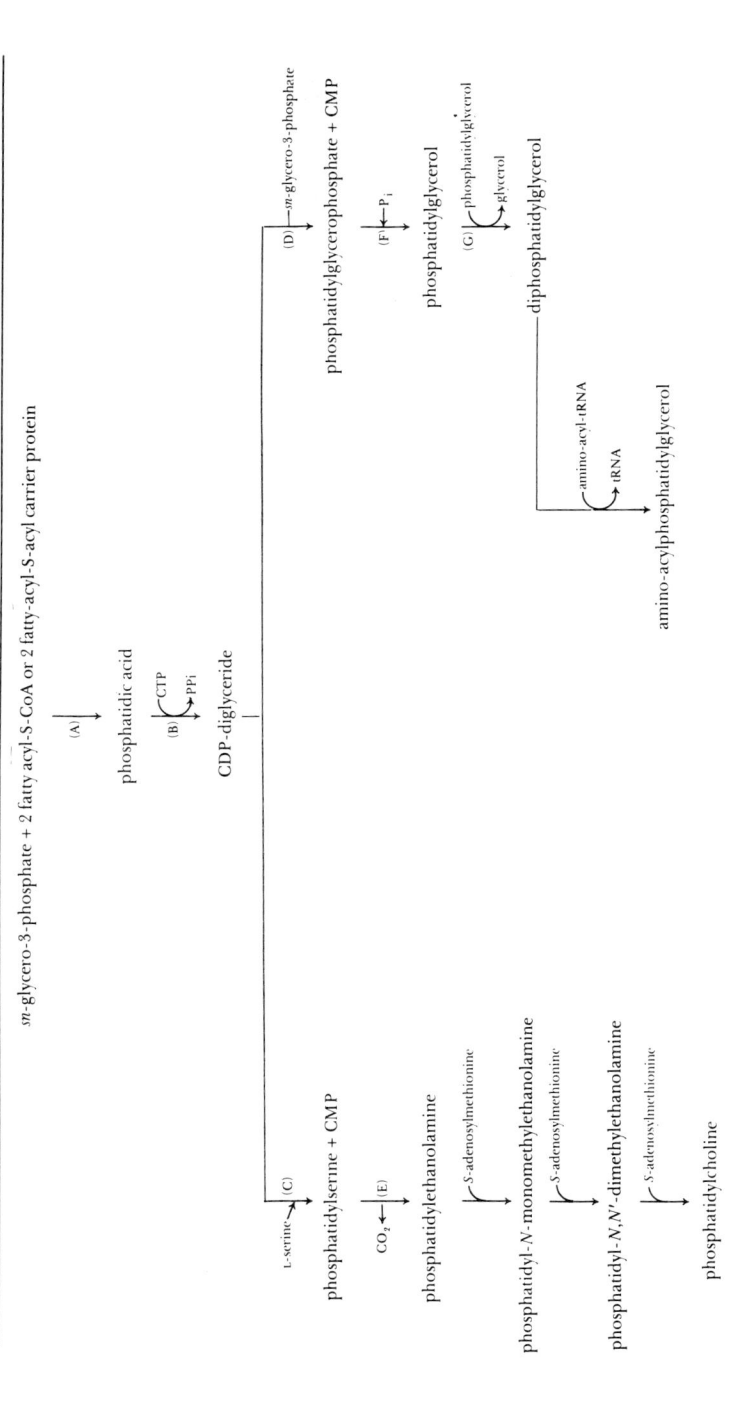

(A) ↓

phosphatidic acid

(B) ⤹ CTP
    ⤸ PPi

CDP-diglyceride ─────────────────────────── (D) → sn-glycero-3-phosphate

                                        phosphatidylglycerophosphate + CMP

L-serine ⤹ (C)

                                    (F) ⤸ Pi

phosphatidylserine + CMP

                                    phosphatidylglycerol

CO_2 ⤹ (E)
                                    (G) ⤹ phosphatidylglycerol
                                         ⤸ glycerol

phosphatidylethanolamine

                                    diphosphatidylglycerol

⤹ S-adenosylmethionine

phosphatidyl-N-monomethylethanolamine

⤹ S-adenosylmethionine   ⤹ amino-acyl-tRNA
                               ⤸ tRNA

phosphatidyl-N,N'-dimethylethanolamine   amino-acylphosphatidylglycerol

⤹ S-adenosylmethionine

phosphatidylcholine

(A) indicates glycero-3-phosphate acyltransferases, (B) indicates phosphatidic acid cytidyl transferase, (C) indicates L-serine : CMP phosphatidyl transferase, (D) indicates glycero-3-phosphate : CMP phosphatidyl transferse, (E) indicates phosphatidylserine decarboxylase, (F) indicates phosphatidylglycerolphosphate phosphatase, and (G) indicates cardiolipin synthetase.

Fig. 1. Microbial phospholipid biosynthesis pathways

unsaturated fatty acids in the 2-position (Okuyama and Wakil, 1973; Okuyama et al., 1976; Ray et al., 1970; Sinensky, 1971).

The studies of Wakil and his colleagues indicate that 1-acyl-sn-glycero-3-phosphate is the sole mono-acyl intermediate in phosphatidic acid biosynthesis (Okuyama and Wakil, 1973; Okuyama et al., 1976), in contrast to other studies which indicate that the products of the first reaction may be either 1-acyl-sn-glycero-3-phosphate or 2-acyl-sn-glycero-3-phosphate (Ray et al., 1970; Sinensky, 1971; van den Bosch, 1974). Recent work has effected a 20- to-40-fold purification of glycero-3-phosphate acyltransferase from *E. coli* (Snider and Kennedy, 1977). This Triton X-100-extracted enzyme lacked lysophosphatidic acid acyltransferase, required phospholipid for activity, formed 1-acyl-sn-glycero-3-phosphate as the sole product, and demonstrated a preference for saturated fatty-acyl-CoA in the 1-position. These experiments further support the view that 1-acyl-sn-glycero-3-phosphate is the product of the glycero-3-phosphate acyltransferase, the sn-2 position being acylated by the lysophosphatidic acid acyltransferase.

A number of glycero-3-phosphate acyltransferase mutants have been derived from *E. coli*. One class of mutants, designated *pls*A, exhibits a thermolabile glycerophosphate acyltransferase (Cronan et al., 1970). These mutants are temperature-sensitive, with growth and macromolecular synthesis ceasing following a shift-up to the non-permissive temperature (Glaser et al., 1973; Schneider and Kennedy, 1976).

Glaser and his colleagues have demonstrated that *pls*A mutants have a thermolabile adenylate kinase (Glaser et al., 1975). Recent workers were unable to demonstrate increased thermolability of the glycerophosphate acyltransferase prepared from *pls*A strain CV15 (Snider and Kennedy, 1977). These mutants are probably pleiotropic, with effects on phospholipid biosynthesis of an indirect nature. When *pls*A strains are grown at an intermediate temperature, phospholipid biosynthesis can be inhibited without affecting nucleic acid biosynthesis (Ray et al., 1976).

A second class of mutants, designated *pls*B, was isolated as glycero-3-phosphate auxotrophs (Bell, 1974, 1975). These mutant strains exhibited an apparent K_m value for glycero-3-phosphate 10 times greater than that observed in wild-type cells. The mutant acyltransferase exhibited a high pH optimum, decreased stimulation by potassium and magnesium ions, 90 per cent inhibition by Triton X-100 and decreased activity in the presence of palmitoyl-CoA. The *pls*B site appears to be the structural gene for acyltransferase. The *pls*A locus maps at minute 13 on the *E. coli* chromosome (Cronan and Godson, 1972) while the *pls*B locus maps at minute 69 (Cronan and Bell, 1974b).

The study of phosphatidic acid biosynthesis in other micro-organisms has offered additional comparative insight into the specificity of microbial acyltransferases. Membrane preparations derived from *Clostridium butyricum* exhibited an obligatory dependence on fatty acyl–acyl carrier protein in acylation of glycero-3-phosphate (Goldfine *et al.*, 1967). Fatty acyl–CoA thio-esters were inactive as acyl donors, while fatty acyl–acyl carrier protein purified from *E. coli* functioned one third as well as the fatty acyl–acyl carrier protein prepared from *Cl. butyricum*. The major product in the reaction was lysophosphatidic acid.

In contrast, the glycero-3-phosphate acyltransferase of *Sacch. cerevisae* required the fatty acyl–CoA thio-ester (Kuhn and Lynen, 1965). The purified fatty-acid synthetase complex was charged with palmitate and incubated with acyltransferase and radio-active glycero-3-phosphate. Labelled phospholipid was formed only when the reaction mixture was supplemented with coenzyme A.

An apparently anomalous reaction has been described in *E. coli* which catalyses synthesis of phosphatidic acid via a direct phosphorylation of diglyceride with ATP (Pieringer and Kunnes, 1965; Chang and Kennedy, 1967a; Weissbach *et al.*, 1971; Schneider and Kennedy, 1973). The physiological function of diglyceride kinase remains uncertain in view of the overshadowing biochemical and genetic evidence that acylation of glycero-3-phosphate represents the sole pathway for phosphatidic acid biosynthesis in micro-organisms, particularly *E. coli*. The enzyme has been purified approximately 600-fold, and exhibits an absolute requirement for magnesium ions as well as showing stimulation by phospholipids (Schneider and Kennedy, 1976). A preference is exhibited for adenine nucleotides with the purified enzyme catalysing phosphorylation of numerous lipids, including ceramide and several ceramide- and diglyceride-like analogues. Interestingly, incubation of purified diglyceride kinase with [^{14}C]Triton X-100 and [^{32}P]ATP yielded an alkali-stable [^{14}C, ^{32}P]lipid, tentatively identified as [^{14}C-Triton X-100-^{32}P]phosphate.

B. CYTIDINE DIPHOSPHATE DIGLYCERIDE BIOSYNTHESIS

The role of CDP-diglyceride in bacterial metabolism as a key intermediary metabolite in biosynthesis of the common phospholipids is amply documented. The enzyme catalysing synthesis of CDP-diglyceride is CTP:phosphatidic acid cytidyl transferase (PACT), and has been studied in membrane preparations derived from *Acinetobacter* sp. (McCaman and Finnerty, 1968), *Bacillus* PP (Patterson and Lennarz, 1971), *E. coli* (White *et al.*, 1971; Carter, 1968), *Micrococcus lysodeikticus* (DeSiervo and Salton, 1971) and *Salmonella typhimurium* (Bell *et al.*, 1971):

$$\text{phosphatidic acid} + \text{CTP} \xrightarrow{M^{2+}} \text{CDP-diglyceride} + PP_i$$

An obligate divalent cation requirement (usually magnesium) was exhibited by all enzyme systems, as was an exogenous supply of substrates. Potassium-ion stimulation of PACT activity was noted for *Acinetobacter* sp. and *Bacillus* PP, but not for *E. coli*. In all cases, a nonionic detergent such as Cutscum or Triton X-100 was necessary for optimal activity. A comparison of the optimal reaction conditions and kinetic properties of PACT derived from selected microbial sources is given in Table 3. The overall characteristics and properties of PACT derived from selected micro-organisms are remarkably similar. A survey documenting the ubiquitous distribution of PACT in micro-organisms is listed in Table 4 (W. R. Finnerty, unpublished results).

TABLE 3. Comparison of properties of CTP:phosphatidic acid cytidyl transferase from various micro-organisms

	Acinetobacter species	*Bacillus* species	*Escherichia coli*
Metal requirement	10 mM Mg^{2+}	4 mM Mg^{2+}	5–10 mM
pH Optimum	8.0	7.0	6.5
Fatty-acyl CoA esters	inhibition	not reported	not reported
Specific activity (μmoles/mg/h)	7.8	0.51	0.28
Product	CDP-diglyceride	CDP-diglyceride	CDP-diglyceride
Potassium effect	stimulation	stimulation	not reported
Detergent requirement	obligate	obligate	not reported
Value for K_m:			
(a) CTP	9.0 mM	0.05 mM	0.7 mM
(b) phosphatidic acid	3.5 mM	0.09 mM	2.0 mM

TABLE 4. Microbial distribution of CTP:phosphatidic acid cytidyl transferase

Micro-organism	Specific activity (μmoles/mg protein/h)
Acinetobacter species H01-N	7.80
Escherichia coli K-10	0.28
Escherichia coli K-12	0.03
Achromobacter chromococcus	0.50
Azotobacter vinelandii	0.20
Serratia marcescens	0.07
Paracolonbacter sp.	0.02
Pseudomonas hydrophilus	0.15
Pseudomonas aeruginosa	0.25
Moraxella lwoffi	1.15
Acinetobacter calcoaceticus	1.90
Desulfovibrio gigas	0.85
Desulfovibrio vulgaris	0.40
Desulfovibrio desulfuricans	0.54
Clostridium butyricum	0.18
Haemophilus parainfluenzae	0.27
Neisseria gonorrheae	0.33
Candida lipolytica	0.95
Mycobacterium rhodochrous	0.48
Mycobacterium albus	0.29
Mycobacterium vaccae	0.73
Micrococcus lysodeikticus	0.97

C. PHOSPHATIDYLGLYCEROPHOSPHATE AND PHOSPHATIDYLGLYCEROL BIOSYNTHESIS

Cytidine diphosphate diglyceride serves as the branch-point metabolite for synthesis of phosphatidylglycerol and phosphatidylserine. Phosphatidylglycerol biosynthesis involves two sequential reactions catalysed by glycero-3-phosphate:CMP phosphatidyl transferase (PGP synthetase) and phosphatidylglycerol phosphate phosphatase (PGP phosphatase):

CDP-diglyceride + *sn*-glycero-3-phosphate → phosphatidylglycerol phosphate + CMP

phosphatidylglycerol phosphate → phosphatidylglycerol + P_i

These enzymes are membrane-bound and require Mg^{2+} and a nonionic detergent for maximal activity. Phosphatidylglycerol phosphate is rarely found in lipid extracts of bacteria, while phosphatidylglycerol appears to undergo rapid turnover and degradation during growth of bacteria (Frerman and White, 1967a; White and Tucker, 1969b; DeSiervo, 1969; Makula and Finnerty, 1971; DeSiervo and Salton, 1973).

The enzymes involved in phosphatidylglycerol biosynthesis have been partially purified from membrane preparations of *E. coli* (Chang and Kennedy, 1967a, b) and more recently purified 6000-fold to 85 per cent homogeneity (Hirabayashi *et al.*, 1976). The membrane-bound PGP synthetase was solubilized with Triton X-100 and purified by CDP-diglyceride Sepharose-affinity chromatography (Larson *et al.*, 1976).

The apparent molecular weight of the purified enzyme was over 200,000 with a minimum subunit molecular weight of 24,000. The enzyme (PGP synthetase) exhibited an absolute requirement for Mg^{2+} (K_m, 50 mM), Triton X-100 (0.5–5%), *sn*-glycero-3-phosphate (K_m, 320 µM) and CDP-diglyceride (K_m, 46 µM) or dCDP-diglyceride (K_m, 34 µM). The purified enzyme also catalysed the reverse reaction yielding glycero-3-phosphate and CDP-diglyceride when incubated with CMP and phosphatidylglycerol phosphate.

The phosphatidylglycerol-synthesizing enzymes of *Bacillus licheniformis* have also been partially purified using CDP-diglyceride Sepharose-affinity chromatography (Larson *et al.*, 1976). The activity of PGP phosphatase exhibited no affinity for the column while PGP synthetase was adsorbed and subsequently eluted with buffer containing 1 mM CDP-diglyceride, 0.5 per cent Triton X-100, 0.2 M KCl and 50 mM $MgCl_2$. The partially purified PGP synthetase required Mg^{2+} ions, and nonionic detergent for maximal activity, while PGP phosphatase required only nonionic detergent. Normal saturation kinetics were determined for PGP synthetase in the presence of 0.2 mM CDP-diglyceride yielding an apparent K_m for *sn*-glycero-3-phosphate of 0.15 mM. The enzyme exhibited an apparent K_m value of 0.18 mM for PGP.

D. O-AMINO-ACYL PHOSPHATIDYLGLYCEROL BIOSYNTHESIS

The amino-acid esters of phosphatidylglycerol have been described from a variety of Gram-positive micro-organisms. These lipo-amino acids were first described from *Cl. perfringens* and *Staph. aureus* (MacFarlane, 1964). Subsequent characterizations were obtained from

Streptococcus faecalis, Bacillus subtilis, Bacillus megaterium and *Lactobacillus acidophilus* (MacFarlane, 1966; Lennarz, 1966). A number of amino acids have been characterized as forming lipo-amino acid complexes, including L-lysine, L-ornithine, L-arginine and L-alanine, with the L-lysine and L-alanine esters predominant. The physiological function of these lipo-amino acids has yet to be established. Biosynthesis of lysylphosphatidylglycerol is a two-step reaction involving activation of the amino acid and its subsequent transfer to phosphatidylglycerol:

$$\text{lysine} + \text{tRNA} \xrightarrow{\text{ATP}} \text{lysyl-tRNA}$$

$$\text{lysyl-tRNA} + \text{phosphatidylglycerol} \rightarrow \text{lysylphosphatidylglycerol} + \text{tRNA}$$

The first step is catalysed by a soluble enzyme and the second step by a particulate enzyme (Lennarz *et al.*, 1966). The specificity of the reaction was established by preparing S-β-amino-ethylcysteinyl-tRNA^{lys} which served as an active substrate for the particulate enzyme prepared from *Staph. aureus* (Nesbitt and Lennarz, 1968; Gould *et al.*, 1968). The specificity for the amino-acyl group of donor tRNA was demonstrated with alanyl-tRNA in *Cl. welchii* particulate enzyme preparations (Gould *et al.*, 1968). *N*-Acetylalanyl-tRNA and lactyl-tRNA were inactive as substrates for alanylphosphatidylglycerol biosynthesis. Enzyme preparations derived from *Staph. aureus* and *Cl. welchii* exhibited a specificity for the donor tRNA. Amino-ethylcysteinyl-tRNA^{lys} was inactive as a substrate in biosynthesis of lysylphosphatidylglycerol and alanylphosphatidylglycerol.

The metabolism and physiological function of amino-acid esters of phosphatidylglycerol remains unknown. Examination of the polar lipids of *Staph. aureus* during the stationary phase of growth showed accumulation of cardiolipin with a concomitant decrease in phosphatidylglycerol (Short and White, 1971). The lysylphosphatidylglycerol content remained constant as the stationary phase continued.

E. FATTY-ACYL PHOSPHATIDYLGLYCEROL BIOSYNTHESIS

A group of phospholipids occurring in biological systems, and of relatively recent recognition as trace components, are *bis*-phosphatidic acid, semi-lyso-*bis*-phosphatidic acid (fatty-acyl phosphatidylglycerol) and lyso-*bis*-phosphatidic acid. Fatty-acyl phosphatidylglycerol was first detected in the lungs of rabbit, pig and rat (Body and Gray, 1967) and has since been observed in humans with lipid-storage diseases (Rouser *et al.*, 1968; Yamamoto *et al.*, 1971). Lysosomes of rat-liver hepatocytes have been shown to be enriched in acylphosphatidylglycerol (Wherret and Huterer, 1972) as well as cultured baby hamster kidney (BHK) cells (Renkonen *et al.*, 1972; Brotherus and Renkonen, 1974). *Bis*-Phosphatidic acid, acylphosphatidylglycerol and lyso-*bis*-phosphatidic acid comprised 0.02 per cent, 0.1 per cent and 1.6 per cent, respectively, of the total lipid phosphorus of BHK cell monolayers. The stereoconfiguration of lyso-*bis*-phosphatidic acid obtained from BHK cells was shown to have the novel backbone structure of 1-*sn*-glycerophosphoryl-1'-*sn*-glycerol, indicating biosynthesis of this phospholipid to be different from those biological systems using *sn*-glycero-3-phosphate (Brotherus *et al.*, 1974).

Acylphosphatidylglycerol has been tentatively identified from *Mycoplasma* sp. (Plackett *et al.*, 1969, 1970), *Listeria monocytogenes* (Carroll *et al.*, 1968), *Bifidobacterium bifidum* (Exterkate and Veerkamp, 1969) and *Acinetobacter* species H01-N (R. A. Makula, personal communication). The phospholipids of several marine bacteria have been shown to contain cardiolipin or *bis*-phosphatidic acid (DeSiervo and Reynolds, 1975; McAllister and DeSiervo, 1975).

Olsen and Ballou (1971) reported isolation and characterization of acylphosphatidylglycerol from *Salmonella typhimurium*. This new phospholipid was identified as 3-*sn*-phosphatidyl-1'-[3'-acyl]-*sn*-glycerol, and comprised 2 per cent of the total phospholipid. Biosynthesis of this lipid appeared to result from either a direct acylation of phosphatidylglycerol or deacylation of *bis*-phosphatidic acid. Regardless of either mechanism, a highly specific enzyme is implicated in synthesis of acylphosphatidylglycerol.

Formation of *bis*-phosphatidic acid by a particulate preparation derived from *E. coli* has been reported (Benns and Proulx, 1971). Labelled [^{14}C]phosphatidylglycerol was incubated with the particulate fraction, resulting in formation of a product that closely resembled *bis*-phosphatidic acid or a close analogue. The product was formed in the absence of energy donors, and under conditions known to activate a number of non-specific lipases or phospholipases. The possible involvement of either a specific enzyme or a non-specific transacylase reaction has yet to be resolved. A subsequent extension of this investigation established the identity of the product as acylphosphatidylglycerol, although no further clarification of the enzymic basis of the reaction was provided (Cho et al., 1973).

A further examination of the synthesis of acylphosphatidylglycerol in *E. coli* has demonstrated that the product is formed from two molecules of phosphatidylglycerol by a transacylation reaction (Cho et al., 1977). Phosphatidylglycerol acts as both an acyl donor and an acyl acceptor. Phosphatidylethanolamine was additionally shown to participate as an acyl donor. The enzyme, acylphosphatidylglycerol synthetase, was membrane-bound, exhibiting an optimal pH value of 7.0 and an obligate Ca^{2+} requirement (6 mM). Magnesium, manganese, copper and Triton X-100 were inhibitory to enzyme activity. A mechanisn for acylphosphatidylglycerol synthesis was postulated as follows:

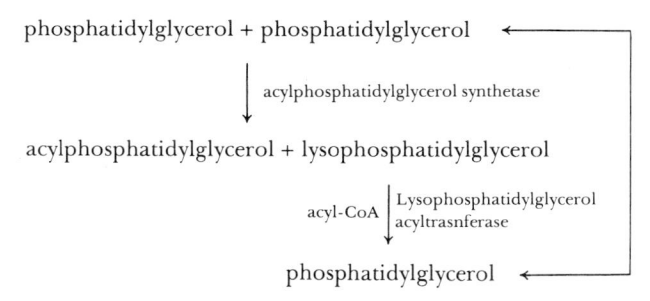

phosphatidylglycerol + phosphatidylglycerol

$\downarrow$ acylphosphatidylglycerol synthetase

acylphosphatidylglycerol + lysophosphatidylglycerol

acyl-CoA $\Big|$ Lysophosphatidylglycerol acyltrasnferase

phosphatidylglycerol

Acylphosphatidylglycerol isolated from *Pseudomonas* BAL-31 and bacteriophage PM2 has been identified as X-3-phosphatidyl-1'-(3'-acyl)-glycerol (Tsukagoshi et al., 1976). This phospholipid accounts for approximately 0.8 per cent of the total phospholipids in both bacterium and virus. Infectious bacteriophage PM2 has been reconstituted by mixing viral proteins, DNA, phosphatidylglycerol and phosphatidylethanolamine. Addition of acylphosphatidylglycerol did not affect the yield of infectious viral particles but, when acylphosphatidylglycerol replaced phosphatidylglycerol, no infectious viral particle resulted.

F. DIPHOSPHATIDYLGLYCEROL (CARDIOLIPIN) BIOSYNTHESIS

Early studies indicated that cardiolipin was formed from CDP-diglyceride and phosphatidylglycerol. Subsequent workers established conclusively that cardiolipin was synthesized in membrane preparations derived from *Micrococcus lysodeikticus* by condensation of two molecules of phosphatidylglycerol yielding cardiolipin (DeSiervo and Salton, 1971):

$$2 \text{ phosphatidylglycerol} \rightarrow \text{cardiolipin} + \text{glycerol}$$

The enzyme catalysing this condensation reaction is cardiolipin synthetase, which is membrane-bound and does not require Mg^{2+} or K^+ ions although Triton X-100 stimulates enzyme activity.

Membranes derived from *Staph. aureus* synthesized cardiolipin from two molecules of phosphatidylglycerol (Short and White, 1972). The enzyme, cardiolipin synthetase, was stimulated by Mg^{2+} and exhibited a low pH optimum (pH 4.4). The ratio of $^{14}C:^{32}P$ in the cardiolipin produced by cardiolipin synthetase was identical to the ratio in double-labelled phosphatidylglycerol (^{14}C in the fatty acids and ^{32}P). Studies with membrane preparations derived from *E. coli* further corroborated synthesis of cardiolipin from two molecules of phosphatidylglycerol (Hirschberg and Kennedy, 1972).

The amount of endogenous CDP-diglyceride in membrane preparations of *E. coli* was determined before and after synthesis of cardiolipin (Tunaitis and Cronan, 1973). No change was detected in CDP-diglyceride, thereby further supporting the reaction mechanism of cardiolipin synthesis from two molecules of phosphatidylglycerol.

Cardiolipin synthetase was shown to be inhibited by several phospholipids (DeSiervo, 1975). Inhibition of cardiolipin synthesis was determined for phosphatidic acid (94%), cardiolipin (72.8%), phosphatidylethanolamine (51.5%), phosphatidylinositol (24.7%), glycolipid (38.1%), glycerol (71.3%) and α-glycerophosphate (0%). Interestingly, the end products of the reaction, namely cardiolipin and glycerol, were among the most effective inhibitors tested.

G. PHOSPHATIDYLSERINE AND PHOSPHATIDYLETHANOLAMINE BIOSYNTHESIS

Biosynthesis of phosphatidylserine is catalysed by CDP-diglyceride: L-serine phosphatidyltransferase (PS synthetase):

$$\text{CDP-diglyceride} + \text{L-serine} \rightarrow \text{phosphatidylserine} + \text{CMP}$$

$$\text{phosphatidylserine} \rightarrow \text{phosphatidylethanolamine} + CO_2$$

Phosphatidylserine is subsequently decarboxylated by phosphatidylserine decarboxylase to phosphatidylethanolamine. Phosphatidylserine represents an obligatory intermediate in phosphatidylethanolamine synthesis in all microbial systems studied to date, and normally represents a minor or trace component of the phospholipid composition of most micro-organisms. Accumulation of phosphatidylserine has been demonstrated in *Megasphaera elsdenii, Veillonella parvula, Selenomonas rumination* var. *lactilytica* and *Anaerovibrio lipolytica* (van Golde et al., 1975). Since phosphatidylserine does not accumulate normally in aerobic or anaerobic bacteria, accumulation of phosphatidylserine in these organisms may be correlated with their ability to ferment lactate. Phosphatidylserine synthetase has been detected only in bacteria, initially being studied in cell-free extracts of *E. coli* (Kanfer and Kennedy, 1964). A later report presented evidence that phosphatidylserine synthetase derived from *E. coli* was predominantly ribosome-bound rather than membrane-bound, and resisted extraction from ribosomes in high ionic-strength buffers (Raetz and Kennedy, 1972). Dissociation of 70 S ribosomes into 50 S and 30 S subunits showed that phosphatidylserine synthetase remained associated with both subunits. The phosphatidylserine synthetase of *Bacillus* species was shown to be membrane-bound (Patterson and Lennarz, 1971). Neither soluble activity nor ribosome-associated activity was observed in this *Bacillus* species.

Studies with *Acinetobacter* species have demonstrated that phosphatidylserine was synthesized by a membrane-bound phosphatidylserine synthetase (W. R. Finnerty, unpublished results). The enzyme exhibited an absolute dependence on CDP-diglyceride, and a partial requirement for a nonionic detergent. The enzyme activity was further stimulated by potassium and magnesium ions, and the product of the reaction was phosphatidylserine. Ribosomes (70 S) and ribosomal subunits (50 S and 30 S) did not exhibit detectable phosphatidylserine synthetase activity.

A further study with *E. coli* has demonstrated a soluble and a ribosomal-bound form of phosphatidylserine synthetase (Ishinago and Kito, 1974). An *E. coli* mutant lacking phosphatidylethanolamine has been isolated, and shown to be the result of a defective phosphatidylserine synthetase (Ohta et al., 1974). This lesion resulted in a significant increase in cardiolipin for this mutant.

Phosphatidylserine synthetase, purified 100-fold from *E. coli* ribosomes, was capable of: (i) exchanging free CMP with the CMP of

CDP-diglyceride, (ii) catalysing an exchange between free serine and the serine of phosphatidylserine, and (iii) catalysing a reverse reaction in the presence of CMP (Raetz and Kennedy, 1974). This enzyme was recently purified to 97 per cent homogeneity with an overall 5500-fold purification (Larson and Dowhan, 1976). Purification was achieved by substrate-specific elution from phosphocellulose using CDP-diglyceride and nonionic detergent. The purified enzyme appeared as a single band in sodium dodecyl sulphate-gel electrophoresis, with an apparent minimum molecular weight of 54,000, and catalysed exchange reactions between CMP and CDP-diglyceride and between serine and phosphatidylserine. Cytidine diphosphate diglyceride was hydrolysed to yield CMP and phosphatidic acid; it could also be replaced by dCDP-diglyceride in all reactions catalysed by the enzyme. Further, the enzyme slowly catalysed synthesis of phosphatidylglycerol phosphate and phosphatidylglycerol when serine was replaced by glycero-3-phosphate or glycerol, respectively.

Mutants of *E. coli* K-12 defective in phosphatidylserine synthetase have been isolated using a rapid autoradiographic screening assay (Raetz, 1975). Four mutants of phosphatidylserine synthetase were characterized as being linked to the *nad* B locus near minute 49, with one (*pss*-8) being temperature-sensitive for growth (Raetz, 1976). When the *pss*-8 mutant was shifted to the non-permissive temperature, the cells stopped dividing and formed long filaments. Incubation at the non-permissive temperature (44°C) for three hours resulted in the content of phosphatidylethanolamine decreasing from 66 per cent to 32 per cent, while phosphatidylglycerol and cardiolipin increased from 34 per cent to 68 per cent.

Molecular cloning has allowed production of strains of *E. coli* carrying copies of specific genes or gene clusters. A collection of *E. coli* strains has been constructed which carry a hybrid ColE1 plasmid into which a fragment of the *E. coli* chromosomal DNA has been inserted (Clarke and Carbon, 1976). The hybrid plasmids are maintained at 10–20 copies per chromosome. Since the fragments were generated by random shearing of chromosomal DNA, the resulting constructed hybrid plasmids would contain most of the *E. coli* genome. Two hybrid plasmids carrying the gene for phosphatidylserine synthetase have been identified from this collection (Raetz *et al.*, 1977). The strains carrying these plasmids overproduced the synthetase by 6- to 15-fold. The overproduced enzyme was found associated with the ribosome fraction. The phospholipid composition of the membrane derived from these overproducing strains was closely similar to that of the wild type.

Decarboxylation of phosphatidylserine results in synthesis of phosphatidylethanolamine through the action of phosphatidylserine decarboxylase. The enzyme is membrane-bound in addition to being significantly more active than phosphatidylserine synthetase. This greater activity of the decarboxylase offers a plausible explanation for the lack of phosphatidylserine in lipid extracts of micro-organisms.

This enzyme has been purified 3600-fold from membranes of *E. coli* (Dowhan *et al.*, 1974). The presence of a nonionic detergent was required for extraction of the enzyme from membranes and all subsequent purification steps. The purified enzyme exhibited an absolute requirement for a nonionic detergent, was specific for phosphatidylserine, and exhibited a normal amino-acid composition.

The isolation and biochemical characterization of *E. coli* strain EH36 carrying a mutation in phosphatidylserine decarboxylase structural gene has been reported (Hawrot and Kennedy, 1975). A subsequent report has established the map position of the *psd*1 locus (Hawrot and Kennedy, 1976).

H. METHYLATED DERIVATIVES OF PHOSPHATIDYLETHANOLAMINE

Several micro-organisms contain one or more methylated derivatives of phosphatidylethanolamine-phosphatidyl-*N*-methylethanolamine, phosphatidyl-*N*,*N*'-dimethylethanolamine and phosphatidylcholine (Goldfine, 1972). Biosynthesis of these derivatives occurs through sequential methylation of phosphatidylethanolamine, with the methyl group donor being S-adenosylmethionine. The first methylation reaction is catalysed by a soluble enzyme forming phosphatidyl-*N*-methylethanolamine. Subsequent methylation reactions are catalysed by a particulate enzyme forming phosphatidyl-*N*,*N*'-dimethylethanolamine and phosphatidylcholine (Kaneshiro and Law, 1964). Not all micro-organisms appear capable of carrying out the three methylation reactions resulting in a complex distribution pattern of methylated derivatives of phosphatidylethanolamine. Early studies failed to document phosphatidylcholine as a phospholipid constituent of micro-organisms. This has not proved to be the case, however, as more careful assessment of bacterial phospholipid composition has demonstrated. The presence of these methylated derivatives, including phosphatidylcholine, varies widely throughout the microbial world. Phosphatidylcholine has been demonstrated in several species of *Rhodopseudomonas, Hyphomicrobium, Agrobacterium, Brucella, Ferrobacillus, Nitrosocystis* and *Gluconobacter*, to name a few. Ikawa (1967) suggested that phosphatidylcholine was present in those bacteria exhibiting an advanced electron-transport system. Hagen *et al.* (1966) suggested a correlation between the presence of phosphatidylcholine and intracytoplasmic membrane systems. Paradoxically, some species within a genus contain phosphatidylcholine while others do not, and no firm positive correlation exists between intracytoplasmic membranes and phosphatidylcholine (Hagen *et al.*, 1966). *Clostridium butyricum* contained only phosphatidyl-*N*-methylethanolamine, while four other clostridial species lacked all three of the methylated derivatives of phosphatidylethanolamine (Goldfine and Ellis, 1964). Similar results have been amply documented for other genera (Yano *et al.*, 1969; Shively and Benson, 1967). A more extensive review of the biosynthesis of methylated derivatives of phosphatidylethanolamine has appeared (Ambron and Pieringer, 1973).

I. ORNITHINE-CONTAINING LIPIDS

A number of ornithine-containing lipids have been isolated and characterized from such micro-organisms as *Mycobacterium bovis* (Laneelle *et al.*, 1963), *Rhodopseudomonas spheroides* (Gorchein, 1964, 1968a, b), *Rhodospirillum rubrum* (DePinto, 1967), *Pseudomonas rubescens* (Wilkinson, 1972), *Streptomyces sioyaensis* (Kawanami *et al.*, 1968; Kimura and Otsuka, 1969), *Brucella melitensis* and *Bordetella pertussis* (Thiele and Schwinn, 1973, 1974), *Thiobacillus thiooxidans* (Knoche and Shively, 1972), *Desulfovibrio gigas* (Makula and Finnerty, 1975), *Actinomyces* sp. (Batrakow *et al.*, 1971) and *Gluconobacter cerinus* (Tahara *et al.*, 1976). These ornithine-containing lipids are widely distributed among bacterial species, and are characterized by a high degree of structural heterogeneity (Table 5). The acyl groups are normal, unsaturated, branched, *iso*, *anteiso*, cyclopropane, 2-hydroxy or 3-hydroxy residues depending on the organism. These structures all contain an amide-linked fatty acid at the α-amino position. The second fatty-acyl constituent is linked in one of three possible ways through: (1) esterification of a fatty alcohol to the carboxyl group of ornithine; (2)

TABLE 5. Structure of ornithine-containing lipids and micro-organisms from which they have been isolated

Structure	Organism

COOH O
| ||
HC–NH–C–CH$_2$–CH–(CH$_2$)$_{12}$–CH$_3$
| |
(CH$_2$)$_3$ O
| |
NH$_2$ C=O
               |
               R
	Desulfovibrio gigas
	Thiobacillus thiooxidans
	Pseudomonas rubescans

O O || || C–O–CH$_2$–CH$_2$–O–C–R | HC–NH–C–R | || (CH$_2$)$_3$ O | NH$_2$	*Mycobacterium bovis* *Brucella melitensis* *Brucella abortus* *Brucella pertussis*
COOH O | || HC–NH–C–R | (CH$_2$)$_3$ | NH$_2$	*Rhodopseudomonas spheroides* *Streptomyces sioyaensis*
O COOH || | C–O–CH–R | HC–NH–C–CH$_2$–CH–R | || | (CH$_2$)$_3$ O OH | NH$_2$	*Actinomyces* strain 660-15

COOH O
| ||
HC–NH–C–CH$_2$–CH–R
| |
(CH$_2$)$_3$ O
| |
NH$_2$ O=C–CH–R
                      |
                      OH
	Acetobacter intermedias
	Acetobacter aceti
	Gluconobacter cerinus
	Gluconobacter suboxydans

esterification via a connector diol to the carboxyl of ornithine; or (3) esterification via the 3-hydroxyl group of the 3-hydroxy fatty acid linked to the α-amino group.

The functional and biological relevance of these lipids remains obscure. However, the lipid does exhibit those amphipathic properties characteristic of membrane polar lipids. An ornithine-containing lipid was shown to be the major polar lipid of *Desulfovibrio gigas* and to be distributed equally between the outer membrane and the cytoplasmic membrane (Makula and Finnerty, 1975). *Desulfovibrio gigas* was further shown to be a phospholipid-deficient organism containing only phosphatidylethanolamine and phosphatidylglycerol (Makula and Finnerty, 1974). A possible structural role for the ornithine-containing lipid of *D. gigas* was suggested in the context of a compensatory role in maintaining the structure of phospholipid-deficient membranes.

An ornithine-containing lipid represented a major polar constituent of nutrient-agar grown *Ps. rubescens* but was undetectable in nutrient broth-grown bacteria (Wilkinson, 1972). Minnikin and Abdolrahimzadeh (1974) demonstrated that, in phosphate-limited cultures of *Ps. fluorescens*, the ornithine-containing lipid was the sole polar lipid.

IV. Catabolism of Microbial Phospholipids

While the enzymology of phospholipid biosynthesis has been studied in some detail, it is only of recent interest that those enzymes involved in phospholipid catabolism have received extensive study. Since the lipid components of most membranes appear in a metabolically dynamic state, catabolism must be considered an important aspect of lipid metabolism. The metabolism of specific lipids to their component parts becomes particularly relevant to the current interest in biomembranes, since the turnover and recycling of portions of lipid molecules could serve to regulate certain membrane functions. Additionally, lipolytic enzymes could further be expected to function in the maintenance, adaptation and regeneration of membranes. Lipolytic enzymes exist for all classes of lipids; however, because of the predominance of phospholipids in microbial membranes, this discussion will be limited to phospholipases.

This group of enzymes includes both carboxyl ester hydrolases and phosphoric diester hydrolases acting principally on phospholipids. The nomenclature of phospholipases has undergone several modifications, and the system in use today will be adhered to throughout this review (Fig. 2). According to this scheme, acyl hydrolases, which act on diacylphospholipids and exhibit specificity

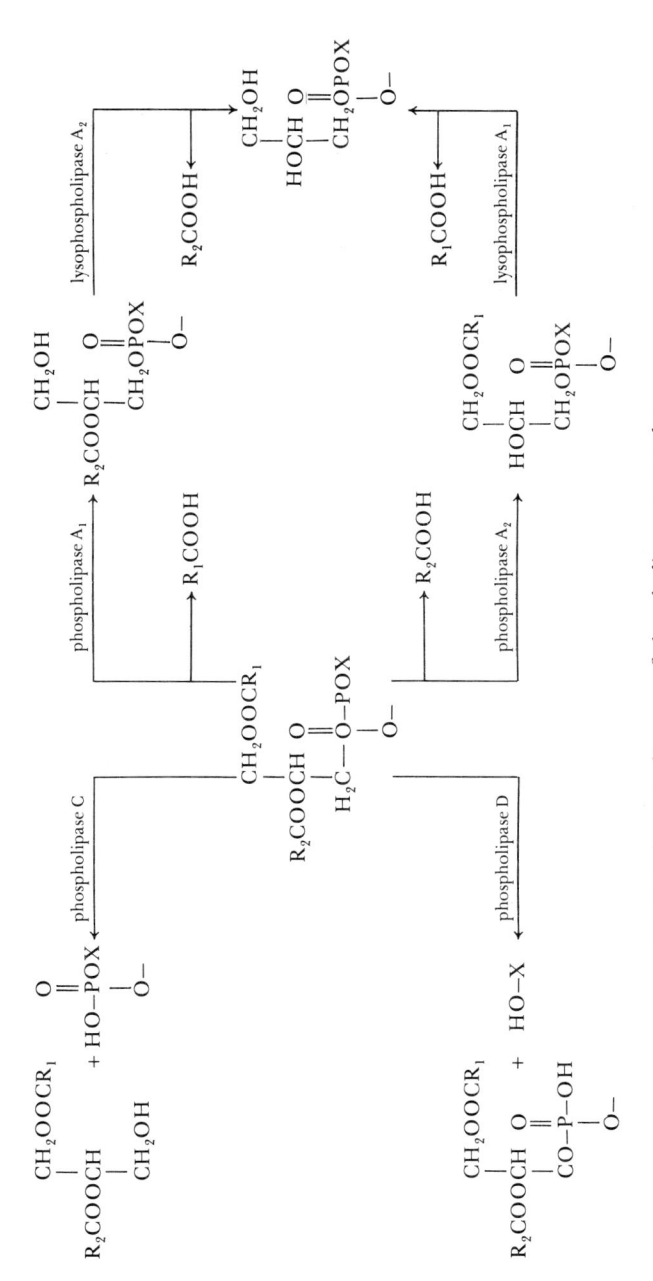

Fig. 2. Schematic diagram of phospholipase nomenclature

for the fatty-acyl ester bonds at the 1 and 2 positions, are classed as phospholipases A_1 and A_2, respectively. Enzymes acting on the respective lysophospholipids (mono-acylphospholipids) are termed lysophospholipases A_1 and A_2. In addition there are the phosphodiesterases. Phospholipase C hydrolyses the ester bond between the glycerol backbone and phosphate, while phospholipase D hydrolyses the ester bond between phosphate and X, with X representing the polar "head group" of the phospholipid.

Those phospholipases that have proved amenable to detailed analysis are the water-soluble phospholipases of various snake venoms and exocrine glands (digestive phospholipases). An extensive discussion of these enzymes has been presented in numerous recent reviews (Bonsen et al., 1972; McMurray and Magee, 1972; van Deenen and deHass, 1966; van den Bosch, 1974) and in a book devoted to lipolytic enzymes (Brockerhof and Jensen, 1974). Kinetic and mechanistic studies of these enzymes have revealed many unusual facets of interfacial catalysis (Verger and deHass, 1976). However, because these phospholipases have evolved with rather specialized functions, one should not expect too many similarities between them and those cellular phospholipases involved in degradation of membrane lipids. In fact, the phospholipases appear to be composed of an extremely heterogeneous group of enzymes varying widely in properties such as molecular weight, pH optimum, metal requirements and heat stability. Additionally, there is evidence that even the snake venom and pancreatic enzymes, which share numerous common characteristics, operate by a different catalytic mechanism (Brockerhof and Jensen, 1974).

In contrast to the advanced state of knowledge of the digestive phospholipases, study of the cellular phospholipases is still in its infancy. These enzymes are of more interest biochemically due to their possible roles in membrane physiology. The last years have seen increased activity in the field, in part, as a consequence of the work of Lands, who demonstrated the existence of enzymes in animal tissues that acylate lysophospholipids (Lands, 1965). A mono acyl-diacyl cycle was postulated for turnover and exchange of the fatty acids of intracellular phospholipids. The phospholipases that are required for such an exchange have since been found in many animal tissues and in micro-organisms. This discussion will necessarily concentrate on the cellular phospholipases of bacteria.

A. PHOSPHOLIPASE A₁

Phospholipase A₁ activity appears to be widely distributed in bacteria, and may represent the predominant lipolytic enzyme of bacteria (Brockerhof and Jensen, 1974; Finnerty and Makula, 1975). By definition, this enzyme should hydrolyse the 1-acyl ester of diacylphospholipids specifically, but such an activity has yet to be conclusively identified in bacteria or higher organisms. The following discussion will show that phospholipases A₁ are generally characterized by a lack of substrate specificity.

A bacterial phospholipase A₁ was first observed in 1967 in *E. coli* (Proulx and van Deenen, 1967) and later confirmed in several other *E. coli* strains (Okuyama and Nojima, 1969; Proulx and Fung, 1969). Activity was characterized by two pH optima (5.0 and 8.4) and was stimulated by calcium ions. Phospholipase A₂ and lysophospholipase activities were also evident. At pH 8.4, phospholipase A₁ activity predominated, especially in the presence of sodium dodecyl sulphate (SDS) which inhibited lysophospholipase activity. Subsequently, Scandella and Kornberg (1971) effected a 5000-fold purification of a phospholipase A₁ from the particulate fraction of *E. coli* B by solubilization with SDS and butanol, fractional precipitation with acetone and SDS-polyacrylamide gel electrophoresis. The purified enzyme was remarkably stable in 3 per cent SDS and tended to aggregate in the absence of detergent. The crude enzyme was substantially stable to heat but became progressively thermolabile during purification. The apparent molecular weight of the final enzyme preparation was 29,000. Substrate specificity was determined using synthetic substrates, for example, 1-palmitoyl-2-oleyl-*sn*-glycero-3-phosphorylethanolamine. Fatty-acid analysis of the lyso derivatives formed indicated that more than 90 per cent of the products were 2-acyllysophospholipids. The enzyme was inactive against the D-isomer of phosphatidylcholine as well as triolein which distinguished it from non-specific lipases. Modification in the acyl chain of the substrate had little effect on the rate of hydrolysis. Although activity in crude extracts was stimulated by detergents, the purified enzyme was equally active against phosphatidylcholine, phosphatidylethanolamine and phosphatidylglycerol in the presence or absence of Triton X-100. Hydrolysis of cardiolipin was stimulated 100-fold by Triton X-100. Interestingly, the rate of hydrolysis of 1-acyllysophosphatidylethanolamine was twice that observed for phosphatidylethanolamine hydrolysis. The enzyme, which was judged greater than 80 per cent pure, thus apparently has both phospholipase A_1 and lysophospholipase A_1 activity.

The presence of a detergent-sensitive as well as detergent-resistant phospholipase A in *E. coli* K-12 was reported by Doi *et al.* (1972). The detergent-resistant phospholipase hydrolysed both phosphatidylethanolamine and phosphatidylglycerol, resulting in accumulation of the respective lysophospholipid. Activity was stimulated by methanol as well as detergent, and was found associated with the crude membrane fraction. In contrast, the detergent-sensitive phospholipase was a soluble enzyme which was thermolabile and showed a specificity for phosphatidyglycerol although lysophosphatidylglycerol did not accumulate as a reaction product. In a later paper, Doi and Nojima (1973) reported that the detergent-resistant phospholipase exhibited both phospholipase A_1 and A_2 activities as well as lysophospholipase activity. Several subsequent reports have indicated that these activities are all present in a single non-specific phospholipase A. Ohki *et al.* (1972) isolated a detergent-resistant phospholipase-deficient mutant by a procedure that revealed colonies unable to undergo autolysis and that crossfed an unsaturated fatty-acid auxotroph. This mutant had lost both phospholipase A_1 and A_2 activities as well as particulate lysophospholipase activity. The loss in activity was not due to synthesis of an inhibitor. The location of this mutation, *pld*A (phospholipid degradation), was found to map between the *ilv* and *met* loci on the *E. coli* chromosome (Abe *et al.*, 1974). Phage Pl *pld*A$^+$ transductant clones of the mutant showed both A_1 and A_2 activities. This simultaneous transduction provided genetic evidence for a single enzyme. Biochemical evidence for a single enzyme was recently reported by Nishijima *et al.* (1977). The detergent-resistant phospholipase A was purified 2000-fold from *E. coli* K-12 by methods similar to those used by Scandella and Kornberg (1971). The final preparation yielded a single peak of activity co-incident with a single protein band on SDS-polyacrylamide gel electrophoresis. Two additional, but very minor, protein bands were observed by urea-acetic acid gel electrophoresis. The final preparation hydrolysed both the 1-acyl and 2-acyl chains of phosphatidylethanolamine or phosphatidylcholine as well as 1-acyl and 2-acyllysophosphatidylethanolamine. The positional specificity of the phospholipase A activity was definitely determined by the presence of radiolabelled lysophospholipids in the hydrolysis products of position-specific radiolabelled diacyl phospholipids. This enzyme appears identical with phospholipase A purified from *E. coli* B by Scandella and Kornberg (1971) with respect to its requirement for Ca^{2+}, pH optimum, phospholipid substrate specificities, behaviour of activity with Triton X-100 and apparent molecular weight. Scandella and Kornberg (1971), however, did not detect phospholipase A_2 activity. This discrepancy was explained by the presence of lysophospholipase A_1 and A_2 activities. Scandella and Kornberg's experiments were carried out in the presence of 0.05% Triton X-100, a condition Nishijima *et al.* (1977) found completely suppressed the lysophospholipase A_2 activity, while lysophospholipase A_1 was still active. Thus, the amount of 1-acyllysophospholipids would be rather small, and their detection in the reaction mixture would be difficult especially in the less sensitive, non-radioactive, method used by Scandella and Kornberg (1971). It should be mentioned that triglyceride lipase activity has been shown to be closely associated with detergent-resistant phospholipase A activity (Doi and Nojima, 1974).

A detailed study of phospholipase and lysophospholipase activities in *E. coli* was reported by Albright *et al.* (1973). This study confirmed the presence of a soluble phosphatidylglycerol-specific phospholipase A as well as both phospholipase A_1 and A_2 activities associated with the outer membrane. Hydrolytic activity toward 1-acyllysophosphatidylglycerol in the outer membrane was ascribed to phospholipase A_1. All of these outer membrane-associated activities showed similar calcium ion and detergent requirements. The pH optima varied somewhat, but it is likely that these activities all reside in the non-specific phospholipase A already referred to.

A second and well characterized bacterial phospholipase A_1 is the highly active enzyme isolated from spores of *Bacillus megaterium* by Raybin *et al.* (1972). The enzyme was obtained in soluble form by extraction of germinated spores with carbonate buffer (pH 10.9) and purified to near homogeneity. The enzyme has a molecular weight of about 26,000, and does not require Ca^{2+} or other metal ions. The pH optimum is 5 or 6.5 depending on the buffer, and the enzyme requires the presence of detergents. The enzyme apparently prefers a substrate-detergent complex bearing a negative charge. Phosphatidylglycerol is hydrolysed equally well in the presence of Triton X-100 (non-ionic) or sodium taurocholate (anionic). However, the zwitterionic substrates phosphatidylethanolamine and phosphatidylcholine were hydrolysed only in the presence of taurocholate. Sodium dodecyl sulphate inhibited activity over 100-fold. The enzyme was specific for the 1-acyl ester of phosphatidylglycerol, with only slight activity toward 1-acyllysophosphatidylglycerol. The specific activity of the sporangial enzyme was almost a thousand times higher than that of the *E. coli* enzyme (Scandella and Kornberg, 1971).

A membrane-bound phospholipase A_1 having lysophospholipase A_1 and A_2 activities was reported in *Mycobacterium phlei* (Ono and Nojima, 1969). Activity in the crude membrane fraction was characterized by pH optima of 4 and 5.1, and was stimulated by Ca^{2+} or Mg^{2+}. Cardiolipin was completely deacylated in extended (20 hour) incubation periods. Phosphatidylcholine and phosphatidylethanolamine were similarly hydrolysed to the corresponding glycerophosphate diesters, phosphatidylcholine being the most rapidly hydrolysed of the three substrates.

The *M. phlei* phospholipase was later purified approximately 500-fold to near homogeneity by solubilization with Triton X-100 (Nishijima *et al.*, 1974). Subsequent treatments included lipid extraction with organic solvents, and chromatography and electrophoresis in the presence of SDS. Two active fractions with molecular weights of 27,000 and 45,000 were observed. The pH optimum for hydrolysis of phosphatidylethanolamine was 7.5. Hydrolysis of phosphatidyglycerol and cardiolipin was observed at acidic pH values (4.5 and 3.7, respectively). The enzyme may thus preferentially hydrolyse nonionic or zwitterionic substrates. The purified enzyme was stimulated by Triton X-100 (seven-fold) and Ca^{2+} (1.2-fold) and was stable but not active in the presence of SDS. It hydrolysed the 1-acyl chains of phosphatidylethanolamine and phosphatidylcholine as well as 1-acyl- and 2-acyllysophosphatidylethanolamine. The enzyme appeared to hydrolyse the two acyl-ester linkages in diacylphospholipids in a stepwise manner, accumulating 2-acyllysophospholipids as intermediates. Many characteristics of this enzyme resemble those of the outer membrane phospholipase of *E. coli*.

Phospholipase A activity has been reported in *Neisseria gonorrhoeae* cell membranes (Senff *et al.*, 1976). Optimal activity was observed in 10 per cent methanol at pH 8.0 to 8.5, in the presence of calcium ions. The activity was both detergent-sensitive and thermolabile. Phosphatidylethanolamine, phosphatidylglycerol and cardiolipin were all hydrolysed, but positional specificity was not determined.

Phospholipase A activity associated with bacteriophage T4 and T4 ghost particles has been reported (Nelson and Buller, 1974). The activity was specific for phosphatidylglycerol, and was stimulated by Triton X-100 and 20 per cent methanol. Activity did not require Ca^{2+} and was inactivated at temperatures above 60°C.

A highly active intrinsic outer-membrane phospholipase A_1 activity derived from *Acinetobacter* sp. has been intensively analysed with respect to product formation (Torregrossa *et al.*, 1977a). Triacyllysocardiolipin and diacyllysocardiolipin were routinely detected in whole-cell lipid extracts of the bacterium, representing from 2–12 per cent of the total cellular phospholipids depending on the age of the culture. Lyso derivatives of cardiolipin have been reported in mycoplasmas and several micro-organisms (Plackett, 1967; Plackett *et al.*, 1970; Smith, 1967; Smith and Koostra, 1967; Langworthy *et al.*, 1976). The phospholipids of *Spiroplasma citri* have been reported to contain 20 per cent triacyllysocardiolipin (Freeman *et al.*, 1976). Exterkate *et al.* (1971) reported lysocardiolipin in nine out of ten strains of bifidobacteria of human intestinal origin.

The structures determined for triacyl- and diacyllysocardiolipin isolated from *Acinetobacter* sp. are shown in Fig. 3. These structures were established through extensive chemical, physical and enzymic procedures (Torregrossa *et al.*, 1977a) and are present as part of the phospholipid complement of *Acinetobacter* sp.

An outer-membrane enzyme activity which converted cardiolipin to triacyl- and diacyllysocardiolipin was described from *Acinetobacter* sp. (Torregrossa *et al.*, 1977b). This outer membrane-bound enzyme, designated cardiolipin phospholipase A_1, was similar to soluble phospholipase A_1 described for spores of *B. megaterium* with respect to pH optimum, heat inactivation and response to calcium ions, EDTA and SDS (Raybin *et al.*, 1972). Cardiolipin phospholipase from *Acinetobacter* sp. was dissimilar to the membrane-bound phospholipase A_1 described for *E. coli* (Scandella and Kornberg, 1971), *B. subtilis* (Kent and Lennarz, 1972) and *M. phlei* (Nishijima *et al.*, 1974), all of which required calcium ions and were heat stable.

The specificity of cardiolipin phospholipase A_1 indicated preferential hydrolysis of the 1-*sn*-acyl ester group of cardiolipin (see Fig. 3), further substantiating the nonequivalent character of the two ends of cardiolipin (Torregrossa *et al.*, 1977b). Figure 3 shows the proposed scheme for hydrolysis of cardiolipin to triacyl- and diacyllysocardiolipin by the outer-membrane phospholipase A_1 of *Acinetobacter* sp. This

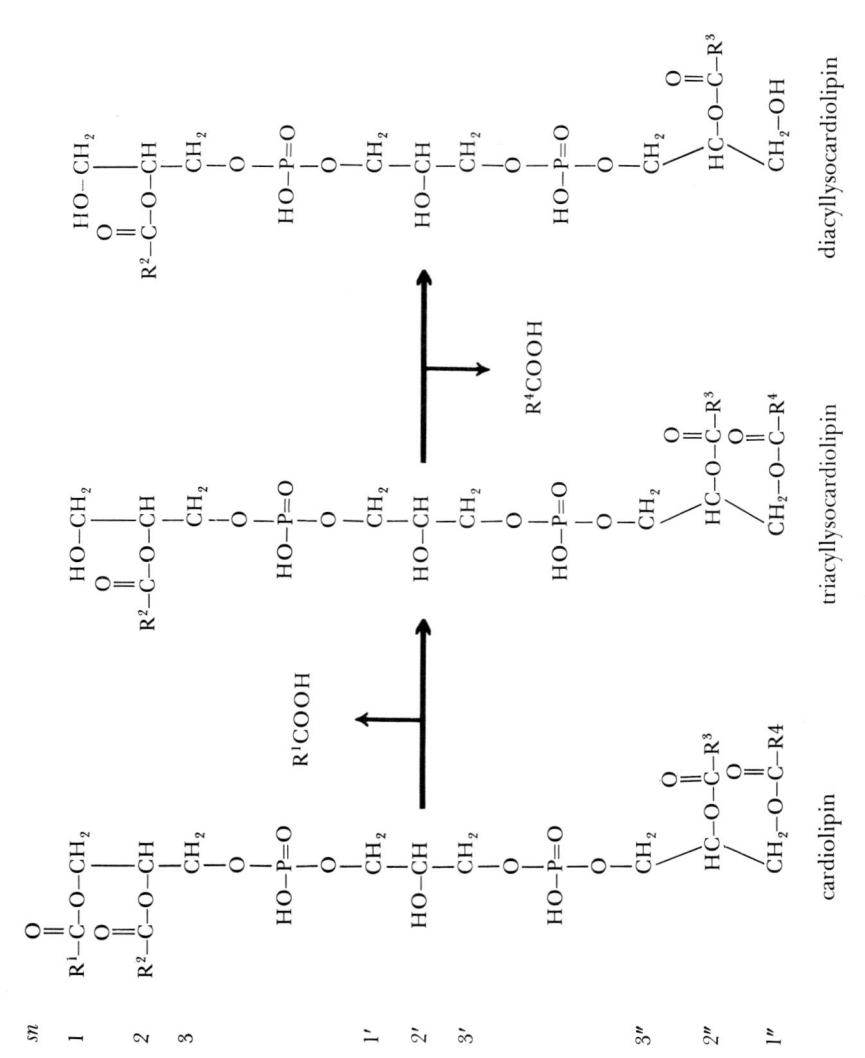

Fig. 3. Hydrolysis of cardiolipin by cardiolipin phospholipase A₁

scheme is based on products formed in *in vitro* studies of cardiolipin hydrolysis by cardiolipin phospholipase, as well as on the structures determined for triacyl- and diacyllysocardiolipin isolated from cellular lipid extracts.

B. PHOSPHOLIPASE A$_2$

To date, there have been few reports of phospholipase A$_2$-specific activity in bacteria. Crude extracts of *E. coli* 015 exhibited phospholipase A$_2$ (Proulx and Fung, 1969); however, this activity could have been due to the nonspecific outer-membrane phospholipase (Nishijima *et al.*, 1977; Scandella and Kornberg, 1971). Bernard *et al.* (1973) reported phospholipase A$_2$-specific activity in both the cytoplasm and the cell membrane of *E. coli* 0118. The two activities may be identical since both exhibited an optimum pH of 8, required Ca^{2+} and detergent, were heat stable and resistant to 40 per cent ethanol. The soluble enzyme was purified about seven-fold, although a considerable loss in activity was observed upon chromatography on cellulose resins. The final preparation had similar characteristics to those already described. The molecular weight of this preparation was estimated to be 600,000, indicating that the activity, though soluble, remained part of a high molecular-weight complex or aggregate. Phospholipase A$_2$ activity was also reported in the outer membrane of *E. coli* by Albright *et al.* (1973); however, it was not established whether this activity was distinct from that expressed by the phospholipase A$_1$ also present. In the studies of Bernard *et al.* (1973), acyl-chain specificity was poorly defined; therefore, the presence of a distinct A$_2$-specific phospholipase in *E. coli* remains to be conclusively demonstrated. In contrast to the A$_1$ phospholipases, the A$_2$ phospholipases of snake venoms, exocrine glands and most mammalian tissues are highly specific for the 2-acyl-ester bond of diacylphospholipids (Brockerhof and Jensen, 1974). Whether bacteria possess analogous enzymes awaits further investigation of additional species.

C. LYSOPHOSPHOLIPASES

Lysophospholipases remove the single fatty-acyl residue from lysophospholipids. Lysophospholipases are universally distributed in cells of animals, plants and micro-organisms (Brockerhof and Jensen, 1974; McMurray and Magee, 1972), and their activity regularly exceeds that of phospholipases A$_1$ and A$_2$. This wide distribution may be the result of their postulated function in protecting the membranes of the organism from the lytic properties of lysophospholipids.

I have already noted that lysophospholipase activity is often found associated with phospholipase A$_1$ activity. In these cases, the two activities may reside in the same enzyme (Abe *et al.*, 1974; Nishijima *et al.*, 1977). True lysophospholipases have, however, been found in the cytoplasmic-membrane and soluble fractions of *E. coli* (Albright *et al.*, 1973; Doi and Nojima, 1975; Doi *et al.*, 1972). These enzymes were active on 1-acyllysophosphatidylethanolamine, were thermolabile and were inhibited by Ca^{2+} and Cutscum (a nonionic detergent). Both activities showed a pH optimum of 9. In addition, lysophospholipase A$_2$ activity which did not require Ca^{2+} or detergent was detected in the cytoplasmic-membrane fraction. A soluble lysophospholipase from *E. coli* K-12 was purified 1500-fold to near homogeneity (Doi and Nojima, 1975). The final enzyme preparation had a molecular weight of 39,000 to 40,000, and hydrolysed 1-acyl- and 2-acyllysophosphatidylethanolamine and 1-acyllysophosphatidylglycerol, but not diacylphospholipids. The purified enzyme also hydrolysed 1-monoglycerides at comparable rates, but it did not hydrolyse di- or triglycerides or *p*-nitrophenyl acetate or palmitate. The activity did not require Ca^{2+}, and detergents were generally inhibitory. This lysophospholipase activity was observed in mutants lacking detergent-resistant and detergent-sensitive phospholipase A (Doi and Nojima, 1975).

A membrane-bound calcium-independent lysophospholipase was reported in *Mycoplasma laidlawii* strain B (van Golde *et al.*, 1971). Highest activity was observed using 1-acyllysophosphatidylglycerol as a substrate. The authors did not determine if phospholipase A was also present in this organism.

D. PHOSPHOLIPASE C

Phospholipase C activity in bacteria normally appears as an extracellular enzyme (Brockerhof and Jensen, 1974; Finnerty and Makula, 1975). The α-toxin of *Cl. perfringens* type A was first identified by MacFarlane and Knight (1941) as phospholipase C (phosphatidylcholine choline-phosphohydrolase), an enzyme which hydrolyses phosphatidylcholine to phosphorylcholine and diglyceride. Since this time, phospholipase C has been studied from *B. cereus* (Chu, 1949;

Zwaal *et al.*, 1971; Otnaess *et al.*, 1972; Doi and Nojima, 1971; van Deenen and deHass, 1966; Haverkate and van Deenen, 1964), *Cl. perfringens* (Takahashi *et al.*, 1974; Sugahara and Ohsaka, 1970; Casu *et al.*, 1971; Stahl, 1973), *Ps. schuylkilliensis* (Arai *et al.*, 1974), *Ps. aureofaciens* (Sonoki and Ikezawa, 1975) as well as other species of *Clostridium, Bacillus, Pseudomonas, Serratia* and *Acinetobacter* (Swartouw and Smith, 1956; Kurioka and Liu, 1967; Esselmann and Liu, 1961; Lehmann, 1971). Phospholipase C derived from these sources was a typical extracellular enzyme. The enzyme purified from *Cl. perfringens* hydrolyses phosphatidylcholine and sphingomyelin, while the enzyme derived from *B. cereus* hydrolyses phosphatidylcholine, phosphatidylethanolamine and phosphatidylglycerol. Enzymes derived from both microbial sources are zinc metalloproteins although phospholipase C purified from *Cl. perfringens* requires calcium as an activator while the *B. cereus* enzyme does not. Phospholipase C purified from *Ps. aureofaciens* likewise did not exhibit an obligate divalent cation requirement, but showed substrate specificity to phosphatidylethanolamine and phosphatidylcholine (Sonoki and Ikezawa, 1975). The presence of intracellular enzyme activity has been reported for *B. cereus* (Koga and Kusaka, 1970), *Cl. perfringens* (Ispolatovskaya, 1971) and *Ps. aureofaciens* (Sonoki and Ikezawa, 1975). Possible explanations for these observations are: (1) an active form of the enzyme, identical to or different from the extracellular form exists within the cell; (2) a form of phospholipase C exists as a membrane-bound enzyme. Further studies will be required to resolve this problem. The enzymes purified to date exhibit several differences in properties such as substrate specificities and metal-ion requirements. A cellular phospholipase C was reported in *E. coli* (Okuyama and Nojima, 1969), but production of O-phosphorylethanolamine from hydrolysis of phosphatidylethanolamine was probably not the result of phospholipase C but rather the result of the combined activities of phospholipase A_1, lysophospholipase and glycerophosphorylethanolamine phosphodiesterase. The last enzyme has been observed in the particulate fraction from *E. coli* (Albright *et al.*, 1973).

E. PHOSPHOLIPASE D

A phospholipase D activity has been described in *Haemophilus parainfluenzae* (Ono and White, 1970) and *E. coli* (Audet *et al.*, 1975). In both organisms, the activity is highly specific for cardiolipin yielding as its products equimolar amounts of phosphatidylglycerol and phosphatidic acid. The *H. parainfluenzae* enzyme has been better characterized, exhibiting an obligate dependence on Mg^{2+}, metal chelators being strongly inhibitory. The activity is apparently membrane-bound, and it is believed to function *in vivo* in turnover of cardiolipin (Tucker and White, 1971; White and Tucker, 1969). Astrachan (1973) studied the bond specificity of phospholipase D, and demonstrated an absolute specificity for the bond between phosphate and the C-3' of cardiolipin's central glycerol. Thus, the enzyme hydrolysed only one of the four phosphate-ester bonds in cardiolipin.

F. REGULATION

Although all of the enzymes necessary for complete degradation of phospholipids have been demonstrated in bacteria (Albright *et al.*, 1973), very little is known about the *in vivo* degradation or regulation of synthesis of bacterial-membrane phospholipids. A physiological role for phospholipases in bacterial lipid metabolism has yet to be conclusively demonstrated despite numerous attempts to do so (Aibara *et al.*, 1972; Bright-Gaertner and Proulx, 1972; Doi *et al.*, 1972). Turnover or degradation of phospholipids is, of course, the suggested role for phospholipases in membrane physiology. However, pulse-chase studies designated to detect turnover of phospholipids and fatty-acyl residues have been difficult to interpret (Mindich, 1973; Siekevitz, 1972). The phosphate and glycerol moieties of phosphatidylethanolamine show little or no turnover in actively growing *E. coli* (Ames, 1968), *Salmonella typhimurium* (Ames, 1968) or *H. parainfluenzae* (White and Tucker, 1969b). Cardiolipin and phosphatidylglycerol exhibited moderate turnover in these organisms (Ames, 1968; Mindich, 1973) as well as in *H. parainfluenzae* (White and Tucker, 1969b). Exchange of phospholipid fatty-acyl residues was extremely slow or undetectable in *E. coli*, *B. subtilis* and *Staphylococcus aureus* (Mindich, 1973); however, acyl-chain turnover has been observed in *H. parainfluenzae* (White and Tucker, 1969b) and in *Acinetobacter* sp. H01-N (Makula and Finnerty, 1971). These data concerning the magnitude and relative rates of phospholipid turnover based on pulse-chase studies are all inherently suspect owing to possible recycling or "salvage" mechanisms. Turnover of phospholipid fatty-acyl residues, for example, could be masked by efficient re-incorporation of labelled fatty acids. In this regard, transacylating enzymes (Siekevitz, 1972) and lysophospholipid acyl transferase (Proulx and van Deenen, 1966; van Deenen, 1966; van den Bosch, 1974) have been demonstrated in various membranes.

A polyglycerophosphatide cycle has been implicated in *E. coli*, relating cardiolipin synthesis and catabolism (Audet *et al.*, 1975). When *E. coli* B was pulsed with ^{14}C-glycerol and chased, a sparing of the phosphatidyl group but not the nonacylated glycerol of phosphatidylglycerol was observed. When energy-depleted cells were restored to an energy-rich medium, cardiolipin was converted to phosphatidylglycerol. Phosphatidic acid produced by the action of phospholipase D on cardiolipin in the presence of a cell-free extract was recycled to polyglycerophosphatide. These cell-free extracts also showed transphosphatidylase activity in the presence of glycerol by degrading cardiolipin to mainly phosphatidylglycerol. The results were interpreted as indicating that cardiolipin synthetase, together with cardiolipin-specific phospholipase D, constitutes a cycle involved in turnover of polyglycerophosphatides in *E. coli*. Since phospholipase D has been shown to be stimulated by ATP (Cole *et al.*, 1974), levels of cellular phosphatidylglycerol and cardiolipin may well be regulated by energy charge.

Mutants deficient in detergent-resistant phospholipase A, or detergent-sensitive phospholipase A, and one deficient in both of these activities, did not differ significantly from the parent strain in growth, composition of phospholipids or turnover of phospholipids, indicating no significant physiological role for either phospholipase A (Doi and Nojima, 1976). It is interesting to note, however, that some residual phospholipase A activity was detectable in these mutants. Additionally, it is conceivable that other phospholipases remained active in the mutants, but were not detected under the assay conditions utilized. Scandella and Kornberg (1971) suggested that the phospholipase A$_1$ of *E. coli* may be responsible for phospholipid breakdown and changes in membrane in integrity which have been observed following phage infection, the addition of antibody and complement, or colicin action.

There is considerable evidence implicating a role for phospholipases in fatty-acid exchange reactions in mammalian tissues. A deacylation-reacylation cycle (diacyl-monoacyl cycle) operates in mammalian tissues to synthesize phospholipids with "custom-tailored" acyl-chain constituents (McMurray and Magee, 1972). Since lysophospholipid acyltransferase has been detected in *E. coli* (Proulx and van Deenen, 1966), all of the enzymes necessary for operation of such a cycle exist in bacteria, but further work is needed to establish *in vivo* operation of the cycle. Studies on turnover of fatty-acyl residues in *E. coli* after a shift-down in temperature have suggested that a diacyl–monoacyl cycle operates *in vivo* to modify the fatty acid composition of phosphatidylglycerol; however, alterations in the fatty-acyl residues of phosphatidylethanolamine apparently resulted from *de novo* synthesis of new phosphatidylethanolamine molecular species (Aibara *et al.*, 1972). Related experiments (Bright-Gaertner and Proulx, 1972) in which *E. coli* was subjected to a cold-induced lag phase did not indicate operation of this cycle. Differences in experimental procedures and conditions could explain this discrepancy.

Heterogeneity in the metabolism of cardiolipin of *H. parainfluenzae* (Tucker and White, 1971; White and Tucker, 1969a) suggested a possible *in vivo* role for cardiolipin-specific phospholipase D. When the phospholipase D was inhibited during growth, cardiolipin content increased and that of phosphatidylglycerol decreased; but when cardiolipin synthesis was inhibited, cardiolipin content decreased, being hydrolysed to phosphatidylglycerol and phosphatidic acid (Tucker and White, 1971).

Regardless of whether they function *in vivo*, phospholipases are potentially autolytic enzymes. This fact alone suggests that their activity must be strictly regulated. To date, there has been but one study regarding the regulation of phospholipases (Kent and Lennarz, 1972). A mutant of *B. subtilis* with an osmotically-fragile protoplast was isolated and found to contain much less phospholipid in its membranes than that found in membranes of protoplasts of the wild type. Subsequently, it was found that extensive phospholipid degradation occurred during conversion of mutant cells to protoplasts. This phospholipid hydrolysis was due to an active membrane-bound phospholipase A_1 and a soluble lysophospholipase. A potent inhibitor of the phospholipase A_1 was found in the wild-type cells. The inhibitor, which was found both in the cytoplasm and associated with the membrane, was a heat-stable protein which irreversibly inactivated the phospholipase A_1 of the mutant. The partially purified inhibitor was not simply a non-specific protease since it had no effect on activity of several enzymes, including three phospholipases from other organisms. The defect in the mutant was assumed to be in production of an active inhibitor which may normally function as a regulator of phospholipase A_1 activity.

Products of phospholipase A activity, namely lysophospholipids, have not been routinely observed in bacterial lipid extracts. In the past, occurrence of trace amounts of lysophospholipids was attributed to chemical or enzymic hydrolysis of phospholipids during extraction. Inability to detect lysophospholipids does not necessarily imply a lack of *in vivo* phospholipase activity, since accumulation of these lytic compounds would be harmful to the cell. Lysophospholipids may, therefore, be rapidly hydrolysed by lysophospholipases or recycled by a lysophospholipid acyltransferase. An increasing number of reports detail the occurrence of lysophospholipids in bacterial lipid extracts for which a physiological role remains unresolved. Lysophospholipids may, however, play an important role in membrane physiology as intermediates in phospholipid turnover or as specific "cofactors" for membrane-bound enzymes.

V. Conclusion

The functional relevance of phospholipids in diverse biological phenomena has gained rapid and increasing importance in cellular physiology. Isolation and characterization of these complex lipids has played a key role in what could be considered as spectacular progress over the past 5–10 years in bridging the gap between cell and molecule, particularly as cell physiology relates to lipid metabolism. As the appreciation of the relationship that exists between the structure of a specific lipid and its function develops, greater attention is being directed to new techniques for determining structure-function interactions resulting in the presentation of more precise questions and the realization of more complete and factual answers. The advances realized in only the past five years have stimulated many new conceptual developments. The asymmetric distribution of phospholipids in membranes as well as the development of affinity chromatography techniques for resolving intrinsic membrane proteins represent significant advances in probing the role of lipids in cellular metabolism. It has become apparent that an understanding of the micro-architecture of the biological membrane is further necessary for understanding the underlying biochemical mechanisms relating to function. As our insight and understanding of lipid–protein interactions develop, the role of lipids in biological processes will undoubtedly assume greater importance. The next 10 years promise the development of many new and revolutionary discoveries as lipid metabolism becomes integrated into the dynamics of cellular metabolism.

REFERENCES

Abe, M., Okamoto, N., Doi, O. and Nojima, S. (1974). *Journal of Bacteriology* **119**, 543.
Aibara, S., Kato, M., Ishinaga, M. and Kito, M. (1972). *Biochimica et Biophysica Acta* **270**, 301.
Ailhaud, G. P. and Vagelos, P. R. (1966). *Journal of Biological Chemistry* **241**, 3866.
Aiyappa, P. S. and Lampen, J. O. (1976). *Biochimica et Biophysica Acta* **448**, 401.
Albright, F. R., White, D. A. and Lennarz, W. J. (1973). *Journal of Biological Chemistry* **248**, 3968.
Ambron, R. T. and Pieringer, R. A. (1973). *In* "Form and Function of Phospholipids", (G. B. Ansell, J. N. Hawthorne and R. M. C. Dawson, eds.), pp. 289–331. Elsevier, New York.
Ames, G. (1968). *Journal of Bacteriology* **95**, 833.
Arai, M., Matsunager, K. and Murao, S. (1974). *Journal of the Agricultural Chemical Society of Japan* **7**, 409.
Astrachan, L. (1973). *Biochimica et Biophysica Acta* **296**, 79.
Audet, A., Cole, R. and Proulx, P. (1975). *Biochimica et Biophysica Acta* **380**, 414.
Barsukov, L. I., Kulikov, V. I. and Bergelson, L. D. (1976). *Biochemical and Biophysical Research Communications* **71**, 704.
Batrakow, S. G., Pilipenko, T. W. and Bergelson, L. D. (1971). *Doklady Academi Nauk S.S.S.R.* **200**, 227.
Bell, R. M., Mavis, R. D., Osborn, M. J. and Vagelos, P. R. (1971). *Biochimica et Biophysica Acta* **249**, 628.
Bell, R. M. (1974). *Journal of Bacteriology* **117**, 1065.
Bell, R. M. (1975). *Journal of Biological Chemistry* **250**, 7174.
Benns, G. and Proulx, P. (1971). *Biochemical and Biophysical Research Communications* **44**, 382.
Bergelson, L. D. and Barsukov, L. I. (1977). *Science, New York* **197**, 224.
Bernard, M. C., Brison, J., Denis, F. and Rosenberg, A. J. (1973). *Biochemie* **55**, 377.
Bevers, E. M., Snoek, G. T., Opdenkamp, J. A. F. and van Deenen, L. L. M. (1977). *Biochimica et Biophysica Acta* **467**, 346.
Bittman, R. and Rottem, S. (1976). *Biochemical and Biophysical Research Communications* **71**, 318.
Bloj, B. and Zilversmit, D. B. (1976). *Biochemistry, New York* **15**, 1277.
Body, D. R. and Gray, G. M. (1967). *Chemistry and Physics of Lipids* **1**, 254.
Bonsen, P. P. M., Pieterson, W. A., Valwerk, J. J. and DeHaas, G. H. (1972). *In* "Current Trends in the Biochemistry of Lipids", (J. Ganguly and R. M. S. Smellie, eds.), pp. 189–202. Academic Press, New York.
Bosch, V. and Braun, V. (1973). *Federation of European Biochemical Societies Letters* **34**, 307.

Bradley, D. E. and Rutherford, E. L. (1975). *Canadian Journal of Microbiology* **21**, 152.
Braun, V. and Sieglin, U. (1970). *European Journal of Biochemistry* **13**, 336.
Braunstein, S. N. and Franklin, R. M. (1971). *Virology* **43**, 685.
Bretscher, M. S. (1972a). *Nature New Biology* **236**, 11.
Bretscher, M. S. (1972b). *Journal of Molecular Biology* **71**, 523.
Bright-Gaertner, E. and Proulx, P. (1972). *Biochimica et Biophysica Acta* **270**, 40.
Brockerhof, H. and Jensen, R. G. (1974). *Lipolytic Enzymes*, pp. 1–285. Academic Press, New York.
Brotherus, J., Renkonen, O., Hermann, J. and Fischer, W. (1974). *Chemistry and Physics of Lipids* **23**, 178.
Brotherus, J. and Renkonen, O. (1974). *Chemistry and Physics of Lipids* **13**, 11.
Carroll, K. K., Cutts, J. H. and Murray, E. G. D. (1968). *Canadian Journal of Biochemistry* **44**, 899.
Carter, J. R., Jr. (1968). *Journal of Lipid Research* **9**, 748.
Cashel, M. and Gallant, J. (1969). *Nature, London* **221**, 838.
Cashel, M. (1975). *Annual Review of Microbiology* **29**, 301.
Casu, A., Pala, V., Monacelli, R. and Nanni, G. (1971). *Italian Journal of Biochemistry* **20**, 166.
Chang, Y.-Y. and Kennedy, E. P (1967a). *Journal of Biological Chemistry* **242**, 516.
Chang, Y.-Y. and Kennedy, E. P (1967b). *Journal of Lipid Research* **8**, 447.
Chang, Y.-Y. and Kennedy, E. P (1967c). *Journal of Lipid Research* **8**, 456.
Chang, Y.-Y. and Kennedy, E. P (1967d). *Journal of Lipid Research* **8**, 456.
Cho, K. S., Benns, G. and Proulx, P. (1977). *Biochimica et Biophysica Acta* **326**, 355.
Chu, H. P. (1949). *Journal of General Microbiology* **3**, 255.
Clarke, L. and Carbon, J. (1976). *Cell* **7**, 91.
Cole, R., Benns, G. and Proulx, P. (1974). *Biochimica et Biophysica Acta* **337**, 325.
Cronan, J. E., Jr. (1974). *Proceedings of the National Academy of Sciences of the United States of America* **71**, 3758.
Cronan, J. E., Jr. (1975). *Journal of Biological Chemistry* **250**, 7074.
Cronan, J. E., Jr. and Bell, R. M. (1974a). *Journal of Bacteriology* **118**, 598.
Cronan, J. E., Jr. and Bell, R. M. (1974b). *Journal of Bacteriology* **120**, 227.
Cronan, J. E., Jr. and Gelman, E. P. (1975). *Bacteriological Reviews* **39**, 232.
Cronan, J. E., Jr. and Godson, G. N. (1972). *Molecular and General Genetics* **116**, 199.
Cronan, J. E., Jr., Ray, T. K. and Vagelos, P. R. (1970). *Proceedings of the National Academy of Sciences of the United States of America* **65**, 737.
Cronan, J. E., Jr. and Vagelos, P. R. (1972). *Biochimica et Biophysica Acta* **265**, 25.
Cronan, J. E., Jr., Weisberg, J. L. and Allen, R. G. (1975). *Journal of Biological Chemistry* **250**, 5835.
Crowfoot, P. D., Esfahani, M. and Wakil, S. J. (1972a). *Journal of Bacteriology* **112**, 1408.
Crowfoot, P. D., Oka, T., Esfahani, M. and Wakil, S. J. (1972b). *Journal of Bacteriology* **112**, 1396.
Cunningham, C. C. and Hager, L. P. (1971a). *Journal of Biological Chemistry* **246**, 1575.
Cunningham, C. C. and Hager, L. P. (1971b). *Journal of Biological Chemistry* **246**, 1583.
Cunningham, C. C. and Hager, L. P. (1975). *Journal of Biological Chemistry* **250**, 7139.
DePinto, J. A. (1967). *Biochimica et Biophysica Acta* **144**, 113.
DeSiervo, A. J. (1967). *Journal of Bacteriology* **100**, 1342.
DeSiervo, A. J. (1975). *Canadian Journal of Biochemistry* **53**, 1031.
DeSiervo, A. J. and Reynolds, J. W. (1975). *Journal of Bacteriology* **123**, 294.
DeSiervo, A. J. and Salton, M. R. J. (1971). *Biochimica et Biophysica Acta* **239**, 280.
DeSiervo, A. J. and Salton, M. R. J. (1973). *Microbios* **8**, 73.

Doi, O. and Nojima, S. (1971). *Biochimica et Biophysica Acta* **248**, 234.
Doi, O. and Nojima, S. (1973). *Journal of Biochemistry, Tokyo* **74**, 667.
Doi, O. and Nojima, S. (1974). *Biochimica et Biophysica Acta* **369**, 64.
Doi, O. and Nojima, S. (1975). *Journal of Biological Chemistry* **250**, 5208.
Doi, O. and Nojima, S. (1976). *Journal of Bacteriology* **80**, 1247.
Doi, O., Ohki, M. and Nojima, S. (1972). *Biochimica et Biophysica Acta* **260**, 244.
Dowhan, W., Wickner, W. T. and Kennedy, E. P. (1974). *Journal of Biological Chemistry* **249**, 3079.
Esselmann, M. T. and Liu, P. V. (1961). *Journal of Bacteriology* **81**, 939.
Endo, A. and Rothfield, L. (1969). *Biochemistry, New York* **8**, 3508.
Exterkate, F. A. and Veerkamp, J. H. (1969). *Biochimica et Biophysica Acta* **176**, 65.
Exterkate, F. A., Otten, B. J., Wassenberg, H. W. and Veerkamp, J. H. (1971). *Journal of Bacteriology* **106**, 824.
Finnerty, W. R. (1977). *Trends in Biochemical Sciences* **2**, 73.
Finnerty, W. R. and Makula, R. A. (1975). *Critical Reviews in Microbiology* **4**, 1.
Fleischer, S., Brierly, G., Klouwen, H. and Slautterback, D. B. (1962). *Journal of Biological Chemistry* **237**, 3264.
Freeman, B. A., Sissenstein, R., McManus, T. T., Woodward, J. E., Lee, I. M. and Mudd, J. B. (1976). *Journal of Bacteriology* **125**, 946.
Frerman, F. E. and White, D. C. (1967a). *Journal of Bacteriology* **94**, 1868.
Frerman, F. E. and White, D. C. (1967b). *Journal of Bacteriology* **94**, 1854.
Frerman, F. E. and White, D. C. (1968). *Journal of Bacteriology* **95**, 2198.
Fulco, A. J. (1974). *Annual Review of Biochemistry* **43**, 215.
Gennis, R. B. and Strominger, J. L. (1976). *Journal of Biological Chemistry* **251**, 1264.
Glaser, M., Bayer, W. H., Bell, R. M. and Vagelos, P. R. (1973). *Proceedings of the National Academy of Sciences of the United States of America* **70**, 385.
Glaser, M., Nulty, W. and Vagelos, P. R. (1975). *Journal of Bacteriology* **123**, 128.
Golden, N. G. and Powell, G. L. (1972). *Journal of Biological Chemistry* **247**, 6651.
Goldfine, H. (1972). *Advances in Microbial Physiology* **8**, 1.
Goldfine, H., Ailhand, G. P. and Vagelos, P. R. (1967). *Journal of Biological Chemistry* **242**, 4466.
Goldfine, H. and Ellis, M. E. (1964). *Journal of Bacteriology* **87**, 8.
Gorchein, A. (1964). *Biochimica et Biophysica Acta* **84**, 356.
Gorchein, A. (1968a). *Biochimica et Biophysica Acta* **152**, 358.
Gorchein, A. (1968b). *Proceedings of the Royal Society, B* **170**, 265.
Gorchein, A., Neuberger, A. and Tait, G. H. (1968). *Proceedings of the Royal Society B* **170**, 311.
Gould, R. M., Thornton, M. P., Liepkalns, V. and Lennarz, W. (1968). *Journal of Biological Chemistry* **243**, 3096.
Green, E. W. and Schaechter, M. (1972). *Proceedings of the National Academy of Sciences of the United States of America* **69**, 2312.
Guarneri, M., Stechmuller, B. and Lehninger, A. L. (1971). *Journal of Biological Chemistry* **246**, 1971.
Hagen, P. O., Goldfine, H. and Williams, P. J. L. (1966). *Science, New York* **151**, 1543.
Hantke, K. and Braun, V. (1973). *European Journal of Biochemistry* **34**, 284.
Haverkate, F. and van Deenen, L. L. M. (1964). *Biochimica et Biophysica Acta* **84**, 106.
Hawkes, S. A., Meechan, J. D. and Bissell, M. J. (1976). *Biochemical and Biophysical Research Communications* **68**, 1226.
Hawrot, E. and Kennedy, E. P. (1975). *Proceedings of the National Academy of Sciences of the United States of America* **72**, 1112.
Hawrot, E. and Kennedy, E. P. (1976). *Molecular and General Genetics* **148**, 271.

Helmkamp, G. M., Harvey, M. S., Wirtz, K. W. A. and van Deenen, L. L. M. (1974). *Journal of Biological Chemistry* **249**, 6382.
Hinckley, A., Muller, A. and Rothfield, L. (1972). *Journal of Biological Chemistry* **247**, 2623.
Hirabayashi, T., Larson, T. J. and Dowhan, W. (1976). *Biochemistry, New York* **15**, 5205.
Hirashima, A., Wang, S. and Inouye, M. (1974). *Proceedings of the National Academy of Sciences of the United States of America* **71**, 4149.
Hirschberg, C. B. and Kennedy, E. P. (1972). *Proceedings of the National Academy of Sciences of the United States of America* **69**, 648.
Ikawa, M. (1967). *Bacteriological Reviews* **31**, 54.
Imai, K. and Brodie, A. F. (1973). *Journal of Biological Chemistry* **248**, 7487.
Inoue, K. and Nojima, S. (1967). *Chemistry and Physics of Lipids* **1**, 369.
Inouye, M., Shaw, J. and Chen, C. (1972). *Journal of Biological Chemistry* **247**, 8154.
Inouye, M. (1974). *Proceedings of the National Academy of Sciences of the United States of America* **71**, 2396.
Ishinaga, M. and Kito, M. (1974). *European Journal of Biochemistry* **42**, 1483.
Ispolatovskaya, M. V. (1971). *In* "Microbial Toxins", (Kadis, S., Montie, T. C. and Ajl, S. J., eds.), pp. 109–158. Academic Press, New York.
Johnson, L. W. and Zilversmit, D. B. (1975). *Biochimica et Biophysica Acta* **375**, 165.
Kamio, Y. and Nikaido, H. (1976). *Biochemistry, New York* **15**, 2561.
Kaneshiro, T. and Law, J. H. (1964). *Journal of Biological Chemistry* **239**, 1705.
Kanfer, J. and Kennedy, E. P. (1963). *Journal of Biological Chemistry* **238**, 2919.
Kanfer, J. N. and Kennedy, E. P. (1964). *Journal of Biological Chemistry* **239**, 1720.
Kawanami, J., Kimura, A. and Otsuka, H. (1968). *Biochimica et Biophysica Acta* **152**, 808.
Kennedy, R. S. and Finnerty, W. R. (1974). *Archives of Microbiology* **102**, 85.
Kennedy, R. S., Finnerty, W. R., Sudarsanan, K. and Young, R. A. (1974). *Archives of Microbiology* **102**, 75.
Kent, C. and Lennarz, W. J. (1972). *Proceedings of the National Academy of Sciences of the United States of America* **69**, 2793.
Kimura, A. and Otsuka, H. (1969). *Agricultural and Biological Chemistry* **33**, 781.
Kito, M. and Pizer, L. I. (1969). *Journal of Bacteriology* **97**, 1321.
Knowles, A. F., Kandrach, A., Racker, E. and Khorana, H. (1975). *Journal of Biological Chemistry* **250**, 1809.
Knoche, H. H. and Shively, J. M. (1972). *Journal of Biological Chemistry* **247**, 170.
Koga, Y. and Kusaka, I. (1970). *European Journal of Biochemistry* **16**, 407.
Kuhn, N. J. and Lynen, F. (1965). *Biochemical Journal* **94**, 240.
Kundig, W. (1974). *Journal of Supramolecular Structure* **2**, 695.
Kundig, W. and Roseman, S. (1971). *Journal of Biological Chemistry* **246**, 1407.
Kurioka, S. and Liu, P. V. (1967). *Journal of Bacteriology* **93**, 670.
Lands, W. E. M. (1965). *Annual Review of Biochemistry* **34**, 313.
Laneelle, M. A., Laneelle, G. and Asselineau, J. (1963). *Biochimica et Biophysica Acta* **70**, 99.
Langworthy, T. A., Mayberry, W. R. and Smith, P. F. (1976). *Biochimica et Biophysica Acta* **431**, 550.
Larson, T. J., Hirabayashi, T. and Dowhan, W. (1976). *Biochemistry, New York* **15**, 974.
Larson, T. J. and Dowhan, W. (1976). *Biochemistry, New York* **15**, 5212.
Lee, N. and Inouye, M. (1974). *Federation of European Biochemical Societies Letters* **39**, 167.
Lehmann, V. (1971). *Acta Pathologica et Microbiologica Scandinavica B* **79**, 61.
Lennarz, W. J. (1966). *Advances in Lipid Research* **4**, 175.
Lennarz, W. J., Nesbitt, J. A. and Reiss, J. (1966). *Proceedings of the National Academy of Sciences of the United States of America* **55**, 934.

Lueking, D. R. and Goldfine, H. (1975). *Journal of Biological Chemistry* **250**, 4911.
McAllister, D. J. and DeSiervo, A. J. (1975). *Journal of Bacteriology* **123**, 302.
McCaman, R. E. and Finnerty, W. R. (1968). *Journal of Biological Chemistry* **243**, 5074.
McMurray, W. C. and Magee, W. L. (1972). *Annual Review of Biochemistry* **41**, 129.
MacFarlane, M. G. and Knight, B. C. J. G. (1941). *Biochemical Journal* **35**, 884.
MacFarlane, M. G. (1964). *Nature, London* **196**, 136.
MacFarlane, M. G. (1966). *Advances in Lipid Research* **2**, 91.
Makula, R. A. and Finnerty, W. R. (1970). *Journal of Bacteriology* **103**, 348.
Makula, R. A. and Finnerty, W. R. (1971). *Journal of Bacteriology* **107**, 806.
Makula, R. A. and Finnerty, W. R. (1974). *Journal of Bacteriology* **120**, 1279.
Makula, R. A. and Finnerty, W. R. (1975). *Journal of Bacteriology* **123**, 523.
Marinetti, G. V. and Love, R. (1976). *Chemistry and Physics of Lipids* **16**, 239.
Mersel, M., Benenson, A. and Doljanski, F. (1976). *Biochemical and Biophysical Research Communications* **70**, 1166.
Merlie, J. P. and Pizer, L. I. (1973). *Journal of Bacteriology* **116**, 355.
Milner, L. S. and Kaback, H. R. (1970). *Proceedings of the National Academy of Sciences of the United States of America* **65**, 683.
Mindich, L. (1973). *In* "Bacterial Membranes and Walls", (L. Leive, ed.), pp. 1–36. Marcel Dekker Inc., New York.
Mindich, L. and Dades, S. (1972). *Journal of Cell Biology* **55**, 32.
Minnikin, D. E. and Abdolrahimzadeh, H. (1974). *Federation of the European Biochemical Societies Letters* **43**, 257.
Munoz, E., Salton, M. R. J., Ng, M. H. and Schor, M. T. (1969). *European Journal of Biochemistry* **7**, 490.
Nelson, E. T. and Buller, C. S. (1974). *Journal of Virology* **14**, 479.
Nesbitt, J. A. and Lennarz, W. (1968). *Journal of Biological Chemistry* **243**, 3088.
Nishijima, M., Akamatsu, Y. and Nojima, S. (1974). *Journal of Biological Chemistry* **249**, 5658.
Nishijima, M., Nakaike, S., Tamori, Y. and Nojima, S. (1977). *European Journal of Biochemistry* **73**, 115.
Nunn, W. D. (1975). *Biochimica et Biophysica Acta* **380**, 403.
Nunn, W. D., Cheng, P. J., Deutsch, R., Tang, C. T. and Tropp, B. E. (1977). *Journal of Bacteriology* **130**, 620.
Nunn, W. D. and Cronan, J. E., Jr. (1974). *Journal of Biological Chemistry* **249**, 3994.
Nunn, W. D. and Cronan, J. E., Jr. (1976a). *Biochemistry, New York* **15**, 2546.
Nunn, W. D. and Cronan, J. E., Jr. (1976b). *Journal of Molecular Biology* **102**, 167.
Nunn, W. D. and Tropp, B. E. (1972). *Journal of Bacteriology* **109**, 162.
Oakley, C. L., Warrack, G. H. and Clarke, P. H. (1947). *Journal of General Microbiology* **1**, 91.
Ohki, M., Doi, O. and Nojima, S. (1972). *Journal of Bacteriology* **110**, 864.
Ohta, A., Shibuya, L., Maruo, B., Ishinaga, M. and Kito, MO. (1974). *Biochimica et Biophysica Acta* **348**, 449.
Okuyama, H. and Nojima, S. (1969). *Biochimica et Biophysica Acta* **176**, 120.
Okuyama, H. and Wakil, S. J. (1973). *Journal of Biological Chemistry* **248**, 5197.
Okuyama, H., Yamada, K., Ikewaza, H. and Wakil, S. J. (1976). *Journal of Biological Chemistry* **251**, 2487.
Olsen, R. W. and Ballou, C. E. (1977). *Journal of Biological Chemistry* **246**, 3305.
Ono, Y. and Nojima, S. (1969). *Biochimica et Biophysica Acta* **176**, 111.
Ono, Y. and White, D. C. (1970). *Journal of Bacteriology* **103**, 111.
Otnaess, A. B., Prydz, H., Bjorklid, E. and Berre, A. (1972). *European Journal of Biochemistry* **27**, 238.

Patterson, P. H. and Lennarz, W. J. (1971). *Journal of Biological Chemistry* **246**, 1062.
Peter, H. W. and Ahlers, J. (1975). *Archives of Biochemistry and Biophysics* **170**, 169.
Pieringer, R. A. and Kunnes, R. S. (1965). *Journal of Biological Chemistry* **240**, 2833.
Plackett, P. (1967). *Annals of the New York Academy of Sciences* **143**, 158.
Plackett, P., Marmion, B. P., Shaw, E. J. and Lemcke, R. M. (1969). *Australian Journal of Experimental Biology and Medical Science* **47**, 171.
Plackett, P., Smith, P. F. and Mayberry, W. R. (1970). *Journal of Bacteriology* **104**, 798.
Polakis, S. E., Guchait, R. B. and Lane, M. D. (1973). *Journal of Biological Chemistry* **248**, 7957.
Proulx, P. and Fung, C. K. (1969). *Canadian Journal of Biochemistry* **47**, 1125.
Proulx, P. and van Deenen, L. L. M. (1966). *Biochimica et Biophysica Acta* **125**, 591.
Proulx, P. and van Deenen, L. L. M. (1967). *Biochimica et Biophysica Acta* **144**, 171.
Racker, E. (1967). *Federation Proceedings. Federation of American Societies for Experimental Biology* **26**, 1335.
Racker, E. and Hinkle, P. C. (1974). *Journal of Membrane Biology* **17**, 181.
Racker, E. and Stoeckenius, W. (1974). *Journal of Biological Chemistry* **249**, 662.
Raetz, C. R. H. (1976). *Journal of Biological Chemistry* **251**, 3242.
Raetz, C. R. H. (1975). *Proceedings of the National Academy of Sciences of the United States of America* **72**, 2274.
Raetz, C. R. H. and Kennedy, E. P. (1972). *Journal of Biological Chemistry* **247**, 2008.
Raetz, C. R. H. and Kennedy, E. P. (1974). *Journal of Biological Chemistry* **249**, 5038.
Raetz, C. R. H., Larson, T. J. and Dowhan, H. (1977). *Proceedings of the National Academy of Sciences of the United States of America* **74**, 1412.
Ray, P. R., White, D. C. and Brock, T. D. (1971). *Journal of Bacteriology* **108**, 227.
Ray, T. H. and Cronan, J. E., Jr. (1975). *Journal of Biological Chemistry* **250**, 8422.
Ray, T. K., Cronan, J. E., Jr. and Godson, G. H. (1976). *Journal of Bacteriology* **125**, 136.
Ray, T. K., Cronan, J. E., Jr., Mavis, R. D. and Vagelos, P. R. (1970). *Journal of Biological Chemistry* **245**, 6442.
Raybin, D. M., Bertsch, L. L. and Kornberg, A. (1972). *Biochemistry, New York* **11**, 1754.
Redwood, W. R., Gibbs, D. C. and Thompson, T. E. (1973). *Biochimica et Biophysica Acta* **31**, 10.
Renkonen, O., Kaariainen, L., Petterson, R. and Oker-Blom, N. (1972). *Virology* **50**, 899.
Rothfield, L. and Romeo, D. (1971). *In* "Structure and Function of Biological Membranes", (L. Rothfield, ed.), pp. 251–284. Academic Press, New York.
Rothman, J. E. and Kennedy, E. P. (1977). *Journal of Molecular Biology* **110**, 603.
Rouser, G., Kritchevsky, G., Yamamoto, A., Knudson, A. G. and Simon, G. (1968). *Lipids* **3**, 287.
Ruettinger, R. T., Olson, S. T., Boyer, R. F. and Coon, M. J. (1974). *Biochemical and Biophysical Research Communications* **57**, 1011.
Salton, M. R. J. (1974). *Advances in Microbial Physiology* **11**, 219.
Sandermann, H., Jr. and Strominger, J. L. (1972). *Journal of Biological Chemistry* **247**, 5123.
Sandermann, H., Jr. and Strominger, J. L. (1971). *Proceedings of the National Academy of Sciences of the United States of America* **68**, 2441.
Sands, J. A. (1976). *Journal of Virology* **19**, 296.
Sands, J. A. and Cadden, S. P. (1975). *Federation of European Biochemical Societies Letters* **58**, 43.
Scandella, C. J. and Kornberg, A. (1971). *Biochemistry, New York* **10**, 4447.
Scandella, C. J., Schindler, H., Franklin, R. M. and Seeling, J. (1974). *European Journal of Biochemistry* **50**, 29.

Schafer, R. and Franklin, R. M. (1975). *Journal of Molecular Biology* **97**, 21.
Schafer, R., Hinnen, R. and Franklin, R. M. (1974). *European Journal of Biochemistry* **50**, 15.
Schiefer, H. G., Gerhardt, V. and Brunner, H. (1975). *Hoppe-Seyler's Zeitschrift für Physiologische Chemie* **356**, 539.
Schneider, E. G. and Kennedy, E. P. (1976). *Biochimica et Biophysica Acta* **441**, 201.
Schneider, E. G. and Kennedy, E. P. (1973). *Journal of Biological Chemistry* **248**, 3739.
Scott, C. C. L. and Finnerty, W. R. (1976). *Journal of Bacteriology* **127**, 481.
Scott, C. C. L., Makula, R. A. and Finnerty, W. R. (1976). *Journal of Bacteriology* **127**, 469.
Senff, L. M., Wagener, W. S., Brooks, G. F., Finnerty, W. R. and Makula, R. A. (1976). *Journal of Bacteriology* **127**, 874.
Sheetz, M. P. and Singer, S. J. (1974). *Proceedings of the National Academy of Sciences of the United States of America* **71**, 4457.
Shively, J. M. and Benson, A. A. (1967). *Journal of Bacteriology* **94**, 1679.
Short, S. A. and White, D. C. (1971). *Journal of Bacteriology* **108**, 219.
Short, S. A. and White, D. C. (1972). *Journal of Bacteriology* **109**, 820.
Siekevitz, P. (1972). *Annual Review of Physiology* **34**, 117.
Sinensky, M. (1971). *Journal of Bacteriology* **106**, 449.
Smith, P. F. (1967). *Annals of the New York Academy of Sciences* **143**, 139.
Smith, P. F. and Koostra, W. L. (1967). *Journal of Bacteriology* **93**, 1853.
Sonoki, S. and Ikezawa, S. (1975). *Biochimica et Biophysica Acta* **403**, 412.
Stahl, W. L. (1973). *Archives of Biochemistry and Biophysics* **154**, 47.
Steiner, S., Burnham, J. C., Conti, S. F. and Lester, R. L. (1970). *Journal of Bacteriology* **103**, 500.
Snider, M. D. and Kennedy, E. P. (1977). *Journal of Bacteriology* **130**, 1072.
Snipes, W., Douthwright, J., Sands, J. and Keith, A. (1974). *Biochimica et Biophysica Acta* **363**, 340.
Sugahara, T. and Ohsaka, A. (1970). *Japanese Journal of Medical Science and Biology* **23**, 61.
Swartouw, H. T. and Smith, H. (1956). *Journal of General Microbiology* **15**, 261.
Taber, H. W. and Morrison, M. (1964). *Archives of Biochemistry and Biophysics* **105**, 367.
Tahara, Y., Kameda, M., Yameda, Y. and Kondo, K. (1976). *Biochimica et Biophysica Acta* **450**, 225.
Takahasi, T., Sugahara, T. and Ohsaka, A. (1974). *Biochimica et Biophysica Acta* **351**, 155.
Teitelbaum, D., Arnon, R., Sela, M., Rabinsohn, Y. and Shapiro, D. (1973). *Immunochemistry* **10**, 775.
Thiele, O. W. and Schwinn, G. (1973). *European Journal of Biochemistry* **34**, 333.
Thiele, O. W. and Schwinn, G. (1974). *Zeitschrift für Allegemeine Mikrobiologie* **14**, 435.
Torregrossa, R. E., Makula, R. A. and Finnerty, W. R. (1977a). *Journal of Bacteriology* **131**, 486.
Torregrossa, R. E., Makula, R. A. and Finnerty, W. R. (1977b). *Journal of Bacteriology* **131**, 493.
Tsukagoshi, N. and Franklin, R. M. (1974). *Virology* **59**, 408.
Tsukagoshi, N., Kania, M. and Franklin, R. M. (1976). *Biochimica et Biophysica Acta* **450**, 131.
Tsukagoshi, N., Petersen, M. H. and Franklin, R. M. (1975). *Nature, London* **253**, 125.
Tucker, A. N. and White, D. C. (1971). *Journal of Bacteriology* **108**, 1058.
Tunaitis, E. and Cronan, J. E., Jr. (1973). *Archives of Biochemistry and Biophysics* **155**, 420.
van Golde, L. M. G., Akkermans-Kruyswijk, W., Franklin-Klein, W., Lankhorst, A. and Prins, R. A. (1975). *Federation of the European Biochemical Societies Letters* **53**, 57.

van Golde, L. M. G., McElhaney, R. N. and van Deenen, L. L. M. (1971). *Biochimica et Biophysica Acta* **231**, 245.
van Deenen, L. L. M. (1966). *Annals of the New York Academy of Sciences* **137**, 717.
van Deenen, L. L. M. and deHaas, G. H. (1966). *Annual Review of Biochemistry* **35**, 178.
van den Bosch, H. (1974). *Annual Review of Biochemistry* **43**, 243.
van den Bosch, H. and Vagelos, P. R. (1970). *Biochimica et Biophysica Acta* **218**, 233.
Verger, R. and deHaas, G. H. (1976). *Annual Review of Biophysics and Bioengineering* **5**, 77.
Weissbach, H., Thomas, E. and Kaback, H. R. (1971). *Archives of Biochemistry and Biophysics* **147**, 249.
Wherret, J. R. and Huterer, S. (1972). *Journal of Biological Chemistry* **247**, 4114.
White, D. A., Albright, F. R., Lennarz, W. J. and Schnaitman, C. A. (1971). *Biochimica et Biophysica Acta* **249**, 636.
White, D. C. (1962). *Journal of Bacteriology* **83**, 851.
White, D. C. (1963). *Journal of Biological Chemistry* **238**, 3757.
White, D. C. and Tucker, A. N. (1969a). *Journal of Bacteriology* **97**, 199.
White, D. C. and Tucker, A. N. (1969b). *Journal of Lipid Research* **10**, 220.
Wilkinson, S. G. (1972). *Biochimica et Biophysica Acta* **270**, 1.
Wirtz, K. W. A. (1974). *Biochimica et Biophysica Acta* **344**, 95.
Yamamoto, A., Adachi, S., Ishikawa, K., Yokomura, T., Kitani, T., Nasu, T., Imoto, T. and Nishikawa, M. (1971). *Journal of Biochemistry, Tokyo* **70**, 775.
Yamamoto, S. and Lampen, J. O. (1975). *Journal of Biological Chemistry* **250**, 3212.
Yamamoto, S. and Lampen, J. O. (1977). *Proceedings of the National Academy of Sciences of the United States of America* **73**, 1457.
Yano, I., Furukawa, Y. and Kusunose, M. (1969). *Journal of Bacteriology* **98**, 124.
Zwaal, R. J. A., Roelofsen, B., Comfurius, P. and van Deenen, L. L. M. (1971). *Biochimica et Biophysica Acta* **233**, 474.

Aspects of Genetic Engineering in Micro-Organisms

S. W. GLOVER

Department of Genetics, Ridley Building, University of Newcastle, Newcastle upon Tyne NE1 7RU

I. Introduction	235
II. Restriction Endonucleases	237
A. Type I Restriction Endonucleases	238
B. Type II Restriction Endonucleases	239
III. End-to-End Joining of DNA Molecules	243
IV. General Methods for Inserting Specific DNA Sequences	244
V. Plasmid Vectors	247
VI. Bacteriophage Lambda Vectors	251
VII. Cloning of Recombinant DNA	254
A. Cloning of Prokaryote genes	254
B. Cloning Yeast DNA	256
C. Cloning Higher Eukaryote DNA	257
D. Screening Methods for Cloned DNA	260
VIII. Potential Biohazards	262
A. Physical Containment	263
B. Biological Containment	264
IX. Acknowledgments	265
References	265

I. Introduction

In undertaking to write this review I had, initially, more than a little sympathy for the statement made by John Abelson in *Science* on the subject of Recombinant DNA, "that many words have been spoken and written about DNA—almost all of them speculative, with little factual basis." While that may have been so a very few years ago, the rate of progress in this field has been so rapid that the reviewer is hard put to it to keep a tally on all that is published. The same issue of

Science (8th April 1977) in which Abelson's remark appeared contains no less than twenty-three original publications on this topic. Late in 1976 a new journal, "*Gene*," appeared devoted entirely to gene cloning and recombinant nucleic acids and it has already (July 1977) completed one volume. Nevertheless, the literature on this subject is widely scattered. In this review I have attempted to bring together some examples of past and present day research on genetic engineering in the narrow context of recombinant DNA. The review is far from definitive and emphasis has been placed on new and developing techniques rather than on attempting to catalogue all that has been reported on this topic.

Many aspects of the subject have been recently reviewed and the readers attention is directed to those sources for current assessments of knowledge in these fields. Restriction endonucleases have been definitely reviewed by Roberts (1976); Curtiss (1976) has reviewed the potential benefits and biohazards of genetic manipulation, and Nathans and Smith (1975) reviewed the role of restriction endonucleases in the analysis and restricting of DNA molecules; Reanney (1978) has written a perceptive review of genetic engineering as an adaptive strategy and previously (Reanney, 1976) reviewed aspects of natural genetic engineering. A review by Murray (1978) on applications of restriction enzymes in biochemistry and genetics is currently in press.

Genetic recombination, in the conventional sense, consists in the breakage and rejoining of DNA molecules and is of fundamental importance to living organisms. Naturally occurring mechanisms which permit genetic recombination to take place are generally thought to be confined by strong taxonomic constraints, because a degree of sequence homology is an essential prerequisite for genetic recombination. This long held view, while still tenable, has been subject to attack in recent years following the discovery of promiscuous plasmids and phages and the widespread occurrence of antibiotic and other resistance markers among diverse groups of bacteria. The importance of insertion (IS) sequences in the DNA of plasmids, viruses, bacteria and eukaryotes which can act as agents to permit exchange of genetic information is increasingly recognized and has been reviewed by Reanney (1978); Cohen (1976); Nevers and Saedler (1977). The recently developed techniques of *in vitro* recombination between DNA molecules are not subject to any of these constraints.

These techniques permit the splicing of DNA molecules of quite diverse origin and, combined with techniques of genetic transformation, permit the introduction of foreign DNA into micro-organisms.

Early work by Jackson et al. (1972) and Lobban and Kaiser (1973) exploited the base pairing between chemically synthesized poly A and poly T tails added to DNA fragments *in vitro* to join together fragments of *Salmonella* P22 bacteriophage DNA (Lobban and Kaiser, 1973) and to join animal virus SV40 to bacteriophage λ DNA after fragmentation with an endonuclease (Jackson et al., 1972). This method still has important applications but has largely been replaced by the use of certain restriction endonucleases which, after cleavage of DNA, generate fragments which have cohesive termini. Thus the fragments produced by *Eco* RI cleavage of DNA have 5'-terminal extensions pAATT and following incubation with DNA ligase may be circularized or dimerized or linked to other fragments (Mertz and Davis, 1972). Bacteriophage T4 coded DNA ligase is also able to join fragments with flush ends, so that even the presence of cohesive ends is no longer a stringent requirement for *in vitro* DNA recombination. The historical developments which led to the development of these techniques have been reviewed by Cohen (1975).

The ability to clone DNA and to study its expression in controlled environments is an exciting development of almost unparalleled importance in modern biology. Opportunities now exist to clone and prepare large quantities of DNA from higher organisms and to study the organization of genes and the systems of controlling their expression which were hitherto unapproachable. These same techniques permit bacteria to be used for the production of many useful enzymes and, eventually, for the production on an industrial scale of other products useful to man.

II. Restriction Endonucleases

Restriction enzymes were first defined as a class of enzymes involved in restriction-modification systems in bacteria recognized initially by their effects on bacteriophage growth in certain strains of *Escherichia coli*. Such restriction-modification systems operate through the activity of two enzymes. First, a restriction endonuclease which recognizes a specific nucleotide sequence within the DNA and as a consequence makes a double strand cleavage of the molecule. The second enzyme is a modification methylase, which recognizes the same nucleotide sequence as the restriction endonuclease and chemically modifies it by methylating a base within the sequence in such a way that the sequence is no longer recognized by the restriction enzyme.

When a DNA molecule enters a bacterial cell the restriction endonuclease "inspects" the molecule and if it contains the appropriate sequence along its length it will cleave the molecule. The modification methylase serves to protect the DNA of the bacterial cell from digestion by its own restriction enzyme and will similarly protect any DNA synthesized in that cell. The two enzymes thus form a pair and provide the enzymological basis of restriction-modification systems.

Restriction endonucleases differ in whether they cleave the DNA at the recognition sequence (Type II), or elsewhere (Type I), and they differ in the type of break produced.

A. TYPE I RESTRICTION ENDONUCLEASES

Genetic studies over a number of years revealed that these enzymes were complex. Mutation in any one of three different genes *hsdS*, *hsdR* and *hsdM* can lead to restriction deficiency. Mutation in *hsdS* and *hsdM* leads to modification and restriction deficiency. Predictions from genetic studies were confirmed when Type I enzymes were isolated and purified. The enzymes have a high molecular weight of about 300,000 daltons; contain three non-identical sub-units and have requirements for Mg^{2+}, ATP and in many cases S-adenosylmethionine, as co-factors for activity. These enzymes recognize specific sites on DNA, but make a limited number of double strand cuts at apparently random sites distant from the recognition sites. The circular DNA molecule which is the replicative form of phage f1 is cut once by the *E. coli* restriction enzyme to yield linear molecules which are resistant to further degradation. The products of restriction cleavage are a heterogenous population of molecules which after denaturation and renaturation can reform circular molecules. These circles are again sensitive to the restriction endonuclease, indicating that the recognition sites are not destroyed by the first cleavage event.

There is evidence that S-adenosylmethionine (SAM) acts as an allosteric effector molecule in the restriction reaction. The enzyme binds SAM rapidly, although activation is slow. The activated enzyme then interacts specifically with specific sites on the DNA in a reaction that requires Mg^{2+} and ATP. It is believed that the state of methylation at the site determines the subsequent action of the enzyme. If the site is unmodified, the enzyme "wanders" along the DNA or a distant part of the DNA molecule is brought into contact with it and cleavage occurs. Cleavage proceeds from a primary single-strand break to a second break on the opposite strand and there is evidence to indicate that the second break may be produced by a different enzyme molecule which may have wandered to the same cleavage site from the opposite direction. One additional unusual feature of the enzymes is that following cleavage the enzyme partly dissociates. One molecular species remains bound to DNA and displays a powerful ATPase activity.

B. TYPE II RESTRICTION ENDONUCLEASES

As Roberts (1976) points out, a majority of enzymes classified as type II restriction endonucleases are not known to be associated with classical restriction-modification systems, although for some of them, viz. *Eco* RI, *Eco* RII (Boyer et al., 1973) and *Hind II* and *Hind III* (Roy and Smith, 1973a, b) the associated methylases have been detected and the restriction-modification systems characterized genetically (Glover and Bannister, 1968; Bannister and Glover, 1970; Boyer, 1971).

The first restriction endonuclease of this class to be isolated was *Hind II* from *Haemophilus influenzae* serotype *d* (Smith and Wilcox, 1970). Subsequently, type II restriction endonucleases have been isolated from all the other *H. influenzae* serotypes except *a* and *e* (see Roberts, 1976). In fact, *H. influenzae* has been the source of nine enzymes of this type. Not only has *H. influenzae* been a rich source of these enzymes, but other *Haemophilus* species have been almost equally productive. Enzymes have been isolated from *H. aegyptius* (*Hae I* Murrey et al. quoted by Roberts, 1976; *Hae II* Roberts et al., 1975; *Hae III* Middleton et al., 1972); *H. aphrophilus* (*Hap I* Roberts, 1976; *Hap II* Takanami, 1974); *H. gallinarum* (*Hga I* Takanami, 1974); *H. haemoglobinophilus* (*H.hg I* Roberts, 1976); *H. haemolyticus* (*H.ha I* Roberts, 1976); *H. parahaemolyticus* (*H.ph I* Kleid et al., 1976); *H. parainfluenzae* (*H.pa I* Sharp et al., 1973; *H.pa II* Sharp et al., 1973) and *H. suis* (*Hsu I* Roberts, 1976). Altogether twenty-two enzymes have been isolated from the twenty-nine strains so far examined. Other genera which have been productive sources for these enzymes include

Bacillus (9 enzymes), *Streptomyces* (8 enzymes), *Moraxella* (7 enzymes), *Xanthomonas* (6 enzymes), *Escherichia* (5 enzymes) and *Anabaena* (4 enzymes). The reader is referred to the comprehensive review by Roberts (1976) for a full listing of the eighty-six enzymes isolated to date. It is of interest to note that restriction endonucleases have been isolated from most of the major groups of prokaryotes, from phototrophic bacteria (*Anabaena*), gliding bacteria (*Myxococcus*), Gram-negative rods and cocci, Gram-positive rods and cocci and *Actinomycetes*. With the possible exception of *Chlamydomonas* (Burton *et al.*, 1976), these enzymes have not been detected in the few eukaryote protozoa, yeasts and fungi that have been examined (Roberts, 1976).

The sequences recognized by these enzymes generally display a characteristic 2-fold rotational symmetry. That is, the nucleotide sequence on one DNA strand read 5' to 3' is the same as the sequence on the complementary strand read 3' to 5'. The majority of type II restriction enzymes have not been purified to homogeneity and their molecular weights and sub-unit composition is not known. One exception is *Eco* RI which Greene *et al.* (1974) have purified to near homogeneity as judged by SDS-polyacrylamide gel electrophoresis. It has a molecular weight of 59,000 daltons and the native enzyme has two sub-units of 29,500 daltons. The recognition sequence is GAATTC (Hedgpeth *et al.*, 1972) and the site of cleavage, indicated by an arrow, G↓AATTC, produces fragments with 5'-tetranucleotide extensions. The oligonucleotide GAATTC is necessary and sufficient for cleavage to occur (Greene *et al.*, 1975), but apparently sequences outside this hexanucleotide influence the rate of cleavage (Thomas and Davis, 1975). A synthetic oligonucleotide TGAATTCA has been synthesized (Greene *et al.*, 1975) and used as a substrate for the modification methylase which methylates at the position indicated by the asterisk in TGAA*TTCA.

While the sub-unit structure of other restriction enzymes is not known, it seems more than likely that the dimeric form of these enzymes is related to their recognition of nucleotide sequences displaying 2-fold rotational symmetry. For practical purposes it is the recognition sequence that is of interest and recognition sequences for many of these enzymes have been determined. From a study of the twenty or so recognition sequences which have been determined, a number of interesting features emerge.

Many enzymes from different sources recognize the same sequence.

The name isoschizomers was coined by Roberts (1976) for these enzymes. For example, the sequence AAGCTT is the recognition sequence for the following enzymes: Bbr I (Bacillus brevis), Chu I (Corynebacterium humiferum), Hin b III (Haemophilus influenzae Rb) Hind III (Haemophilus influenzae Rd) and Hsu I (Haemophilus suis). The tetranucleotide sequence GGCC is recognized by six different enzymes: Blu II (Brevibacterium luteum), Bsu RI (Bacillus subtilis strain R), Hae III (Haemophilus aegyptuis), Hhg I (Haemophilus haemoglobinophilus), Pal I (Providencia alcalifaciens) and Sfa I (Streptococcus faecalis). The tetranucleotide CCGG is recognized by three different enzymes and in this case it is established that the site of cleavage C↓CGG is the same for all three of the enzymes Hap II (Haemophilus aphrophilus), Hpa II (Haemophilus parainfluenzae) and Mno I (Moraxella nonliquefaciens). Two other isoschizomers Xma I (Xanthomonas malvacearum) and Sma I (Serratia marcescens strain Sb) recognize the sequence CCCGGG. However, Xma I produces the cleavage C↓CCGGG, while Sma I produces the cleavage CCC↓GGG.

It is interesting to note that the isolation of isoschizomers is no indication of genetic relatedness. Isoschizomers which have been isolated from Haemophilus (Gram negative) also occur in Corynebacterium (Gram positive). For example, the enzyme Hind III and Chu I are isoschizomers, as are the enzymes Hind II and Chu II.

Two enzymes recently isolated from Diplococcus pneumoniae are interesting in that both recognize the sequence GATC. The enzyme Dpn II cleaves the sequence while the enzyme Dpn I only cleaves the methylated sequence GA*TC. Dpn I is the only specific endonuclease known for which cleavage is dependent upon methylation of the DNA. (Lacks and Greenberg, 1975.)

Several enzymes are apparently able to cleave both double and single-stranded DNA (Godson and Roberts, 1976; Horiuchi and Zinder, 1975; Blakesley and Wells, 1975). While it is possible that such enzymes recognize and cleave recognition site in single-stranded DNA, it seems more probable that they recognize and cleave small duplex regions containing the recognition sequence that may be generated by secondary loop structures within single stranded DNA.

As a result of the large number of restriction enzyme recognition sites now known, it is of interest to note the variation in the type of cleavage produced by these enzymes. Examples are known of enzymes which cleave DNA to produce fragments with 5' single strand extensions, 3' single strand extensions, and flush ends. Examples of enzymes which produces 5' extensions include *Eco* RII (5'-pentanucleotide extensions); *Eco* RI (5'-tetranucleotide extension; *Hin f I* (5'-trinucleotide extension); *Hpa II* (5'-dinucleotide extension). The enzyme *Hae III* produces fragments with flush ends, while the enzyme *Pst I* (*Providencia stuartii*) produces 3'-tetranucleotide extensions and the enzyme *Hha I* (*Haemophilus haemoglobinophilus*) produces 3'-dinucleotide extensions.

At least two enzymes are known for which no specificity has been demonstrated at the site of cleavage. These enzymes are *Hph I* (*Haemophilus parahaemolyticus*) and *Mbo II* (*Moraxella bovis*). However, DNA cleaved by these enzymes does yield specific fragments indicating precise points of cleavage. No common features of nucleotide sequence occur at the sites of cleavage but a common feature does occur eight nucleotides distant from the cleavage site (Kleid *et al.*, 1976). Apparently *Hph I* recognizes the pentanucleotide sequence GGTGA and cleaves DNA at a site eight nucleotides from this sequence. *Mbo II* recognizes the pentanucleotide sequence GAAGA and cleaves DNA also at a site eight nucleotides distant.

Of the sixteen possible tetranucleotide palindromes that can be written from four bases, enzymes have been isolated which recognize six of these, viz. AGCT (*Alu I, Arthrobacter luteus*); GATC (*Mbo I, Moraxella bovis*); GCGC (*Haemophilus haemolyticus*) GGCCC (*Hae III, Haemophilus aegyptuis*); CCGG (*Hpa II, Haemophilus parainfluenzae*) and TCGA (*Taq I, Thermus aquaticus* strain YT1). By modifying cleavage reaction conditions the enzyme *Eco* RI cleaves a seventh sequence (AATT) of this set. Roberts (1976) states that three other enzymes have been isolated which may also cleave tetranucleotides of this set, but their recognition sequences are not yet known.

As will be discussed in a following section, restriction enzymes which cleave DNA to produce fragments with flush ends can be utilized in genetic engineering, but the enzymes of greatest value are those that create fragments with cohesive ends. To date, fourteen such enzymes are known, *Bam I* (G↓GATCC); *Bgl II* (A↓GATCT); *Eco* RI (G↓AATTC); *Eco* RII (↓CCAGG and ↓CCTGG); *Hae II* (PuGCGC↓Py) *Hha I* (GCG↓C); *Hind III* (A↓AGCTT); *Hinf I* (G↓ANTC); *Hpa II* (C↓CGG); *Pst I* (CTGCA↓G); *Taq I* (T↓CGA); *Xma I* (C↓CCGGG); *Mbo I* (↓GATC); *Eco* RI (↓AATT under certain conditions), and *Sal I* an enzyme from *Streptococcus albus* strain G which produces cohesive termini although the recognition sequence is not yet known (Roberts, 1976; Hamer and Thomas, 1976).

Quite dramatically, restriction endonucleases have become essential tools in many areas of molecular biology. Major advances have been made possible in nucleotide sequence analysis as a result of the availability of specific DNA fragments generated by restriction enzymes (reviewed by Salser, 1974; Murray and Old, 1974; Wu *et al.*, 1975; Nathans and Smith, 1975). One of the most productive applications of restriction endonucleases has been in the physical mapping of chromosomes, especially in the case of viral chromosomes consisting of double stranded DNA. Where the viral genome is single stranded, physical maps have been determined by restriction endonuclease cleavage of double stranded DNA replicative intermediates (reviewed by Roberts, 1976; Nathans and Smith, 1975). Equally, restriction enzymes have seen rapid application in the physical mapping of bacterial and other small plasmid DNA molecules including mitochondrial DNA. Roberts (1976) lists over twenty-six viruses and plasmids for which physical maps have been produced by these methods. Further applications have been made in techniques for the isolation of specific genes and in the analysis of the organization of eukaryote DNA (reviewed by Nathans and Smith, 1975; Roberts, 1976).

In following sections the applications of restriction endonucleases to the construction and cloning of *in vitro* recombinant DNA molecules will be reviewed.

III. End-to-End Joining of DNA Molecules

Two general methods are available for joining DNA molecules end-to-end.

In the first method, complementary single stranded termini are added enzymatically to the 3'-ends of the molecules to be joined (Jackson *et al.*, 1972; Lobban and Kaiser, 1973). Initially it is necessary to remove a small number of nucleotides from the 5'-end of each DNA strand with λ exonuclease. The terminal, single stranded 3'-ends are then extended by adding dAMP residues to one set of molecules and dTMP to the other set, using the terminal deoxynucleotidyl transferase reaction (Chang and Bollum, 1971). Molecules with oligo dA termini can then be joined end-to-end with molecules carrying oligo dT termini, relying on complementary base pairing. The joined molecules are then covalently linked by the combined action of exonuclease III (to remove unpaired single strands), DNA polymerase I (to ensure the completeness of the double stranded join), and DNA ligase (to seal the single strand gaps).

The method now more generally employed in recombinant DNA studies is the joining of DNA fragments by the cohesive ends produced by restriction endonucleases which make staggered breaks in DNA. Any DNA fragments produced by cleavage with a given enzyme of this type will have complementary single stranded ends and can be joined end-to-end to any other DNA fragments produced by the same enzyme. Such fragments may also circularize via these cohesive ends (Mertz and Davis, 1972). The adjacent 5'-phosphoryl and 3'-hydroxyl ends can then be linked covalently using DNA ligase (Richardson, 1969).

The procedure developed by Blattner *et al.* (1977) for joining DNA molecules differs significantly from that generally employed. The conditions for ligation of fragments were chosen to favour the production of giant concatenates from the fragments rather than, as usual, to limit ligation to produce dimers. The vector Charon phage λ DNA was first annealed to join its ends. The vector and target DNAs were digested with the chosen restriction enzyme. To generate recombinant DNA molecules, the vector and target DNA digests were then mixed in proportions adjusted to keep the concentration of target DNA fragments below that of vector DNA and the total DNA concentration was kept high, to ensure that the formation of recombinant molecules competed effectively with monomolecular cyclization of the shorter target fragments. The mixture was then incubated with T4 DNA ligase to join the molecules covalently. Viable phages were then obtained by transfection of either *E. coli* spheroplasts (Henner, Kleber and Benzinger, 1973) or calcium treated *E. coli* (Mandel and Higa, 1970), or by *in vitro* packaging into λ capsids (Kaiser, Syvanen and Masuda, 1975; Hohn and Hohn, 1974; Sternberg *et al.*, 1976). It is assumed that λ-sized recombinant molecules are extracted from the concatenated DNA adsorbed by the λ packaging function, or that circles are first generated by recombination in the transformed cells.

IV. General Methods for Inserting Specific DNA Sequences

Two groups of workers have devised general methods for inserting specific DNA sequences into cloning vehicles (Bahl *et al.*, 1976a;

Scheller et al., 1977a). Cohesive ends are often lacking in many DNA molecules of interest and hence these molecules cannot be cloned by conventional methods into appropriate vehicles using restriction endonucleases. Molecules obtained by random cleavage of DNA, complementary DNA (cDNA) and many restriction endonuclease fragments fall into this category. The methods employed to achieve this end are essentially similar. First decadeoxyribonucleotides were synthesized by the improved phosphotriester method (Itakura et al., 1973; Itakura et al., 1975a; Katagiri et al., 1975; Itakura et al., 1975b; Stawinsky et al., 1976). Successful synthesis was achieved for d(C–C–G–G–A–T–C–C–G–G) the *Bam I* linker sequence and d(A–C–A–A–G–C–T–T–G–T) the *Hind III* linker sequence by Bahl et al. (1976a), and for d(C–C–G–A–A–T–T–C–G–G) the *Eco* RI linker sequence, as well as the *Bam I* and *Hind III* sequence, by Scheller et al. (1977a). It is interesting to note that the *Hind III* decamer, which contains the recognition sequence AAGCTT, is not cleaved by *Hsu I* despite the fact that *Hind III* and *Hsu I* are isoschizomers (Roberts, 1976) and therefore recognize identical sequences. However, *Hsu I* does cleave the sequence when it forms part of a longer thirty base pair fragment of DNA. The same decamer also contains the *Alu I* sequence AGCT, but in this instance *Alu I* does cleave the short fragment. The decamer containing the *Eco* RI sequences and the decamer containing the *Bam I* sequence are cleaved by their respective endonucleases, but the *Bam I* decamer which contains the *Hpa II* recognition sequence (C–C–G–G) at each end is not cleaved by *Hpa II* presumably because these sequences are too close to the ends of the short fragment. Four methods for joining DNA molecules are in common use. First, end-to-end joining of cohesive ends generated by restriction endonucleases (Chang and Cohen, 1974); second, adding poly dA and poly dT tails to fragments (Jackson et al., 1972); third, adding poly dC and poly dG tails to fragments (Rougeon et al., 1975); and, lastly, blunt-end ligation using T4 ligase (Backman et al., 1976), based on the ability of T4 ligase efficiently to join blunt-ended DNA duplexes under certain conditions (Sgaramelia et al., 1976).

Bahl et al. (1976a) joined, by end-to-end T4 mediated ligation, the *Bam I* decamer to a synthetic *lac* operator, a twenty-one base-pair long DNA duplex (Bahl et al., 1976b), to produce a fragment which bears two *Bam I* decamers flanking the *lac* operator. This larger fragment was then subjected to *Bam I* restriction endonuclease cleavage to produce a *lac* promoter flanked by *Bam I* generated cohesive ends, which was then inserted into the plasmid pMB9 (Rodriques et al., 1976) which had been digested with *Bam I* to produce linear molecules with cohesive ends. The resulting hybrid *lac⁻* pMB 9 DNA was then used to transform competent *E. coli*, employing the tetracycline resistance marker of pMB9 for selection. Transformed clones were shown, by the capacity to over-produce β-galactosidase (Marians et al., 1976), to contain functional extra copies of the *lac* promoter. Hybrid *lac*-pMB9 DNA was isolated after chloramphenicol stimulated amplification and shown *in vitro* to possess appropriate *lac* repressor binding properties.

Scheller et al. (1977a) reported the use of linker sequences to clone both eukaryotic DNA sequences and *c* DNA sequences. The *Eco* RI decamer was ligated to repetitive DNA from sea urchin sperm and the fragments generated by *Eco RI* were cloned in plasmid RSF2124 which has an Eco RI site within its colicin gene (So et al., 1975). The hybrid RSF2124–sea urchin DNA can be detected after transformation by screening for clones which fail to produce colicin. The *Bam I* decamer is reported in Scheller et al. (1977b) to have been linked to *c* DNA and cloned in the plasmid pB313 which has ampicillin and tetracycline resistance genes with a *Bam I* site in the tetracycline gene (J. Shine and H. M. Goodman, personal communication, quoted by Scheller et al., 1977a).

An important problem to be overcome in the use of chemically synthesized linker molecules to clone DNA segments is the efficient transcription and translation of inserted DNA sequences. Scheller et al. (1977b) were careful to use a linker molecule so designed that, after linking, three extra nucleotides were introduced on either side of the inserted *lac* operator. This leaves unchanged the reading frame for the plasmid-directed protein synthesis. Despite a report that both the *Hind III* and *Bam I* restriction endonuclease targets in pMB9 are located in the tetracycline resistance gene (Rodrigues et al., 1976), the insertion of twenty-seven nucleotides (twenty-one from the *lac* promoter plus three on each side from the cleaved *Bam I* decamer) into the plasmid, did not change the tetracycline resistance of bacteria transformed by the hybrid plasmid. As Scheller et al. (1977b) point out, either the *Bam I* site is not within the tetracycline resistance gene, or the insertion of twenty-seven nucleotides does not alter the activity of the protein this gene codes.

Even if the ends of a DNA fragment are not blunt-ended, these general methods can be used after first digesting the DNA fragment with the *Aspergillus* nuclease 51 (Ando, 1966; Ghangas and Wu, 1975) to produce even-ended molecules.

Scheller *et al.* (1977a) suggest a further extension of this general method, applicable to almost any DNA molecule. One DNA molecule is chosen for the ease with which it can be selected. It could then be linked to a second DNA molecule "X" bearing the same cohesive ends as the selected DNA molecule, both sets of molecules having been generated by restricted endonuclease cleavage of chemically synthesized linkers attached to the ends. These two molecules could then be joined by their cohesive ends and T4 ligation and inserted into a suitable vehicle. The presence of DNA molecule "X" in cells transformed with the hybrid vehicle could then be monitored by screening for the activity coded for by the selected DNA molecule.

A rather different method was used by Sadler *et al.* (1977) to clone chemically synthesized *lac* operators. They synthesized single-stranded fragments of the *lac* operator which, when annealed, generated the central portion of the *lac* operator flanked by 5' pAATT single-strand ends complementary to the single-strand sequences produced by *Eco* RI endonuclease. The synthetic *lac* operator was then inserted into *Eco* RI cleaved pMB9 plasmid DNA and used to transform *E. coli*. About 10% of the tetracyline resistant transformants showed evidence of the presence of functional *lac* operator. Functional *lac* operator has thus been cloned independently by five groups, namely, Marians *et al.* (1976); Bahl *et al.* (1976a); Heyneker *et al.* (1976); Sadler *et al.* (1977) and Scheller *et al.* (1977a).

V. Plasmid Vectors

The *E. coli* plasmid Col E1 and its many derivatives have been used extensively as cloning vehicles for foreign DNA. Col E1 is a relatively small, covalently closed circular DNA molecule some 4.2×10^6 daltons in size (Bazaral and Helinski, 1968). It is present in multiple copies (20–30 molecules per cell) in *E. coli* under normal conditions of growth (Clewell and Helsinki, 1972). When bacteria harbouring Col E1 are grown in the presence of chloramphenicol, the Col E1 plasmid is amplified many fold until it reaches a condition in which about 45% of the total DNA in the culture is attributable to Col E1 DNA (Clewell and Helinski, 1972; Clewell, 1972). Consequently by using this amplification step, very high yields of both Col E1 DNA and also any

DNA fragment that can be inserted into the plasmid may be readily obtained (Hershfield et al., 1974). Since Col E1 and many of its derivatives contain a non-essential region, a single site sensitive to the restriction endonuclease *Eco* RI, *Eco* RI-digested DNA can be inserted into the plasmid after cleavage with the enzyme. Recombinant plasmids constructed by insertion of foreign DNA at the *Eco* RI site can be introduced into *E. coli* by transformation of calcium treated bacteria. Their presence can readily be detected since they no longer produce the antibiotic protein colicin E1 (Hershfield et al., 1974).

Col E1 is a non-conjugative plasmid, that is, it is unable to promote its own transfer from one bacterium to another by cell–cell contact. This feature is of value in the biological containment of recombinant DNA. However, the conjugal transfer of Col E1 is promoted by many conjugative (self-transmissible) plasmids, although they vary considerably in their ability to promote Col E1 transfer (Kahn and Helinski, 1964). Recently, it has been shown that addition of the *Eco* RI fragment 6 from the *E. coli* sex factor F to pSC101 (a non-conjugative plasmid) increases the frequency of its transmission by two conjugative plasmids R1–19 and Col VB (Crisona and Clark, 1977). These increases are believed to result from addition of an origin of conjugational transfer to the plasmid derived from the sex factor fragment. Several derivatives of Col E1 have been developed for use as cloning vehicles. A mini-Col E1 (pVH51) is a spontaneously isolated plasmid (2.1×10^6 daltons) from which about one-half of Col E1 DNA has been deleted. It retains the essential features of the vector, having a single *Eco* RI site, and is amplified after chloramphenicol treatment to an even greater extent than the parental Col E1 plasmid. As mentioned, mobilization of non-conjugative plasmids by conjugative plasmids is a disadvantageous feature in biological containment and it is of interest that pVH51 is mobilized much less readily than Col E1 (Hershfield et al., 1976). Several low molecular weight derivatives of Col E1-like plasmids that carry antibiotic resistance genes have been constructed and are now extensively used as cloning vehicles (Betlach et al., 1976; Rodriguez et al., 1976).

Kanamycin resistant derivatives of Col E1 and mini-Col E1 have been constructed by *in vitro* insertion of a 4.5×10^6 dalton kanamycin resistance fragment into the single Eco RI site and subsequent removal of one of the two *Eco RI* sites generated by the *in vitro* recombination. These plasmids have two useful properties for gene cloning experiments. The parental Col E1 plasmid lacks a readily selectable marker, so the addition of kanamycin resistance provides such a marker. Since one *Eco RI* site has been removed from the Col E1-*Ran* plasmid, the remaining site provides a means for introducing DNA segments into the plasmid without loss of the kanamycin resistance marker. A restriction map of this plasmid pCR1 (Covey, Richardson and Carbon, 1976) has recently been published (Armstrong *et al.*, 1977). It contains restriction targets for at least six endonucleases, some of which cut the molecule at one site only.

Insertion of a gene conferring kanamycin resistance into Col E1 and its derivatives and its value as a selective marker for recombinant plasmids, highlights the value of naturally occurring and specifically constructed R factors as cloning vehicles. Cohen *et al.* (1973) isolated the first R factor derivative useful as a cloning vehicle, pSC101. This small plasmid was isolated following mechanical shear of a large plasmid carrying genes for multiple antibiotic resistance. It was only a fraction of the size of the parental R factor but still carried the genes essential for replication in *E. coli* and a gene conferring tetracycline resistance. It carries a single target for the enzyme *Eco* RI, which lies outside the regions controlling the replication functions and tetracycline resistance. This plasmid has been used to clone the DNA from several higher organisms including ribosomal genes from *Xenopus laevis* (Morrow *et al.*, 1974), fragments carrying histone genes for the sea urchin (Kedes *et al.*, 1975), and fragments of mouse mitochondrial DNA (Chang *et al.*, 1975). Plasmid pSC101 was subsequently modified *in vitro* by introducing into it a segment of DNA derived from another plasmid conferring kanamycin resistance. The recombinant plasmid obtained, now conferred resistance to both tetracycline and kanamycin. In a further development, Chang and Cohen (1974) constructed a chimeric plasmid from pSC101 and the *Staphylococcus aureus* plasmid pI258. The *S. aureus* plasmid was known to have several *Eco* RI targets and conferred resistance to ampicillin. *Eco* RI digests of pSC101 and pI258 were ligated and from transformed *E. coli* hosts bacteria resistant to both tetracycline and ampicillin were selected. From these doubly resistant bacteria a new molecular species of DNA was isolated which had molecular characteristics of the staphylococcal plasmid as well as characteristics derived from the *E. coli* plasmid.

Another plasmid also known to carry determinants for resistance to tetracycline and ampicillin and derived from pSC101, is pBR313 (H. W. Boyer and H. M. Goodman, personal communication). this plasmid has the very useful feature that it contains single targets for at least four different restriction endonucleases, and three of these targets are located within the region of the plasmid controlling tetracycline resistance. Thus this plasmid can be used to clone DNA fragments prepared by digestion with four different restriction endonucleases. Additionally, the location of restriction targets within the region determining tetracycline resistance facilitates the isolation of recombinant plasmids resulting from appropriate restriction and ligation reactions. After transformation of *E. coli*, plasmid carrying cells can be recognized and selected for by their resistance to ampicillin. Bacteria harbouring recombinant plasmids can be distinguished by the fact that they are no longer resistant to tetracycline.

Some of the advantages of plasmid vectors and the distinct advantages of phage vectors have been combined by Backman *et al.* (1977) in the construction of plasmid vectors which have unique sites for several different restriction endonucleases and which additionally carry the bacteriophage lambda gene CI, which directs the synthesis of λ repressor protein and confers immunity to λ infection on bacteria which harbour it. With one of these, namely plasmid pKB158, Backman *et al.* (1977) were able to compare the efficiency of selection using either the λCI gene or the tetracycline resistance gene on the same plasmid. In the experiment reported, selection for tetracycline resistance was about twice as efficient as selection for λ immunity after transformation of *E. coli* with *Eco* RI-digested plasmid DNA.

Consideration has been given by Levin and Stewart (1977) to the probability of establishing chimeric plasmids carried by disarmed host bacteria in natural populations of bacteria. These estimates are low and are certainly much lower for non-conjugative plasmids. However, many non-conjugative plasmids can be mobilized by conjugative plasmids and interbacterial transfer of *E. coli–Drosophila melanogaster* DNA recombinant plasmids has been detected (Hamer, 1977). These non-conjugative plasmid pSC101–*D. melanogaster* recombinants were mobilized by the *Salmonella* sex factors Col I and Col Ib. Of forty-seven such recombinants studied, forty-six were mobilized at the same or slightly lower frequency than the parental plasmids, whereas one was mobilized about one thousand times less efficiently.

VI. Bacteriophage Lambda Vectors

Bacteriophage λ is a particularly valuable cloning vehicle which combines a number of properties important in the selection of a vector for cloning DNA segments. Phage λ has been intensively studied over a period of many years and has been well characterized both physically and genetically (Hershey, 1971). As a vector it offers a number of features which contribute to its biological containment. Strains sensitive to phage λ are relatively rare in nature (R. Davis quoted in Federal Register, 1976) and λ particles apparently do not survive in the human intestine (K. Murray and W. Szybolski quoted in Federal Register, 1976). While the establishment of λ as a stable lysogen is a very frequent event, in normal circumstances most of the λ vectors currently in use lack one or more of the att^+, int^+ and CI^+ functions essential for lysogeny (Thomas *et al.*, 1974; Murray and Murray, 1974; Rambach and Tiollais, 1974). Bacteriophage λ unable to establish lysogeny may, however, be converted into a plasmid state at a low frequency (Signer, 1969). At about the same low frequency, λ lacking the int^+ and att^+ functions can also lysogenize *E. coli* (Manly *et al.*, 1969; Gottesman and Weisberg, 1971; Shimada *et al.*, 1972). However plasmid-bearing and true λ lysogens are efficiently killed by the routine chloroform treatment used in the preparation of phage lysates (Adams, 1959). Although λ sensitive strains are rare in nature, many *E. coli* strains carry related lambdoid phages (Jacob and Wollman, 1956; Parkinson, 1971), thus the avenue of escape should not be neglected.

As a lytic cloning vector, phage λ offers the special advantage that almost one-third of its DNA can be replaced without the phage losing functions necessary for lytic development. There are limits, set by the amount of DNA that can be packaged by λ, to the length of DNA that can be inserted. The amount of DNA that can be packaged appears to be at the upper limit about 105% of the normal genome and at the lower limit about 80%. Thus a λ deletion lacking 20% of the genome can accept an insertion amounting to 25% of the length of the λ DNA molecule. If we recall that the λ genome, containing about 50 genes, consists of approximately 50,000 base pairs, these limits do not present major obstacles to many studies of recombinant DNA.

Several groups of investigators have constructed derivatives of λ specifically adapted for DNA cloning (Murray and Murray, 1974;

Rambach and Tiollais, 1974; Enquist et al., 1976; Williams et al., 1977; Thomas et al., 1974; Murray and Murray, 1975). The main object of these genetic and biochemical manipulations of λ is to construct phages which lack restriction endonuclease targets in the essential parts of the genome and have only one target in the dispensible centrally placed one-third of the λ DNA. This then permits, by insertion, the addition of foreign DNA fragments. Lambda phage have also been constructed which have two restriction targets in the non-essential region and these phages allow insertion of foreign DNA by replacement rather than addition. In addition to manipulating restriction targets, λ phages have been constructed to clone DNA fragments in particular size classes and to clone DNA fragments produced by more than one restriction enzyme (Murray, Murray and Brammar, 1975). Phage λ particles have been constructed so that by insertion or, more particularly, by replacement of part of the λ genome with foreign DNA the plaque type of the phage is altered. This procedure offers the important advantage that after transfection with the recombinant λ phages direct screening can be used to distinguish between parental phage and the required recombinant phages.

Advantage has been taken of the detailed knowledge available about the control gene expression in λ to construction phages which permit the insertion of foreign DNA in such a way that it comes under the control of the very efficient left-ward promoter P_L, and in such phage the λ N gene product permits RNA polymerase to override transcriptional stop signals (Franklin, 1974). Regulatory mutants in the λ gene Q can be used to prevent cell lysis which depends on the products of the late λ genes S and R and also to prevent the expression of the "late" λ genes which codes for head and tail proteins. Foreign genes inserted into such phages close to the left-ward promoter P_L may thus be subject to enhanced expression. The cells containing such recombinant λ phages are not lysed, so that recovery of these foreign DNA gene products is greatly increased. One startling example of the application of these methods is the construction of a λ phage carrying the *E. coli* tryptophan synthetase operon, so that, following infection of *E. coli*, the products of this operon comprise a very high proportion of the total cell protein. (Moir and Brammar, 1976.)

A series of phages, designated Charon phages after the mythological boatman of the River Styx, has been constructed by Blattner et al. (1977). The phages in this series possess virtually all of the properties referred to previously. Charons 8 and 9 have the largest cloning capacity, approaching the theoretical maximum for a non-defective λ vector, and can accommodate up to almost 25,000 base pairs of DNA; others are useful for cloning much smaller DNA fragments. Charon phages have been constructed that are suitable for cloning DNA fragments produced by *Eco* RI, *Hind III* and *Sst I*, and certain of these phages contain sites for two of these enzymes and can therefore be used to clone DNA digested with pairs of enzymes. The advantage of such phages and the use of double digests is that heterogeneously terminated vectors and fragments prevent self-cyclization from taking place. Several of the Charon phages have been so constructed that a lower limit is placed on the size of the cloned DNA they can accept. When the non-essential part of the phage has been deleted the resulting shortened DNA falls below the lower limit for successful packaging, thus these vector molecules can only be converted into viable phage by the insertion of an additional piece of DNA. Thus the ability of the recombinant λ to form a plaque is a mark of successful cloning. Three indicator systems have been employed in the Charon phages as indicators of successful cloning. These are phages which following insertion produce clear rather than turbid plaques (Charons 6 and 7); phages which following insertion into the λ *red* gene are unable to plaque on *pol A* strains (Charon 12) and phages which following insertion into the *lac Z* gene of λ *lac 5* derivatives produce colourless rather than coloured plaques on *lac*⁻ indicator bacteria on appropriate media (Charons 2 and 16). In many of the Charon phages the inserted DNA fragment comes under the transcriptional control of λ P_L or under the control of the *lac* promoter in λ *lac* phages.

Another group, Leder *et al.* (1977), have produced a λ derivative of particular use in cloning DNA fragments carrying specific unique mammalian genes. This λ derivative, constructed from a λ vector previously described by Thomas *et al.* (1974), incorporates amber mutations in gene E (the major head protein of phage λ), in gene W (a protein required for the joining of λ heads and tails during maturation) and in gene S (a protein required for a host cell lysis). The phage can be propagated by lytic infection of host strains carrying appropriate suppressors or as a lysogen since it carries the λ integration–excision system intact as well as the temperature sensitive repressor λCI857. The phage carries both *Eco* RI and *Sst I* targets and can thus accept both *Eco* RI and *Sst I* generated fragments up to about

10,000 base pairs in length, and has been successfully used to clone mouse ribosomal RNA sequences as well as ribosomal gene fragments from chicken DNA (McClements and Skalka, 1977).

Experiments have been carried out by Cameron and Davis (1977) which indicate that insertion of yeast DNA segments into λ, or the insertion of *E. coli* DNA into λ, either tends to inhibit the growth of the phage or have little to no effect on λ growth. In no case did a foreign DNA fragment confer an advantage.

VII. Cloning of Recombinant DNA

A. CLONING PROKARYOTE GENES

Although the expression of eukaryotic DNA in prokaryotic cells provides the most dramatic use of *in vitro* recombination and cloning to cross genetic barriers, the well defined and genetically isolated Gram-negative and Gram-positive prokaryotes provide an interesting system in which to study the replication, regulation and expression of cloned genes. The first example of the successful translation of a gene from a Gram-positive organism in a Gram-negative organism was the demonstration by Chang and Cohen (1974) that a gene coding for a β lactamase from *Staphyloccus aureus* was fully expressed, and could be maintained indefinitely, in a chimeric plasmid in *E. coli*. More recently, Duncan *et al.* (1977) reported that a gene from the *B. subtilis* bacteriophage ϕ3T (*thyP3*, which codes for *thymidylate synthetase*) functions in both *E. coli* and *B. subtilis* thymine auxotrophs and converts them to prototrophy. The ϕ3T bacteriophage DNA was digested with *Eco* RI and the fragments ligated to an *Eco* RI digest of the *E. coli* plasmid pMB9. Thy⁻ *E. coli* was transformed with the recombinant DNA and a chimeric plasmid pCD1 was isolated from a transformed clone which is able to transform *E. coli* to tetracycline resistance and to thymine independence simultaneously. This success, taken together with the recognition that several *Bacillus* species harbour small plasmids (Lovett and Branucci, 1974, 1975; Lovett *et al.*, 1976), opens the way for the cloning of foreign DNA in *B. subtilis* as an alternative to *E. coli*.

Further examples of the use of the versatile *E. coli* plasmid Col E1 as a cloning vehicle include its use to clone the *E. coli* gene (*sbc B*) which codes for exonuclease I (Vapnek et al., 1976) and its use to clone the *trp*, *ara* and *leu* operons from *E. coli* by Clarke and Carbon (1975). The same plasmid was used by Collins et al. (1976) to clone both the *trp* and the *gal* operons from *Eco* RI digests of unfractionated total *E. coli* DNA. The efficiency of *in vitro* recombination followed by transformation in recovering clones containing the selected DNA segment can be assessed from the fact that Collins et al. (1976) reported that ten to twenty clones containing selected DNA segments were obtained per µg of digested DNA.

A particular pertinent example of the use of gene cloning as a highly effective means to amplify the number of copies of a specific gene is provided by the work of Cameron et al. (1975). They constructed a hybrid bacteriophage lambda into which was inserted the mutant *E. coli* gene coding for DNA ligase (*lop-11*) which leads to a five-fold overproduction of the ligase (Gottesman et al., 1973). More recently, the same group of investigators (Panasenko et al., 1977) have modified this recombinant lambda DNA molecule to produce a vector which, under the appropriate conditions, generates the production of an amount of DNA ligase that is five hundred-fold greater than that produced by uninfected *lig*$^+$ strains and sufficient to account for about 5% of the total cellular protein of *E. coli*. The value of this vector in simplifying the purification of DNA ligase is obvious and should make it substantially easier to prepare large quantities of pure DNA ligase for *in vitro* recombination experiments, as well as for investigations into its structure and mechanism of action.

New light has recently been shed on what is arguably one of the earliest examples of variation in bacteria *viz.* phase variation in *Salmonella* (Andrewes, 1922). Flagella antigens in *Salmonella* are specified by two genes, H1 and H2, and phase variation is the result of the ability of *Salmonella* strains to switch from expression of one gene to expression of the other under the control of a genetic element linked to H2 (Stocker, 1949; Lederberg and Iino, 1956). A variety of mechanisms have been proposed over the years to account for phase variation. Zeig et al. (1977) now present evidence that the expression of H2 is controlled by the inversion of a DNA sequence adjacent to H2 so that in one orientation H2 is switched on whilst in the opposite orientation H2 is switched off. This work was made possible by the molecular cloning of segments of *Salmonella* DNA which contain these genetic loci and the analysis by heteroduplex techniques of the cloned segments.

B. CLONING YEAST DNA

A collection of yeast hybrid circular DNA molecules has been constructed *in vitro* by Ratzkin and Carbon (1977) using the poly dA–poly dT end-to-end joining method. Each hybrid circle contained one molecule of poly dT-tailed DNA of the *E. coli* plasmid Col E1 converted to linear DNA by *Eco* RI endonuclease digestion and annealed to a poly dA-tailed fragment of *Saccharomyces cerevisiae* DNA prepared by shearing total yeast DNA to an average size of 8×10^6 daltons. The hybrid Col E1-yeast DNA was used to transform *E. coli* and Col E1 resistant clones were selected. Sufficient clones were selected to ensure, as far as possible, that the hybrid plasmid population represented the entire yeast genome. From this population, hybrid Col E1-yeast DNA plasmids were selected on the basis of their ability to complement auxotrophic *E. coli* strains.

Plasmid pYe leu 10 is able to complement several different point and deletion mutations in the *E. coli* or *Salmonella typhimurium* gene *leuB*. Several other plasmids appear to be specific suppressors of the *leuB6* mutation in *E. coli* strains C600. An *E. coli hisB* gene deletion is complemented by hybrid plasmid pYe his2. From these results, it is apparent that complementation of bacterial mutations by DNA segments derived from yeast is not a particularly rare event. If these results can be confirmed by isolation and characterization of yeast gene products from *E. coli* strains carrying hybrid plasmids, the major problem of full expression through transcription and translation of eukaryote DNA in prokaryotes will, in this instance, have been solved.

Functional genetic expression of yeast DNA in *E. coli* has also been demonstrated by Struhl *et al.* (1976). Using a modified bacteriophage lambda vector, *Eco* RI-digested *Saccharomyces cerevisiae* DNA was used to transfect *E. coli*. The *att⁻ int⁻* recombinant phage was successfully integrated in the chromosome of *hisB⁻* strains to make double lysogens which were able to grow in the absence of histidine.

In a major piece of work recently reported, Beckman *et al.* (1977) isolated some four thousand *E. coli* clones containing yeast DNA segments inserted into the bacterial plasmid pBR313 (Bolivar *et al.*, 1977). Using an *in situ* bacterial colony hybridization technique with individual radioactively labelled yeast transfer RNA species, one hundred and seventy-five clones were identified as carrying yeast tRNA genes. The analysis of these clones is not yet completed, but there is evidence that some clones hybridize with more than one purified tRNA species and some clones show evidence of tRNA gene clustering, however most do not. Further analysis should reveal much about the organization of tRNA genes on the chromosomes of yeast.

C. CLONING HIGHER EUKARYOTE DNA

The frequency and presence in multiple copies of ribosomal RNA genes, and the ready availability of highly purified RNA probes for their detection, has provided a convenient system for cloning higher eukaryotic DNA (Morrow et al., 1974); *Euglena gracilis* DNA (Lomax et al., 1977); DNA of the chicken, *Gallus domesticus* (McClements and Skalka, 1977), in either plasmid or phage λ vectors.

The first eukaryotic gene to be inserted by *in vitro* recombination into a bacterial plasmid was the gene coding for β-globin. Almost simultaneously, four different laboratories reported the production of recombinants containing β-globin DNA (Rougeon et al., 1975; Efstratiadis et al., 1976; Rabbitts, 1976; Higuchi et al., 1976b). All of these groups of investigators used reverse transcription of isolated globin in RNA to make complementary DNA as starting material for the cloning experiments. Rougeon et al. (1975) used a 300–400 nucleotide long cDNA preparation to synthesize double-stranded globin DNA and after cloning and amplification, obtained a 100–150 base-pair long segment of globin DNA. This represents only about one-fifth of the length of the original mRNA. This result pin points certain limitations of this method. For example, the inserted DNA can only contain sequences present in the isolated mRNA, any additional non-transcribed sequences in the original eukaryote gene will be lost and, secondly, the method relies on the efficient reverse transcription of the mRNA. This latter limitation appears to have been overcome by Efstratiadis et al. (1975) who successfully obtained a full length copy of globin mRNA which was used to produce double-stranded globin DNA with the aid of DNA polymerase I (Efstratiadis et al., 1976). This synthetic gene was then inserted into the plasmid PMB9 and used to transform *E. coli* (Maniatis et al., 1976). The synthetic gene is 500 base pairs long and only 80 nucleotides shorter than the original β-globin in RNA. Interestingly it contains the full coding region plus 40 and 110 nucleotide sequences from the untranslated regions at the 3' and 5' termini. Maniatis et al. (1976) characterized the synthetic gene and the inserted globin DNA after amplification in *E. coli* by analysis of restriction enzyme cleavage maps of the DNAs. Using eight different enzymes with known recognition sequences, restriction maps could be compared with the cleavage pattern predictable from what is known of the nucleotide sequence of β-globin in RNA.

The group of heritable diseases known as β-thalassemias are characterized by an inherited reduction or a total absence of β-globin chain synthesis, although recent reports (Ramirez *et al.*, 1976) indicated that β-thalassemias may be far more complex and heterogenous in respect of their precise molecular bases. The α-thalassemias similarly are characterized by a heritable absence of α-globin chains, due to a deletion of the DNA coding for the α-chain (Ottolenghi *et al.*, 1974; Taylor *et al.*, 1974). Recently, Kan *et al.* (1975) synthesized radioactively labelled α-globin cDNA from α-globin mRNA and hybridized it to extracted foetal DNA in a successful attempt to diagnose a suspected case of α-thalassemia. Even more serious health problems are posed by the diseases associated with faulty β-globin chain synthesis. There is encouragement from the progress already made in cloning and amplifying the β-globin gene that useful diagnostic probes may be made available as a result of applying these techniques to human disease problems. It may be noted here that Liu *et al.* (1977), in addition to characterizing rabbit β-globin gene-containing clones (Higuchi *et al.*, 1976a, b; Browne *et al.*, 1977), now report characterization of bacterial plasmids containing rabbit α-globin gene sequences. In a series of papers, Weissman and his collaborators have almost completed the sequencing of the human β-globin mRNA and of human α-globin mRNA (Marotta *et al.*, 1977a, b; Cohen-Solal *et al.*, 1977). They report (Wilson *et al.*, 1977) the cloning in *E. coli* plasmids of cDNA complementary to human foetal haemoglobin HbF mRNA.

Important progress has been reported recently towards the cloning of eukaryote DNA sequences in the absence of specific screening procedures. (Ullrich *et al.*, 1977.) This work relies on the already established approach of copying mRNA into double-stranded cDNA using reverse transcriptase and then inserting the cDNA copy into a bacterial plasmid vector. This method has already been used successfully to clone eukaryote genes such as the globin gene whose mRNA can be obtained in highly purified form (Efstratiadis *et al.*, 1976). The procedure is to introduce the hybrid plasmid–eukaryote gene into *E. coli* by transformation and to identify transformed clones containing the required DNA sequence by hydridization with a specific probe. In the case of the globin gene experiments such a probe is available in the form of purified globin mRNA.

Ullrich *et al.* (1977), in attempting to isolate the rat DNA sequences coding for insulin, had no such specific probe available because pure rat insulin mRNA had not been isolated. The remarkable success achieved, clearly depended on the development of several ingenious techniques which may have general application in exploitation of *in vitro* DNA recombination to produce on an industrial scale a variety of eukaryote proteins. To obtain presumptive insulin cDNA, Ullrich *et al.* (1977) enriched pancreatic preparations for those specialized cells which synthesize insulin. Total poly-adenylated mRNA was isolated from these cells and cDNA prepared using reverse transcriptase. Following digestion with the restriction endonuclease *Hae III*, the cDNA yielded prominent fragments of 180 and 80 nucleotides compared with an undigested length of about 450 nucleotides. The cDNA prepared was, in fact, a single DNA strand with a hairpin loop. The loop was digested with S1 nuclease and converted to a linear double stranded DNA molecule with blunt ends. To insert the molecule into a vector, the double-stranded cDNA was first linked by end-to-end joining to synthetic decanucleotides containing the *Hind III* recognition sequence, using the T4 ligase reaction. The smaller *Hae III* fragments of the presumptive rat insulin cDNA were similarly linked to the synthetic *Hind III* recognition sequences. Insertion into the plasmid pMB9 was then achieved by *Hind III* digestion of these DNA segments and annealing by their cohesive ends followed by ligation. To reduce recyclization of the pMB9 plasmid DNA, digestion with *Hind III* endonuclease was followed by digestion with alkaline phosphatase to remove 5' phosphate groups. Under these circumstances, only those plasmid molecules which have annealed to foreign DNA segments with terminal 5'-phosphates can be covalently closed by DNA digase. The hybrid plasmids were then used to transform *E. coli* and transformants selected by tetracycline resistance—the plasmid marker. Since no specific probe for rat insulin DNA sequences was available, Ullrich *et al.* (1977) prepared crude DNA preparations from transformed clones and made *Hae III* digests in an attempt to find those hybrid plasmids which carried the 450, 180 and 80 nucleotide fragments characteristic of the cDNA. Hybrid plasmids were found which yielded *Hae* digest fragments of 410, 180 and 80 nucleotides. Identification of the cloned cDNA as the sequence which coded for insulin was made possible by direct sequence analysis which showed that the cloned DNA coded for the known amino acid sequence of rat insulin.

It is frequently the case in eukaryotes that the immediate product of translation of mRNA is a precursor polypeptide which is processed in several steps to produce the functional protein. The immediate product of translation of insulin mRNA is preproinsulin. The hybrid plasmids isolated by Ullrich *et al.* (1977) contain between them the whole of the proinsulin coding sequence and part of the *N*-terminal prepeptide. However none of them contains the important sequences from the 5′ untranslated end of the mRNA. However, it seems from this work that general principles have been defined that should permit the cloning of selected eukaryote gene sequences that can be transcribed and translated in *E. coli*. It seems unlikely that accurate post-translational processing can be accomplished in bacteria and eventual success in the production of eukaryote gene products may have to rely on *in vitro* manipulation of the primary translation products.

D. SCREENING METHODS FOR CLONED DNA

The isolation of large quantities of DNA corresponding to a single gene is considerably aided if complementary RNA or DNA probes are available to screen clones produced by recombinant DNA techniques. Screening methods have been developed for recombinant plasmids by Kedes *et al.* (1975) and Grunstein and Hogness (1975). Similar methods have also been developed for screening recombinant bacteriophages by Jones and Murray (1975), Yu *et al.* (1976), Sanzey *et al.* (1976) and Kramer *et al.* (1976).

The most rapid of these methods, developed by Grunstein and Hogness (1975) is based on the technique pioneered by Olivera and Bonhoeffer (1974) for growing the lysing bacteria on membrane filters. The limiting factor of such methods is the need to pick and spot plaques or colonies onto filters for further growth before screening. Benton and Davis (1977) have improved the rapidity of screening by the use of λgt recombinant phage. Plaques of this phage contain enough phage DNA for detectable hybridization to complementary labelled nucleic acid and this DNA can be fixed to nitrocellulose filters by direct contact. In this way it is claimed that 2×10^4 plaques can be screened per hour. The method has been successfully applied to detect plaques produced by recombinant λgt-yeast rRNA genes using ^{32}P-labelled total yeast ribosomal RNA as the probe.

Hybridization *in situ* has also been developed as a method to detect recombinant SV40 virus carrying specific sequences of non-viral DNA (Villarreal and Berg, 1977). Plaques produced by a λ-SV40 recombinant were constructed and cloned in monkey kidney cells with the aid of a superinfecting virus, necessary because the λDNA segment replaces SV40 genes coding for virion capsid proteins. After transfer to nitrocellulose filters, the presence of λ DNA was detected using a ^{32}P-labelled λ DNA probe specific for the segment of λ DNA included in the λ-SV40 hybrid.

Assays for specific functions encoded in cloned genes have been developed by Hershfield *et al.*, 1974; Clarke and Carbon, 1975; and Struhl *et al.*, 1976. These assays include methods for detection of functions which can complement mutational defects in the metabolic pathways of auxotrophic bacterial hosts, for recombinant DNA, and for structural defects such as those involved in mutant flagellar genes (Silverman *et al.*, 1976). The identification of a cloned gene product through expression of its function is, however, limited by the difficulty of predicting its precise function in the new environment of the host strain used for cloning. Additionally, the requirement for functional expression can preclude identification of clones encoding proteins which have to be processed or assembled in specific ways before function can be detected. These methods also preclude identification of genes coding for proteins for which there are no convenient assays and virtually all eukaryotic proteins for which there are no counterparts in *E. coli*. Recently, assays for successful translation of cloned eukaryote DNA sequences based upon detection of new protein bands in acrylamide gels have been reported by Boyer (1976) in the analysis of the histone genes of *Drosophila melanogaster*. However, positive identification of these proteins will require subsequent confirmation by other methods. Skalka and Shapiro (1976) have developed a direct method for detecting cloned gene translation which is independent of the expression of protein functions. The method is based on specific antigen–antibody complex formation within a vector phage plaque or surrounding a vector-containing bacterial colony. The lambda phage

($\lambda plac5$) developed by Rambach and Tiollais (1974) was successfully used to clone the β-galactosidase gene of *E. coli* and recombinant phage lambda plaques were detected by immunoprecipitation using anti-β-galactosidase antibody. Colonies of *E. coli* lysogenic for recombinant phage were similarly detected by immunoprecipitation after either heat-induction of prophage or *in situ* lysis with lysozyme. It is likely that these methods will see wide application in the future.

VIII. Potential Biohazards

The potential biohazards associated with recombination DNA research led to a series of discussions and ultimately to the publication simultaneously in *Science* and *Nature* of a letter by a committee led by Paul Berg (Berg *et al.*, 1974a, b, c). In this letter, Berg and his colleagues assessed some of the potential risks attendant on recombinant DNA research and requested that scientists defer certain types of experiments and use considerable caution in the conduct of other types of experiments. The latter generated extensive discussion and debate among scientists and others and led to the convening of an international meeting at the Asilomar Conference Center, Pacific Grove, California in February 1975 (Berg *et al.*, 1975a, b, c). At this meeting, provisional guidelines were suggested for the conduct of recombinant DNA research and the recommendation was made that certain types of experiments should be deferred. These provisional guidelines have now been fully codified for U.S. scientists in the NIH Guidelines for Research Involving Recombinant DNA Molecules (1976). A revised version of these Guidelines is to be published in 1977. Similar guidelines have been produced by committees in other countries; for the U.K., see Ashby (1975), and Williams (1976). Draft Guidelines have been published in Canada and are shortly to be published in U.S.S.R., France, The Netherlands, Federal Republic of Germany, Japan, Australia and Israel. Other nations may propose adoption of guidelines published in the U.S.A. or U.K. At the international level, ICSU has established a Committee on Genetic Experimentation (COGENE) whose purposes and objectives include consideration of potential biohazards.

The principal argument is that the construction of recombinant DNA molecules and their introduction into micro-organisms might create a novel organism which, if it was inadvertently released from a laboratory, would present a biohazard to man and/or his environment. The counter argument is that such a micro-organism, newly constructed, would not have been subject to the forces of natural selection and burdened with the maintenance and expression of additional genetic information would not compete effectively with naturally occurring strain in nature. A second counter argument is that genetic engineering experiments have already been tried during the course of evolution (so-called natural genetic engineering) and potentially hazardous combinations have not been successful and therefore have not survived.

The published guidelines for recombinant DNA research encompass two-line defence mechanisms; first, physical containment, and second, biological containment.

A. PHYSICAL CONTAINMENT

As in conventional work with micro-organisms, the primary defence against the escape of a pathogen has to be the skills of the personnel who work in the laboratory. The procedures and precautions which should be taken by such laboratory workers are fully documented (Collins and Hartley, 1974; Darlow, 1969; Godber, 1975; Hellman *et al.*, 1973; Lenette *et al.*, 1974; Shapton and Board, 1972; Steere, 1971) and several Governmental Organizations in the U.S. and elsewhere have published laboratory safety guides.

Given trained personnel, a laboratory engaged in recombinant DNA research will require appropriate physical containment equipment such as microbiological safety cabinets, laminar flow hoods, properly installed autoclaves and in some cases negative-pressure laboratories with special facilities. The physical containment facilities appropriate to recombinant DNA research experiments differ depending on the potential risk associated with particular experiments. The NIH guidelines in the U.S. provide detailed descriptions of physical containment facilities for each of four different categories of experiment ranging from P1 to P4. The U.K. guidelines similarly, though in less detail, define physical containment facilities appropriate to four categories of experiment. In Canada, rather more categories are defined.

B. BIOLOGICAL CONTAINMENT

A new concept in microbiological safety has been introduced in relation to the control of potential biohazards associated with recombinant DNA research. Of course, microbiologists have always, and for obvious reasons, preferred to work with non-pathogens rather than pathogens, with attenuated rather than non-attenuated strains wherever possible. The new concept is the deliberate genetic manipulation of micro-organisms to reduce the probability of their survival and perpetuation in the environment.

There are two quite distinct elements in organisms constructed by *in vitro* recombination. There is the microbial host strain and the cloning vector.

For plasmid cloning vectors the use of non-conjugative plasmids is clearly to be preferred to the use of conjugative plasmids. Among non-conjugative plasmids those that are inefficiently mobilized by other plasmids confer an additional advantage. Plasmid-cloning vectors can be genetically manipulated so that transmission to other hosts is blocked, or transmission is lethal and replication of the plasmid can be rendered dependent upon particular host strains likely to be rare in nature (see Plasmid vectors above). For viral cloning vectors, replication can be rendered dependent upon particular host strains, replication can be made dependent upon a narrow range of temperature and a virus can be chosen which has a very narrow host range so that it is unlikely to encounter a susceptible host outside the laboratory (see Bacteriophage lambda vectors above).

The host strain may also be manipulated genetically in a variety of ways to reduce its potential as a biohazard when used to clone foreign DNA. The organism in most common use is *E. coli* which is prevalent in the intestinal tracts of both warm-blooded and cold-blooded animals and occurs in sewers, rivers, lakes and contaminated estuaries as well as on urban and agricultural land (Cooke, 1974). Curtiss (1976) has endeavoured to construct a number of *E. coli* strains that would be suitable as host strains for cloned DNA. By a combination of thirteen genetic manipulations he has obtained a strain of *E. coli* which incorporates many failsafe features. The strain cannot synthesize the murein layer of its cell wall except under special laboratory conditions, and it is unable to synthesize the mucopolysaccharide colanic acid which facilitates survival in the absence of murein from the cell wall. It is sensitive to bile salts and detergents, and undergoes thymineless death degrading its DNA in the absence of thymine or thymidine. It is resistant to several *E. coli* transducing phages and sensitive to ultraviolet radiation. It is defective as a recipient in matings with donor strains carrying certain conjugative plasmids and is unable to function as a recipient in matings with donor strains carrying certain other conjugative plasmids. Taken together, these features contribute very substantially towards the construction of a completely disarmed host bacterium for recombinant DNA experiments (Curtiss *et al.*, 1977; Leder *et al.*, 1977).

Young (1976) has been prominent among those advocating the advantages of *B. subtilis* rather than *E. coli* as a host system for cloning foreign DNA. Among the advantages of *B. subtilis* is the belief that it is non-pathogenic and its circular chromosome map is well defined with about two hundred genes well mapped (Young and Wilson, 1975; Lepesant-Kejzlarova *et al.*, 1975). The organism is important in the fermentation industry and, unlike *E. coli*, it lacks endotoxins in the cell wall and offers a source of single cell protein. Genetic manipulation is easily accomplished through high frequency transformation and several generalized transducing phages are known. The major disadvantage of *B. subtilis,* sporulation, which readily permits its survival in otherwise adverse environments can be overcome by the introduction of asporogenic deletion mutations. However, knowlege of its genetics and physiology is inferior to that of *E. coli* and while there are many *B. subtilis* phages in use, knowledge of them is primitive compared with bacteriophage λ. Little is known about *B. subtilis* plasmids, and gene amplification in *B. subtilis* cannot be carried out using specialized transducing phages at the present time.

IX. Acknowledgments

I wish to express my thanks to Dr. Monique Glover in the preparation of material for this review and to Professor B. W. Holloway for criticizing the manuscript. My thanks are also due to Mrs. Julie Waugh for her patient typing.

REFERENCES

Abelson, J. (1977). *Science, New York* **196**, 159.
Adams, M. H. (1959). "Bacteriophages." Interscience, New York.
Ando, T. (1966). *Biochimica et Biophysica Acta* **114**, 158.

Andrewes, F. W. C. (1922). *Journal of Pathology and Bacteriology* **25**, 515.
Armstrong, K. A., Hershfield, V. and Helinski, D. R. (1977). *Science, New York* **196**, 172.
Ashby, E. (1975). "Report of the Working Party on the Experimental Manipulation of the Genetic Composition of Micro-organisms." *HMSO Cmnd 5880*, HMSO, London.
Backman, K., Ptashne, M. and Gilbert, W. (1976). *Proceedings of the National Academy of Sciences of the United States of America* **73**, 4174.
Backman, K., Hawley, D. and Ross, M. J. (1977). *Science, New York* **196**, 182.
Bahl, C. P., Marians, K. J., Wu, R., Stawinsky, J. and Narang, S. A. (1976a). *Gene* **1**, 81.
Bahl, C. P., Wu, R., Itakura, K., Katagiri, N. and Narang, S. A. (1976b). *Proceedings of the National Academy of Sciences of the United States of America* **73**, 91.
Bannister, D. and Glover, S. W. (1970). *Journal of General Microbiology* **61**, 63.
Bazaral, M. and Helinski, D. R. (1968). *Journal of Molecular Biology* **36**, 185.
Beckman, J. S., Johnson, P. F. and Abelson, J. (1977). *Science, New York* **196**, 205.
Benton, W. D. and Davis, R. W. (1977). *Science, New York* **196**, 180.
Berg, P., Baltimore, D., Boyer, H. W., Cohen, S. N., Davis, R. W., Hogness, D. S., Nathans, D., Roblin, R. O., Watson, J. D., Weissman, S. and Zinder, N. D. (1974a). *Science, New York* **185**, 303.
Berg, P., Baltimore, D., Boyer, H. W., Cohen, S. N., Davis, R. W., Hogness, D. S., Nathans, D., Roblin, R. O., Watson, J. D., Weissman, S. and Zinder, N. D. (1974b). *Proceedings of the National Academy of Sciences of the United States of America* **71**, 2593.
Berg, P., Baltimore, D., Boyer, H. W., Cohen, S. N., Davis, R. W., Hogness, D. S., Nathans, D., Roblin, R. O., Watson, J. D., Weissman, S. and Zinder, N. D. (1974c). *Nature, London* **250**, 175.
Berg, P., Baltimore, D., Brenner, S., Roblin, R. O. and Singer, M. F. (1975a). *Science, New York* **188**, 991.
Berg, P., Baltimore, D., Brenner, S., Roblin, R. O. and Singer, M. F. (1975b). *Proceedings of the National Academy of Sciences of the United States of America* **72**, 1981.
Berg, P., Baltimore, D., Brenner, S., Roblin, R. O. and Singer, M. F. (1975c). *Nature, London* **225**, 442.
Betlach, M. C., Hershfield, V., Chow, L., Brown, W., Goodman, H. M. and Boyer, H. W. (1976). *Federation Proceedings of the American Society for Experimental Biology* **35**, 2037.
Blakesley, R. W. and Wells, R. D. (1975). *Nature, London* **257**, 421.
Blattner, F. R., Williams, W. G., Blechl, A. E., Denniston-Thompson, K., Faber, H. E., Furlong, L.-A., Grunwald, D. J., Kiefer, D. O., Moore, D. D., Schumm, J. W., Sheldon, E. L. and Smithies, O. (1977). *Science, New York* **196**, 161.
Bolivar, F., Rodriguez, R. L., Betlach, M. C. and Boyer, H. W. (1977). In "DNA Insertion Elements, Plasmids and Episomes" (A. I. Bukhari, J. A. Shapiro and S. L. Adhya, eds.), p. 684, Cold Spring Harbor Laboratory, New York.
Boyer, H. W. (1971). *Annual Review of Microbiology* **25**, 153.
Boyer, H. W., Chow, L. T., Dugaiczyk, A., Hedgpeth, J. and Goodman, H. M. (1973). *Nature New Biology* **244**, 40.
Boyer, H. (1976). In "Proceedings of the Tenth Miles International Symposium". Raven Press, New York.
Browne, J., Paddock, G. V., Liu, A., Clarke, P., Heindell, H. and Salser, W. (1977). *Science, New York* **195**, 389.
Burton, W. G., Roberts, R. J., Myers, P. A. and Sager, R. A., (1976). *Federation Proceedings of the American Society for Experimental Biology* **35**, 1588.
Cameron, J. R. and Davis, R. W. (1977). *Science, New York* **196**, 212.
Cameron, J. R., Panasenko, S. M., Lehman, I. R. and Davis, R. W. (1975). *Proceedings of the National Academy of Sciences of the United States of America* **72**, 3416.

Chang, A. C. Y. and Cohen, S. N. (1974). *Proceedings of the National Academy of Sciences of the United States of America* **71**, 1030.
Chang, A. C. Y., Lansman, R. A., Clayton, D. A. and Cohen, S. N. (1975). *Cell* **6**, 231.
Chang, L. M. S. and Bollum, F. J. (1971). *Biochemistry, New York* **10**, 536.
Clarke, L. and Carbon, J. (1975). *Proceedings of the National Academy of Sciences of the United States of America* **72**, 4361.
Clewell, D. B. (1972). *Journal of Bacteriology* **110**, 667.
Clewell, D. B. and Helinski, D. R. (1972). *Journal of Bacteriology* **110**, 1135.
Cohen, S. N. (1975). *Scientific American* **233**, 24.
Cohen, S. N. (1976). *Nature, London* **263**, 731.
Cohen, S. N., Chang, A. C. Y., Boyer, H. W. and Helling, R. B. (1973). *Proceedings of the National Academy of Sciences of the United States of America* **70**, 3240.
Cohen-Solal, M., Forget, B. G., Prensky, W., Marotta, C. A. and Weissman, S. M. (1977). *Journal of Biological Chemistry* **252**, 5032.
Collins, C. H. and Hartley, E. G. (1974). *In* "The Prevention of Laboratory Acquired Infections". *Public Health Laboratory Service, Monograph Series No. 6*. HMSO, London.
Collins, C. J., Jackson, D. A. and DeVries, A. J. (1976). *Proceedings of the National Academy of Sciences of the United States of America* **73**, 3838.
Cooke, E. M. (1974). "*Escherichia coli* and Man." Churchill-Livingston, London.
Covey, C., Richardson, D. and Carbon, J. (1976). *Molecular and General Genetics* **145**, 155.
Crisona, N. J. and Clark, A. J. (1977). *Science, New York* **196**, 187.
Curtiss, R. (1976). *Annual Review of Microbiology* **30**, 507.
Curtiss, R., Pereira, D. A., Clark, J. E., Hsu, J. C., Goldschmidt, R., Hall, S. I., Moody, R., Maturin, L. and Inoue, M. (1977). *In* "Proceedings of the Tenth Miles International Symposium". Raven Press, New York.
Darlow, H. M. (1969). *In* "Methods in Microbiology" (J. R. Norris and D. W. Ribbons, eds.), p. 169. Academic Press, London.
Duncan, C. H., Wilson, G. A. and Young, F. E. (1977). *Gene* **1**, 153.
Efstratiadis, A., Kafatos, F. C., Maxam, A. M. and Maniatis, J. (1975). *Cell* **4**, 367.
Efstratiadis, A., Kafatos, F. C., Maxam, A. M. and Maniatis, T. (1976). *Cell* **7**, 279.
Enquist, L., Tiemeier, D., Leder, P., Weisberg, R. and Sternberg, N. (1976). *Nature, London* **259**, 596.
Federal Register (1976). *Recombinant DNA Research* **41**, 113–259.
Franklin, N. C. (1974). *Journal of Molecular Biology* **89**, 33.
Ghangas, G. S. and Wu, R. (1975). *Journal of Biological Chemistry* **250**, 4601.
Glover, S. W. and Bannister, D. (1968). *Biochemical and Biophysical Research Communications* **30**, 735.
Godber, G. (1975). "Report of the Working Party on the Laboratory Use of Dangerous Pathogens." *HMSO Cmnd 6054*, HMSO, London.
Godson, G. N. and Roberts, R. J. (1976). *Virology* **73**, 561.
Gottesman, M. M. and Weisberg, R. (1971). *In* "The Bacteriophage Lambda" (A. D. Hershey, ed.), p. 113. Cold Spring Harbor, New York.
Gottesman, M. E., Hicks, M. L. and Gilbert, M. (1973). *Journal of Molecular Biology* **77**, 531.
Greene, P. J., Betlach, M. C., Boyer, H. W. and Goodman, H. M. (1974). *Methods in Molecular Biology* **7**, 87.
Greene, P. J., Poonian, M. S., Nussbaum, A. L., Tobias, L., Garfin, D. E., Boyer, H. W. and Goodman, H. M. (1975). *Journal of Molecular Biology* **99**, 237.
Grunstein, M. and Hogness, D. (1975). *Proceedings of the National Academy of Sciences of the United States of America* **72**, 3961.

Hamer, D. H. (1977). *Science, New York* **196**, 220.
Hamer, D. H. and Thomas, C. A. (1976). *Chromosoma* **49**, 243.
Hedgpeth, J., Goodman, H. M. and Boyer, H. W. (1972). *Proceedings of the National Academy of Sciences of the United States of America* **69**, 3448.
Hellman, A., Oxman, M. N. and Pollack, R. (eds.) (1973). "Biohazards in Biological Research." Cold Spring Harbor, New York.
Henner, W. D., Kleber, I. and Benzinger, R. (1973). *Journal of Virology* **12**, 741.
Hershey, A. D. (1971). "The Bacteriophage Lambda." Cold Spring Harbor, New York.
Hershfield, V., Boyer, H. W., Yanofsky, C., Lovett, M. and Helinski, D. R. (1974). *Proceedings of the National Academy of Sciences of the United States of America* **71**, 3455.
Hershfield, V., Boyer, H. W., Chow, L. and Helinski, D. R. (1976). *Journal of Bacteriology* **126**, 447.
Heyneker, H. L., Shine, J., Goodman, H. M., Boyer, H. W., Rosenberg, J., Dickerson, R. E., Narang, S. A., Ikakura, K., Lin, S. and Riggs, A. D. (1976). *Nature, London* **263**, 748.
Higuchi, R., Paddock, G. V., Wall, R. and Salser, W. (1976a). *Federation Proceedings of the American Society for Experimental Biology* **35**, 1369.
Higuchi, R., Paddock, G. V., Wall, R. and Salser, W. (1976b). *Proceedings of the National Academy of Sciences of the United States of America* **73**, 3146.
Hohn, B. and Hohn, T. (1974). *Proceedings of the National Academy of Sciences of the United States of America* **71**, 2372.
Horiuchi, K. and Zinder, N. D. (1975). *Proceedings of the National Academy of Sciences of the United States of America* **72**, 2555.
Itakura, K., Bahl, C. P., Katagiri, N., Michniewicz, J. J., Wightman, R. H. and Narang, S. A. (1973). *Canadian Journal of Chemistry* **51**, 3649.
Itakura, K., Katagiri, N., Bahl, C. P., Wightman, R. M. and Narang, S. A. (1975a). *Journal of the American Chemical Society* **97**, 7327.
Itakura, K., Katagiri, N., Narang, S. A., Bahl, C. P., Marians, K. J. and Wu, R. (1975b). *Journal of Biological Chemistry* **250**, 4592.
Jackson, D. A., Symons, R. H. and Berg, P. (1972). *Proceedings of the National Academy of Sciences of the United States of America* **69**, 2904.
Jacob, F. and Wollman, E. L. (1956). *Annales de l'Institut Pasteur, Paris* **91**, 486.
Jones, K. W. and Murray, K. (1975). *Journal of Molecular Biology* **96**, 455.
Kahn, P. and Helinski, D. R. (1964). *Journal of Bacteriology* **88**, 1573.
Kaiser, D., Syvanen, M. and Masuda, T. (1975). *Journal of Molecular Biology* **91** 175.
Kan, Y. W., Dozy, A. M., Varmus, H. E., Taylor, J. M., Holland, J. P., Lie-Injo, L. E., Ganesan, J. and Todd, D. (1975). *Nature, London* **255**, 255.
Katagiri, N., Itakura, K. and Narang, S. A. (1975). *Journal of the American Chemical Society* **97**, 7332.
Kedes, L. H., Cohen, S. N., Houseman, D. and Chang, A. C. Y. (1975). *Nature, London* **255**, 533.
Kleid, D., Humayun, Z., Jeffrey, A. and Ptashne, M. (1976). *Proceedings of the National Academy of Sciences of the United States of America* **73**, 293.
Kramer, R. A., Cameron, J. R. and Davis, R. W. (1976). *Cell* **8**, 227.
Lacks, S. and Greenberg, B. (1975). *Journal of Biological Chemistry* **250**, 4060.
Leder, P., Tiemeier, D. and Enquist, L. (1977). *Science, New York* **196**, 175.
Lederberg, J. and Iino, T. (1956). *Genetics* **41**, 744.
Lenette, E. H., Spaulding, E. H. and Truant, J. P., eds. (1974). "Manual of Clinical Microbiology." American Society for Microbiology, Washington D.C.

Lepesant-Kejzlarova, J., Lepesant, J. A., Walle, J., Billault, A. and Dedonder, R. (1975). *Journal of Bacteriology* **121**, 823.
Levin, B. R. and Stewart, F. M. (1977). *Science, New York* **196**, 218.
Liu, A. Y., Paddock, G. V., Heindell, H. C. and Salser, W. (1977). *Science, New York* **196**, 192.
Lobban, P. E. and Kaiser, A. D. (1973). *Journal of Molecular Biology* **78**, 453.
Lomax, M. I., Helling, R. B., Hecker, L. I., Schwartbach, S. D. and Barett, W. E. (1977). *Science, New York* **196**, 202.
Lovett, P. S. and Branucci, M. G. (1974). *Journal of Bacteriology* **120**, 408.
Lovett, P. S. and Branucci, M. G. (1975). *Journal of Bacteriology* **124**, 484.
Lovett, P. S., Duvall, E. J. and Keggins, K. M. (1976). *Journal of Bacteriology* **127**, 817.
Mandel, M. and Higa, A. (1970). *Journal of Molecular Biology* **53**, 159.
Maniatis, T., Ku, S. G., Efstratiadis, A. and Kafatos, F. C. (1976). *Cell* **8**, 163.
Manly, K. F., Signer, E. R. and Radding, C. M. (1969). *Virology* **39**, 137.
Marians, K. J., Wu, R., Stawinsky, J., Hozumi, T. and Narang, S. A. (1976). *Nature, London* **263**, 744.
Marotta, C. A., Forget, B. G., Cohen-Solal, M., Wilson, J. T. and Weissman, S. M. (1977a). *Journal of Biological Chemistry* **252**, 5019.
Marotta, C. A., Wilson, J. T., Forget, B. G. and Weissman, S. M. (1977b). *Journal of Biological Chemistry* **252**, 5040.
McClements, W. and Skalka, A. M. (1977). *Science, New York* **196**, 195.
Mertz, J. E. and Davis, R. W. (1972). *Proceedings of the National Academy of Sciences of the United States of America* **69**, 3370.
Middleton, J. H., Edgell, M. H. and Hutchinson, C. A. (1972). *Journal of Virology* **10**, 42.
Moir, A. and Brammar, W. J. (1976). Cited by Murray, K. *In* "Biochemical Manipulation of Genes", *Endeavour XXXV* 129.
Morrow, J. F., Cohen, S. N., Chang, A. C. Y., Boyer, H. W., Goodman, H. M. and Helling, R. B. (1974). *Proceedings of the National Academy of Sciences of the United States of America* **71**, 1743.
Murray, K. (1978). *Proceedings of the Royal Society Series B* (In Press).
Murray, K. and Murray, N. E. (1975). *Journal of Molecular Biology* **98**, 551.
Murray, K., Murray, N. E. and Brammar, W. J. (1975). *Proceedings of the Xth Federation of European Biochemical Societies Meeting* 38, p. 193, North-Holland-American Elsevier.
Murray, K. and Old, R. W. (1974). *Progress in Nucleic Acid Research and Molecular Biology* **14**, 117.
Murray, K. Morrison, A., Cooke, H. W. and Roberts, R. J. (1977). (Unpublished observations) cited by Roberts, R. J. (1976).
Murray, N. E. and Murray, K. (1974). *Nature, London* **251**, 476.
Nathans, D. and Smith, H. O. (1975). *Annual Reviews of Biochemistry* **44**, 273.
Nevers, P. and Saedler, H. (1977). *Nature, London* **268**, 109.
Olivera, B. M. and Bonhoeffer, F. (1974). *Nature, London* **250**, 513.
Ottolenghi, S., Lanyon, W. G., Paul, F., Williamson, R., Weatherall, D. J., Clegg, J. B., Pritchard, J., Pootrakul, S. and Boon, W. H. (1974). *Nature, London* **251**, 389.
Panasenko, S. M., Cameron, J. R., Davis, R. W. and Lehman, I. R. (1977). *Science, New York* **196**, 188.
Parkinson, J. S. (1971). Cited by Hershey, A. D. and Dove, W. *In* "Bacteriophage Lambda" (A. D. Hershey, ed.), p. 792. Cold Spring Harbor, New York.
Rabbitts, T. H. (1976). *Nature, London* **260**, 221.
Rambach, A. and Tiollais, P. (1974). *Proceedings of the National Academy of Sciences of the United States of America* **71**, 3927.

Ramirez, F., O'Donnell, J. V., Marks, P. A., Bank, A., Musumeci, S., Pizzarelli, G., Russo, G., Lupis, B. and Gambino, R. (1976). *Nature, London* **263**, 471.
Ratzkin, B. and Carbon, J. (1977). *Proceedings of the National Academy of Sciences of the United States of America* **74**, 487.
Reanney, D. C. (1976). *Bacteriological Reviews* **40**, 552.
Reanney, D. C. (1978). *Microbiological Reviews* (In press).
Richardson, C. C. (1969). *Annual Review of Biochemistry* **38**, 795.
Roberts, R. J. (1976). *Critical Reviews in Biochemistry* **3**, 123.
Roberts, R. J., Breitmeyer, J. B., Tabachnik, N. F. and Myers, P. A. (1975). *Journal of Molecular Biology* **91**, 121.
Rodriguez, R. H., Bolivar, F., Goodman, H. M., Boyer, H. W. and Betlach, M. (1976). "ICH-UCLA Symposia on Molecular and Cellular Biology" (P. Nierlich and R. J. Rutter, eds.) Vol. 5, p. 471.
Rougeon, F., Kourilsky, P. and Mach, B. (1975). *Nucleic Acids Research* **2**, 2365.
Roy, P. H. and Smith, H. O. (1973a). *Journal of Molecular Biology* **81**, 427.
Roy, P. H. and Smith, H. O. (1973b). *Journal of Molecular Biology* **81**, 445.
Sadler, J. R., Tecklenburg, M., Bety, J. L., Goeddel, D. V., Yansura, D. G. and Caruthers, M. H. (1977). *Gene* **1**, 305.
Salser, W. A. (1974). *Annual Review of Biochemistry* **43**, 923.
Sanzey, B., Mercereau, O., Ternynck, T. and Kourilsky, P. (1976). *Proceedings of the National Academy of Sciences of the United States of America* **73**, 3394.
Scheller, R. H., Dickerson, R. E., Boyer, H. W., Riggs, A. D. and Itakura, K. (1977a). *Science, New York* **196**, 177.
Scheller, R. H., Thomas, T., Lee, A., Niles, W., Klein, R., Britton, R. and Davidson, E. (1977b). *Science, New York* **196**, 197.
Sgaramella, V., Van de Sanche, J. H. and Khorana, H. G. (1976). *Proceedings of the National Academy of Sciences of the United States of America* **73**, 1468.
Shapton, D. A. and Board, R. G. eds. (1972). "Safety in Microbiology." Academic Press, London.
Sharp, P. A., Sugden, B. and Sambrook, J. (1973). *Biochemistry, New York* **12**, 3055.
Shimada, K., Weisberg, R. A. and Gottesman, M. E. (1972). *Journal of Molecular Biology* **63**, 483.
Signer, E. R. (1969). *Nature, London* **233**, 158.
Silverman, M., Matsumura, P., Draper, R., Edwards, S. and Simon, M. (1976). *Nature, London* **261**, 248.
Skalka, A. and Shapiro, L. (1976). *Gene* **1**, 65.
Smith, H. O. and Wilcox, K. W. (1970). *Journal of Molecular Biology* **51**, 379.
So, M., Gill, R. and Falkow, H. S. (1975). *Molecular and General Genetics* **142**, 239.
Stawinsky, J., Hozumi, T., Narang, S. A., Bahl, C. P. and Wu, R. (1976). *Canadian Journal of Chemistry* **54**, 670.
Steere, N. V. ed. (1971). "Handbook of Laboratory Safety." Chemical Rubber Company, Cleveland, U.S.A.
Sternberg, N., Tiemeier, D. and Enquist, L. (1976). *Gene* **1**, 255.
Stocker, B. A. D. (1949). *Journal of Hygiene* **47**, 398.
Struhl, K., Cameron, J. R. and Davis, R. W. (1976). *Proceedings of the National Academy of Sciences of the United States of America* **73**, 1471.
Takanami, M. (1974). *Methods in Molecular Biology* **7**, 113.
Taylor, J. M., Dozy, A., Kan, Y. W., Varmus, H. E., Lie-Injo, L. E., Ganesan, J. and Todd, D. (1974). *Nature, London* **251**, 392.
Thomas, M. and Davis, R. W. (1975). *Journal of Molecular Biology* **91**, 315.

Thomas, M., Cameron, J. R. and Davis, R. W. (1974). *Proceedings of the National Academy of Sciences of the United States of America* **71**, 4579.

Ullrich, A., Shine, J., Chirgwin, J., Pictet, R., Tischer, E., Rutter, W. J. and Goodman, H. M. (1977). *Science, New York* **196**, 1313.

Vapnek, D., Alton, N. K., Basett, C. L. and Kushna, S. R. (1976). *Proceedings of the National Academy of Sciences of the United States of America* **73**, 3492.

Villarreal, L. P. and Berg, P. (1977). *Science, New York* **196**, 183.

Williams, R. (1976). Report of the Working Party on the Practice of Genetic Manipulation. *HMSO Cmnd 6600*. HMSO, London.

Williams, B. G., Moore, D. D., Schumm, J. W., Grunwald, D. J., Blechl, A. E. and Blattner, F. R. (1977). *In* "Genetic Alteration: Impact of Recombinant Molecules in Genetic Research" (R. Beers, ed.) p. 196. Raven Press, New York.

Wilson, J. T., Forget, B. G., Wilson, L. B. and Weissman, S. M. (1977). *Science, New York* **196**, 200.

Wu, R., Bambara, R. and Jay, E. (1975). *Critical Reviews in Biochemistry* **2**, 455.

Young, F. E. (1976). *Federal Register* **41**, 209.

Young, F. E. and Wilson, G. A. (1975). *In* "Spores VI" (P. Gerhardt, R. N. Costilow and H. L. Sadoff, eds.), p. 596. American Society for Microbiology, Washington D.C.

Yu, V., Ilyin, N. A., Tchurikov, G. and Georgiev, P. (1976). *Nucleic Acids Research* **3**, 2115.

Zeig, J., Silverman, M., Hilmen, M. and Simon, M. (1977). *Science, New York* **196**, 170.

AUTHOR INDEX

Numbers in italic are those pages on which References are listed

A

Abdolrahimzadeh, H., 210, *230*
Abe, M., 127, *168*, 214, 220, *226*
Abelson, J., 235, 256, *265*, *266*
Adachi, S., 202, *233*
Adams, M. H., 251, *265*
Adler, H. I., 117, 157, 158, 159, 162, *168*
Adman, R., 91, *101*
Adolph, E. F., 110, *168*
Ahlers, J., 185, 186, *231*
Aibara, S., 222, 224, *226*
Ailhaud, G. P., 194, 197, *226*, *228*
Aiyappa, P. S., 185, 187, *226*
Akamatsu, Y., 214, 215, 216, 217, *230*
Akkermans-Kruyswijk, W. 205, *232*
Albertini, A. M., 152, *175*
Albino, A., 128, 155, *175*
Albright, F. R., 180, 198, 215, 219, 220, 221, 222, *226*, *233*
Allen, J. S., 162, *168*
Allen, R. G., 162, *168*, 189, *227*
Alton, N. K., 255, *271*
Ambron, R. T., 208, *226*
Allison, D. P., 159, *168*
Ames, B. N., 22, *64*
Ames, G., 193, 222, *226*
Amy, N. K., 7, 8, 9, 10, 57, *62*
Anderson, B. M., 7, 8, 9, 10, *62*, *63*
Anderson, C., 95, 97, *101*
Ando, T., 247, *265*
Andrewes, F. W. C., 255, *266*
Antoine, A. D., 12, 13, 17, 48, *62*

Arai, M., 221, *226*
Araki, 77, *101*
Archibald, A. R., 107, 140, 159, *168*
Armitage, J., 160, *174*
Armstrong, K. A. 249, *266*
Arnon, R., 182, *232*
Arst, H. N., Jr, 17, 18, 40, 43, 44, 45, 46, *63*
Ashby, E., 262, *266*
Asmus, A., 146, 155, *174*
Asselineau, J., 208, *229*
Astrachan, L., 222, *226*
Aten, J. A., 152, *175*
Atkins, J., 95, 97, *101*
Audet, A., 221, 223, *226*
Autissier, F., 145, *171*

B

Backman, K., 245, 250, *266*
Bahl, C. P., 244, 245, 247, *266*, *268*, *270*
Baltimore, D., 262, *266*
Ballou, C. E., 76, *101*, 202, *230*
Bambara, R., 243, *271*
Bank, A., 258, *270*
Bannister, D., 239, *266*, *267*
Barett, W. E., 257, *269*
Barlour, S. D., 163, *169*
Barner, H. D., 153, *169*
Barsukov, L. I., 181, 182, 183, *226*
Barth, P. T., 112, *173*

273

AUTHOR INDEX

Bartnicki-Garcia, S., 70, 71, 72, 73, 74, 75, 76, 77, 78, 79, 87, 93, *102, 103*
Basett, C. L., 255, *271*
Batrakow, S. G., 208, *226*
Bayer, W. H., 190, 196, *228*
Bayne-Jones, S., 110, *168*
Bazaral, M., 247, *266*
Beachey, E. H., 143, *168*
Beaufils, A. M., 107, 140, 151, 159, *170*
Beckman, B. E., 107, 159, *170*
Beckman, J. S., 256, *266*
Begg, K. J., 112, 123, 124, 125, 130, 131, 143, 146, 147, 148, 152, 153, 167, *169*
Bell, R. M., 180, 190, 196, 197, 198, *226, 227, 228*
Benenson, A., 182, *230*
Benns, G., 202, 203, 223, *226, 227*
Ben-Sasson, S., 120, *170*
Benson, A. A., 208, *232*
Benton, W. D., 260, *266*
Benzer, S., 91, *104*
Benzinger, R., 244, *268*
Berg, F. M., 147, 148, *175*
Berg, P., 237, 243, 245, 261, 262, *266, 268, 271*
Bergelson, L. D., 181, 182, 183, 208, *226*
Bergman, K., 99, *102*
Bernard, M. C., 219, *226*
Berre, A., 221, *230*
Bertsch, L. L., 215, 217, *231*
Betlach, M., 246, 248, *270*
Betlach, M. C., 240, 248, 256, *266, 267*
Bety, J. L., 247, *270*
Bevers, E. M., 185, 187, *226*
Bhatti, A. R., 130, *169*
Billault, A., 265, *269*
Binding, H., 100, *102*
Bissell, M. J., 182, *228*
Bisset, K. A., 146, *169*
Bittman, R., 183, *226*
Bjorklid, E., 221, *230*
Bjorksten, K., 131, *175*
Blakeslee, A. F., 99, *102*
Blakesley, R. W., 241, *266*
Blanco, M., 165, *169*
Blattner, F. R., 244, 252, *266, 271*
Blechl, A. E., 244, 252, *266, 271*
Bleecken, S., 112, *169*
Bloj, B., 182, *226*
Bloom, G. D., 125, 155, 159, *169, 173*
Bloxham, D. P., 84, *102*

Blumberg, G., 126, *171*
Blumberg, P. M., 154, 157, *172*
Board, R. G., 263, *270*
Bockrath, R. C., 161, *173*
Body, D. R., 202, *226*
Boguslawski, G., 91, *102*
Bolivar, F., 246, 248, 256, *266, 270*
Bollum, F. J., 243, *267*
Bonhoeffer, F., 260, *269*
Bonsen, P. P. M., 212, *226*
Boon, W. H., 258, *269*
Boos, W., 134, *175*
Boothby, D., 151, 154, *171*
Boquet, P., 125, *173*
Borgia, P., 76, 82, 83, 100, *102*
Bosch, V., 184, *226*
Bouvier, F., 165, *169*
Boyer, H., 261, *266*
Boyer, H. W., 239, 240, 245, 246, 247, 248, 249, 256, 257, 261, 262, *266, 267, 268, 269, 270*
Boyer, R. F., 185, 187, *231*
Boylen, R. J., 157, *169*
Bradley, D. E., 188, *227*
Brammar, W. J., 252, *269*
Branucci, M. G., 254, *269*
Braun, V., 139, 154, 155, 157, 158, *169, 170, 184, 185, 226, 227, 228*
Braunstein, S. N., 187, *227*
Breitmeyer, J. B., 239, *270*
Bremer, H., 129, *169*
Brenner, S., 107, 148, 149, *171*, 262, *266*
Bretscher, M. S., 181, *227*
Brierly, G., 184, *228*
Bright-Gaertner, E., 222, 224, *227*
Briles, E. B., 136, *169*
Brinkley, S. B., 127, *169*
Brison, J., 219, *226*
Britton, R., 246, *270*
Brock, T. D., 189, *231*
Brockerhof, H., 212, 213, 219, 220, *227*
Brodie, A. F., 185, *229*
Brody, S., 75, *102*
Brooks, G. F., 216, *232*
Brostrom, M. A., 127, *169*
Brotherus, J., 202, *227*
Brown, C. M., 2, *63*
Browne, J., 258, *266*
Brown, W., 248, *266*
Brunner, H., 182, *232*
Bruns, G., 91, *102*

AUTHOR INDEX

Buller, C. S., 217, *230*
Burdett, I. D. J., 107, 125, 154, 155, 157, *169*, *174*
Burgeff, H., 99, *102*
Burke, P. V., 99, *102*
Burman, L. G., 154, 155, *169*, *173*
Burnham, J. C., 191, *232*
Burton, W. G., 240, *266*
Butlin, G., 127, 163, *171*

C

Cabib, E., 78, *103*
Cadden, S. P., 188, *231*
Cambier, H. Y., 14, *63*
Cameron, J. R., 251, 252, 253, 254, 255, 256, 260, 261, *266*, *268*, *269*, *270*
Capaldo, F. N., 163, *169*
Carbon, J., 206, *227*, 249, 255, 256, 261, *267*, *270*
Carroll, K. K., 202, *227*
Carter, J. R., Jr. 180, 198, *227*
Caruthers, M. H., 247, *270*
Cashel, M., 190, *227*
Cass, K. H., 25, *65*
Casu, A., 221, *227*
Cerdá-Olmedo, E., 99, *102*
Chan, J. K., 7, *63*
Chan, L., 139, *172*
Chang, A. C. Y., 245, 249, 254, 257, 260, *267*, *268*, *269*
Chang, H. C. P., 18, 21, *63*
Chang, L. M. S., 243, *267*
Chang, Y-Y., 180, 197, 200, *227*
Chastellier, C., de, 139, *169*
Chen, C., 184, *229*
Cheng, P. J., 192, *230*
Chirgwin, J., 258, 259, 260, *271*
Cho, K. S., 203, *227*
Chow, L., 248, *266*, *268*
Chow, L. T., 239, *266*
Chu, E., 127, *172*
Chu, H. P., 220, *227*
Chung, K. L., 135, 140, 143, *169*
Churchward, G. G., 133, 134, *169*
Clark, A. J., 248, *267*
Clark, D. J., 125, 127, 132, 152, 153, 161, 162, *169*, *171*, *174*
Clark, J. B., 131, *169*
Clark, J. E., 265, *267*

Clarke, L., 206, *227*, 255, 261, *267*
Clarke, P., 258, *266*
Clarke, P. H., 221, *230*
Clark-Walker, G. D., 72, 84, *102*, *103*
Claus, G. W., 131, *176*
Clayton, D. A., 249, *267*
Clegg, J. B., 258, *269*
Clewell, D. B., 247, *267*
Clivio, A., 152, *175*
Coapes, H. E., 107, 140, 159, *168*
Coddington, A., 11, 12, 13, 16, 17, 18, 38, 42, 47, 48, *63*, *64*
Cohen, A., 162, *168*
Cohen, S. N., 236, 237, 245, 249, 254, 257, 260, 262, *266*, *267*, *268*, *269*
Cohen, S. S., 153, *169*
Cohen-Solal, M., 258, *267*, *269*
Cole, R., 221, 223, *226*, *227*
Cole, R. M., 107, 135, 140, 143, 159, 162, *168*, *169*, *172*, *174*
Collins, C. H., 263, *267*
Collins, C. J., 255, *267*
Collins, J., 112, *173*
Collins, J. F., 116, 117, 119, 162, *169*, *175*
Comfurius, P., 221, *233*
Conti, S. F., 191, *232*
Cook, K. A., 50, *63*
Cooke, E. M., 264, *267*
Cooke, H. W., 239, *269*
Coon, M. J., 185, 187, *231*
Cooper, S., 111, 112, 116, 119, 120, 125, 129, 150, *169*, *170*
Cota-Robles, E., 70, *102*
Cotton, F. A., 6, *63*
Cove, D. J., 11, 16, 17, 18, 19, 20, 21, 38, 40, 41, 42, 43, 44, 45, 46, 48, 49, 50, 60, *63*, *64*, *65*
Covey, C., 249, *267*
Coyne, S. I., 162, *172*, *174*
Cress, D. E., 150, *171*
Crisona, N. J., 248, *267*
Cronan, J. E., Jr. 188, 189, 191, 196, 197, 204, *227*, *230*, *231*, *232*
Crowfoot, P. D., 191, *227*
Cummins, J. E., 133, *173*
Cunningham, C. C., 185, 186, *227*
Cunningham, W. P., 107, 159, *170*
Curtiss, R., 236, 264, 265, *267*
Cutler, R. G., 128, *169*
Cuzin, F., 107, 148, 149, *171*
Cutts, J. H., 202, *227*

D

Dades, S., 181, *230*
Dales, S., 145, *173*
Daneo-Moore, L., 151, 154, *171*, *175*
Darlow, H. M., 263, *267*
David, C. N., 99, *102*
Davidson, E., 246, *270*
Davies, J., 54, 55, 57, *65*
Davies, M. C., 151, *170*
Davis, B. D., 139, *172*
Davis, R. W., 237, 240, 244, 251, 252, 253, 254, 255, 256, 260, 261, 262, *266*, *268*, *269*, *270*, *271*
Dean, A. C. R., 129, *169*
Debanne, M. T., 55, 56, 57, *65*
Dedonder, R., 265, *269*
Degnen, S. T., 150, 152, *169*
Delbrück, M., 99, *102*
Denhardt, D. T., 161, *172*
Denis, F., 219, *226*
Dennis, P. P., 129, 132, *169*
Dennison, D. S., 99, *102*
Denniston-Thompson, K., 244, 252, *266*
De Vries, A. J., 255, *267*
De Pinto, J. A., 208, *227*
Der Vartanian, D., 29, *64*
De Servio. A. J., 180, 189, 190, 198, 200, 202, 204, *227*, *230*
Deutsch, R., 192, *230*
De Voe, I. W., 130, *169*
Devoret, R., 165, *169*
De Vries, J., 14, *64*
Dickerson, R. E., 245, 246, 247, *268*, *270*
Dix. D. E., 152, *169*
Dio, O., 214, 215, 220, 221, 222, 223, *226*, *228*, *230*
Doljanski, F., 120, *170*, 182, *230*
Donachie, W. D., 112, 120, 123, 124, 125, 128, 130, 131, 143, 146, 147, 148, 152, 153, 162, 167, *169*, *171*, *175*
Douthwright, J., 188, *232*
Dowhan, H., 206, *231*
Dowhan, W., 200, 206, 207, *229*
Downey, R. J., 11, 12, 13, 16, *63*, *64*
Dozy, A., 258, *270*
Dozy, A. M., 258, *268*
Draper, R., 261, *270*
Dring, G. J., 156, *169*
Dugaiczyk, A., 239, *266*
Ducan, C. H., 254, *267*
Dunn, E., 46, *65*

Dunn-Coleman, N. S., 60, 61, *63*
Duvall, E. J., 254, *269*
Dyer, J. C., 18, 21, *63*

E

Eberle, H., 111, 150, *169*
Echlin, G., 127, *170*
Ecker, R. E., 129, 132, *169*, *170*
Edelmann, P. L., 127, *170*
Edgell, M. H., 239, *269*
Edwards, S., 261, *270*
Efstratiadis, A., 257, 258, *267*, *269*
Ehrenstein, G., von. 91, *104*
Ellis, M. E., 208, *228*
Elmer, G. W., 74, *102*
Elmros, T., 156, *176*
Endo, A., 185, *228*
Engberg, B., 130, *170*
Enquist, L., 244, 252, 253, 265, *267*, *268*, *270*
Ensign, J. C., 131, *172*
Epstein, E., 73, 81, 82, 84, *103*
Erickson, R., 14, 15, 16, 19, *64*
Erickson, R. H., 6, 7, 35, 36, *65*
Errington, F. P., 110, 114, 116, 117, *170*, *173*
Esfahani, M., 191, *227*
Esselmann, M. T., 221, *228*
Evans, H. J., 2, 7, 38, *64*
Evans, J. E., 128, *169*
Exterkate, F. A., 202, 217, *228*

F

Faber, H. E., 244, 252, *266*
Falkow, H. S., 246, *270*
Fan, D. P., 107, 159, *170*
Fangman, W. L., 153, *171*
Federal Register, 251, *267*
Fielding, P., 145, *175*
Filip, C. C., 162, *168*
Fincham. J. R. S., 49, 50, *64*
Finnerty, W. R., 180, 189, 193, 194, 198, 200, 208, 210, 212, 216, 217, 220, 222, *228*, *229*, *230*, *232*
Fischer, W., 202, *227*
Fisher, T. L., 7, *63*
Fisher, W. D., 117, 162, *168*
Fitz-James, P., 156, *170*
Flawia, M. M., 81, *102*
Fleischer, S., 184, *228*
Flores-Garreon, A., 77, 83, *102*, *103*

Focht, W. J., 12, *63*
Fontana, R., 157, *174*
Forbes, E., 46, *65*
Fores-Carreon, A., 77, *103*
Forget, B. G., 258, *267*, *269*, *271*
Forsberg, C. W., 157, 158, 159, *170*
Foster, K. W., 99, *102*
Fox, C. F., 145, 146, *175*, *176*
Francombe, W. H., 138, 156, *173*
Frank, H., 146, 155, *174*
Frank, M. E., 114, 120, 130, *171*
Frankland, P. F., 110, *270*
Franklin, N. C., 252, *267*
Franklin, R. M., 181, 187, 188, 203, *227*, *231*, *232*
Franklin-Klein, W., 205, *232*
Frazier, W. A., 12, 13, 14, 17, *63*
Freedman, M. L., 134, *172*
Freeman, B. A., 217, *228*
Frehel, C., 107, 140, 151 159, *170*
Frelat, G., 125, *173*
Frerman, F. E., 192, 200, *228*
Friedman, M., 93, *102*
Fugiwara, T., 157, *170*
Fukuda, A., 150, *170*
Fukui, S., 157, *170*
Fulco, A. J., 189, *228*
Fung, C. K., 213, 219, *231*
Furlong, L.-A., 244, 252, *266*
Furukawa, Y., 208, *233*

G

Gaffar, A., 130, *175*
Galizzi, A., 152, *175*
Gallant, J., 153, *176*, 190, *227*
Gambino. R., 258, *270*
Gan, Y. Y., 107, 159, *173*
Ganelin, V. L., 19, *63*, *64*
Ganesan, J., 258, *268*, *270*
Ganthner, F., 100, *102*
Gardner, A. D., 154, *170*
Garfin, D. E., 240, *267*
Garner, C. D., 10, *63*
Garrett, R. H., 3, 4, 5, 7, 8, 9, 10, 11, 12, 16, 17, 18, 20, 21, 22, 23, 24, 25, 26, 27, 28, 29, 30, 31, 32, 33, 34, 35, 36, 37, 38, 39, 42, 46, 48, 49, 50, 51, 57, 58, 59, 60, 62, *63*, *64*, *65*
Gauger, W. L., 99, *102*
Gelman, E. P., 189, *227*
Gennis, R. B., 184, 185, *228*

Georgiev, P., 260, *271*
Gerhardt, V., 182, *232*
Gesteland, R., 95, 97, *101*
Gette, W., 92, *102*
Ghangas, G. S., 247, *267*
Gibbs, D. C., 186, *231*
Giesbrecht, P., 151, *170*
Gilbert, M., 255, *267*
Gilbert, W., 245, *266*
Gill, R., 246, *270*
Glaser, D. A., 132, *176*
Glaser, L., 139, *172*
Glaser, M., 190, 196, *228*
Glick, M. C., 131, *170*
Glover, S. W., 239, *266*, *267*
Godber, G., 263, *267*
Godson, G. H., 196, *231*
Godson, G. N., 197, *227*, 241, *267*
Goeddel, D. V., 247, *270*
Goldberg, M., 92, *102*
Golden, N. G., 190, *228*
Goldfine, H., 180, 191, 197, 207, 208, *228*, *230*
Goldschmidt, R., 265, *267*
Golecki, J. R., 143, *173*
Gonnans, F., 91, *104*
Goodell, E. W., 99, *102*, *155*, *159*, *170*
Goodman, H. M., 239, 240, 246, 247, 248, 249, 257, 258, 259, 260, *266*, *267*, *268*, *269*, *270*, *271*
Goodwin, B. C., 128, *170*
Gorchein, A., 191, 208, *228*
Gottesman, M. E., 251, 255, *270*
Gottesman, M. M., 251, *267*
Gottlieb, P. 120, *176*
Gould, R. M., 201, *228*
Gray, G. M., 202, *226*
Gray, T. R. G., 131, *172*
Green, E. W., 145, 146, *170*, 181, *228*
Greenawalt, J. W., 161, *176*
Greenbaum, P., 23, 24, 25, 33, 36, 60, *63*, *65*
Greenberg, B., 241, *268*
Greene, P. J., 240, *267*
Greenbaum, P., 25, 36, *63*
Greenwood, D., 154, 155, *170*
Gross, J. D., 163, 164, *170*, *176*
Grossman, N., 125, 127, 151, *174*
Grover, N. B., 120, *170*
Grover, N. R., 121, 123, 130, 168, *170*
Groves, D. J., 161, 162, *174*

H

Grunstrom, T., 125, *173*
Grunstein, M., 260, *267*
Grunwald, D. J., 244, 252, *266*, *271*
Guarneri, M., 182, *228*
Guchait, R. B., 191, *231*
Gumpert, J., 159, *169*
Gunstaphson, R. A., 162, *168*

Haas, G. H., de, 212, 221, *226*, *233*
Hackenbeck, R., 133, *170*
Haga, J. Y., 154, 155, *171*
Hagen, P. O., 208, *228*
Hager, L. P., 185, 186, *227*
Hahn, B., 157, 158, *170*
Hahn, J. J., 135, 140, *169*
Haidle, C. W., 73, *102*
Hale, C. M. F., 146, *169*
Hall, B., 91, 93, *101*, *102*
Hall, S. I., 265, *267*
Hallock, L. L., 162, *174*
Hamadi, K., 131, 132, *171*
Hamer, D. H., 243, 250, *268*
Hampe, A., 120, *175*
Hancock, R., 151, 156, *170*
Hankinson, O., 43, 60, *63*
Hantke, K., 185, *228*
Harada, B., 153, *176*
Hardesty, B., 97, *102*
Hardigree, A. A., 117, 157, 158, 159, 162, *168*, *171*
Hartley, E. G., 263, *267*
Harvey, M. S., 182, *229*
Harvey, R. J., 111, 113, 115, 116, 117, 119, 120, *170*, *172*
Hash, J. H., 151, *170*
Hastings, Wilson, T., 160, *173*
Haverkate, F., 221, *228*
Hawirko, R. Z., 135, 140, 143, *169*
Hawkes, S. A., 182, *228*
Hawley, D., 245, 250, *266*
Hawrot, E., 207, *228*
Hayakawa, K., 157, *172*
Hecker, L. I., 257, *269*
Hedgpeth, J., 239, 240, *266*, *268*
Heindell, H., 258, *266*
Heindell, H. C., 258, *269*
Heisenborg, M., 99, *102*
Held, W., 92, *102*

Helinski, D. R., 247, 248, 249, 261, *266*, *267*, *268*
Helling, R. B., 249, 257, *267*, *269*
Hellio, R., 139, 167, *169*, *174*
Hellman, A., 263, *268*
Helmkamp. G. M., 182, *229*
Helmstetter, C. E., 111, 112, 116, 119, 120, 121, 125, 127, 128, 129, 151, 152, *169*, *170*, *173*, *174*
Hempstead, P. G., 164, *170*
Henner, W. D., 244, *268*
Henning, U., 157, 158, *170*
Herbert, D., 129, *170*
Herman, R. K., 129, *169*
Hermann, J., 202, *227*
Hermoso, J., 92, *103*
Hershey, A. D., 251, *268*
Hershfield, V., 248, 249, 261, *266*, *268*
Herskowitz, I., 95, *102*
Hewitt, E. J., 2 22, *63*
Heyneker, H. L., 247, *268*
Hicks, M. L., 255, *267*
Higa, A., 244, *269*
Higgins, M. L., 107, 135, 136, 148, 149, 151, 154, 156, *170*, *171*, *175*
Highton, P. J., 156, *171*
Higuchi, R., 257, 258, *268*
Hill, W. E., 153, *171*
Hilmen, M., 255, *271*
Hinckley, A., 185, *229*
Hinkle, P. C., 185, *231*
Hinnen, R., 181, *232*
Hirabayashi, T., 200, *229*
Hirashima, A., 184, *229*
Hirota, Y., 114, 127, 138, 139, 145, 150, 160, 161, 162, 163, 164, 165, 167, *171*, *172*, *174*
Hirsch, P., 150, 151, *171*, *173*
Hirschberg, C. B., 204, *229*
Hjalmarsson, K., 130, *170*
Hobbs, D. G., 156, *171*
Hocking, D., 99, *102*
Hoffman, B., 132, *171*
Hoffman, H., 114, 120, 130, *171*
Hofschneider, P. H., 157, *171*
Hogness, D., 260, *267*
Hogness, D. S., 262, *266*
Hohman, R. J., 131, *171*
Hohn, B., 244, *268*
Hohn, T., 244, *268*

AUTHOR INDEX

Holland, I. B., 133, 134, *169*
Holland, J. P., 258, *268*
Holme, T., 129, *171*
Hood, J. R., 110, 113, 114, 116, 117, *174*
Hopper, A., 93, *102*
Horiuchi, K., 241, *268*
Hotta, S., 131, 132, *171*
Houseman, D., 249, 260, *268*
Howe, W. E., 163, 164, 165, *171*
Hozumi, T., 245, 246, 247, *269*, *270*
Hsu, J. C., 265, *267*
Hughes, R. C., 140, 142, 151, 167, *171*
Humayun, Z., 239, 242, *268*
Hurst, A., 156, *169*
Husain, I., 131, *171*
Hutchinson, C. A., 239, *269*
Huterer, S., 202, *233*
Hyde, M. R., 10, *63*
Hynes, M. J., 60, *63*

I

Iba, H., 150, *170*
Iino, T., 255, *268*
Ikawa, M., 208, *229*
Ikewaza, H., 196, *230*
Ikezawa, S., 221, *232*
Ilyin, N. A., 260, *271*
Imahori, K., 157, *172*
Imai, K., 185, *229*
Imoto, T., 202, *233*
Inderlied, C., 84, 86, *102*
Ingram, J. M., 130, *169*
Inoue, K., 182, *229*
Inoue, M., 265, *267*
Inouye, M., 125, 163, 165, *171*, *186*, 184, 185, *229*
Isaac, P. K., 135, 140, 143, *169*
Ishikawa, K., 202, *233*
Ishinaga, M., 205, 222, 224, *226*, 229, *230*
Ispolatovskaya, M. V., 221, *229*
Itakura, K., 245, 246, 247, *266*, *268*, *270*
Ito, S., 160, *173*

J

Jackson, D. A., 237, 243, 245, 255, *267*, *268*
Jacob, F., 107, 114, 127, 138, 145, 148, 149, 150, 160, 162, 163, *171*, *172*, 251, *268*
Jacob, G. S., 7, 9, 10, 24, 36, *63*

James, R., 154, 155, *171*
Jay, E., 243, *271*
Jeffrey, A., 239, 242, *268*
Jensen, R. G., 212, 213, 219, 220, *227*
Jobaggy, A., 72, 75, *104*
Jones, G., 15, *64*
Johnson, L. W., 182, *229*
Johnson, P. F., 256, *266*
Jones, K. W., 260, *268*
Jones, N. C., 125, 128, *171*
Jong, M. A., de, 108, 121, 123, 125, 152, *176*

K

Kaariainen, L., 202, *231*
Kaback, H. R., 185, 197, *230*, *233*
Kafatos, F. C., 257, 258, *267*, *269*
Kahn, P., 248, *268*
Kaiser, A. D., 237, 243, *269*
Kaiser, D., 244, *268*
Kaltschmidt, E., 97, *102*
Kameda, M., 208, *232*
Kamin, H., 29, 30, 31, *64*
Kamio, Y., 182, *229*
Kamirvo, T., 154, 157, *172*
Kan, Y. W., 258, *268*, *270*
Kandrach, A., 185, *229*
Kaneshiro, T., 207, *229*
Kanfer, J., 193, *229*
Kanfer, J. N., 205, *229*
Kania, M., 203, *232*
Karamata, D., 164, *170*
Katagiri, N., 245, *266*, *268*
Kato, M., 222, 224, *226*
Kawanami, J., 208, *229*
Kedes, L. H., 249, 260, *268*
Keener, S., 162, *172*
Keggins, K. M., 254, *269*
Keith, A., 188, *232*
Keleman, M. V., 142, *171*
Kelly, C., 110, *171*
Kendall, D. G., 112, *171*
Kennedy, E. P., 180, 183, 193, 196, 197, 200, 204, 205, 206, 207, *227*, *228*, *229*, *231*, *232*
Kennedy, R. S., 193, *229*
Kent, C., 217, 224, *229*
Kepes, A., 145, *171*
Ketchum, P. A., 10, 12, 13, 14, 16, 17, 19, 59, *63*, *64*
Khachatourians, G. G., 162, *171*

Khorana, H., 185, *229*
Khorana, H. G., 245, *270*
Kiefer, D. O., 244, 252, *266*
Kimura, A., 208, *229*
Kinghorn, J. R., 2, 17, 45, 46, *64*, *65*
Kinsky, S. C., 3, 42, *64*
Kipnis, D. M., 80, *103*
Kitani, T., 202, *233*
Kito, M., 190, 194, 205, 222, 224, *226*, *229*
Kito, M. O., 205, *230*
Kjeldgaard, N. O., 128, 130, *174*
Kleber, I., 244, *268*
Kleid, D., 239, 242, *268*
Klein, R., 246, *270*
Kleppe, K., 149, *172*
Kline, B. C., 150, *171*
Klouwen, H., 184, *228*
Knight, B. C. J. G., 220, *230*
Knocne, H. H., 208, *229*
Knowles, A. F., 185, *229*
Knudson, A. G., 202, *231*
Kobayashi, G., 91, *102*
Koch, A. L., 106, 110, 112, 113, 114, 116, 117, 118, 119, 121, 126, *171*, *174*
Koga, Y., 221, *229*
Kojima, M., 131, 132, *171*
Kokaisl, G., 132, *169*
Kolenbrander, P. E., 131, *171*
Kondo, K., 208, *232*
Konecki, D., 97, *102*
Koostra, W. L., 217, *232*
Koppes, L., 108, 121, 123, 125, 152, *176*
Korch, C., 149, *172*
Kornberg, A., 213, 214, 215, 216, 217, 219, 223, *231*
Kourilsky, P., 245, 257, 260, *270*
Koyama, T., 114, 136, *172*, *176*
Kramer, G., 97, *102*
Kramer, R. A., 260, *268*
Krebs, E. G., 82, *104*
Kretovich, V. L., 19, *63*, *64*
Krisch, R. E., 161, *172*
Kritchevsky, G., 202, *231*
Krotski, D. M., 162, *176*
Krulwich, T. A., 131, *172*
Ku, S. G., 257, *269*
Kubitschek, H. E., 110, 113, 114, 115, 118, 120, 126, 129, 132, 134, 153, 168, *172*

Kuhn, D. A., 114, *172*
Kuhn, N. J., 197, *229*
Kulikov, V. I., 182, 183, *226*
Kundig, W., 185, *229*
Kunnes, R. S., 197, *231*
Kurioka, S., 221, *229*
Kusaka, I., 221, *229*
Kushna, S. R., 255, *271*
Kusunose, M., 208, *233*
Kvetkas, M. J., 161, *172*

L

Lacks, S., 241, *268*
Lafferty, M. A., 20, 21, 22, 23, 24, 25, 26, 32, 33, 36, 58, *64*
Lamberti, A., 14, *64*
Lampen, J. O., 185, 187, *226*, *233*
Lands, W. E. M., 212, *229*
Lane, H. E. D., 161, *172*
Lane, M. D., 191, *231*
Laneelle, G., 208, *229*
Laneelle, M. A., 208, *229*
Langworthy, T. A., 217, *229*
Lankhorst, A., 205, *232*
Lansman, R. A., 249, *267*
Lanyon, W. G., 258, *269*
Lara, S. L., 78, *102*
Lardy, H. A., 84, *102*
Lark, K. G., 111, 150, *169*
Larsen, A., 73, 79, 80, 81, *102*
Larson, T. J., 200, 206, *229*, *231*
Laurent, S. J., 127, *172*
Law, J. H., 207, *229*
Lazdunski, C., 157, 158, *172*
Leder, P., 252, 253, 265, *267*, *268*
Lederberg, J., 255, *268*
Lee, A., 246, *270*
Lee, D. K., 12, 13, 17, *64*
Lee, I. M., 217, *228*
Lee, J-P., 29, *64*
Lee, K-Y., 14, 15, 16, 19, *64*
Lee, N., 184, *229*
Leffler, S., 92, *103*
Le Gall, J., 29, *64*
Lehman, I. R., 255, *266*, *269*
Lehmann, V., 221, *229*
Lehninger, A. L., 36, *64*, 182, *228*
Leinweber, F. J., 21, 32, 33, *64*, *65*
Leive, L., 139, *172*
Lemeke, R. M., 202, *231*

Lenette, E. H., 263, *268*
Lennarz, W., 201, *228*, *230*
Lennarz, W. J., 180, 198, 201, 205, 215, 217, 219, 220, 221, 222, 224, *226*, *229*, *230*, *233*
Lepesant, J. A., 265, *269*
Lepesant-Kejzlarova, J., 265, *269*
Lester, R. L., 191, *232*
Lev, M., 132, *172*
Levin, B. R., 250, *269*
Lewis, C. M., 49, 50, *64*
Lewis, J., 95, 97, *101*
Lewis, N. J., 19, 20, *64*
Lie-Injo, L. E., 258, *268*, *270*
Liepkalns, V., 201, *228*
Lin, E. C. C., 114, 138, 145, 150, *172*
Lin, S., 247, *268*
Linnett, P., 154, 157, *172*
Lippman, E., 71, 77, 93, *102*
Lin, A., 258, *266*
Liu, A. Y., 258, *269*
Liu, P. V., 221, *228*, *229*
Lobban, P. E., 237, 243, *269*
Lockhart, W. R., 129, *175*
Lodish, H., 92, 95, 97, *102*
Lomax, M. I., 257, *269*
London, I., 91, *102*
Loomis, W., 94, *103*
Lorian, V., 156, *172*
Losada, M., 20, 23, 59, 60, *64*
Love, R., 182, *230*
Lovett, J., 93, *103*
Lovett, M., 248, 261, *268*
Lovett, P. S., 254, *269*
Lowell, N., 131, 158, *173*
Lueking, D. R., 191, *230*
Lund, F., 154, *172*
Lupis, B., 258, *270*
Luscombe, B. M., 131, *175*
L'vov, N. P., 19, *63*, *64*
Lynen, F., 197, *229*

M

Maaløe, O., 128, 130, *174*
Mabbs, F. E., 10, *63*
Mach, B., 245, 257, *270*
Madansky, C. H., 14, *63*
Magee, P., 93, *102*
Magee, W. L., 212, 219, 224, *230*

Makula, R. A., 189, 193, 200, 208, 210, 213, 216, 217, 220, 222, *228*, *230*, *232*
Mandel, M., 244, *269*
Mandelstam, J., 151, *172*, *174*
Maniatis, J., 257, *267*
Maniatis, T., 257, 258, *267*, *269*
Manis, J. J., 150, *171*
Manly, K. F., 251, *269*
Marians, K. J., 244, 245, 246, 247, *266*, *268*, *269*
Marinetti, G. V., 182, *230*
Marks, P. A., 259, *270*
Marmion, B. P., 202, *231*
Marotta, C. A., 258, *267*, *269*
Marr, A. G., 111, 113, 115, 116, 117, 119, 120, 129, 130, *170*, *172*, *175*
Martin, D. T. M., 152, *169*
Martin, H. H., 157, *171*, *172*
Martin, J. T., 135, *175*
Martin, L., 116, *175*
Martin, R. G., 22, *64*
Maruo, B., 205, *230*
Masuda, T., 244, *268*
Mateles, R. I., 130, *174*
Matney, T. S., 127, *172*
Matsuhashi, M., 114, 136, *172*, *176*
Matsuhashi, S., 154, 157, *172*
Matsumura, P., 261, *270*
Matsunager, K., 221, *226*
Matsuzawa, H., 157, *172*
Maturin, L., 265, *267*
Mauck, J., 139, *172*
Mavis, R. D., 180, 196, 198, *226*, *231*
Maxam, A. M., 257, 258, *267*
May, F., 15, *64*
May, J. W., 143, *172*
Mayall, B. H., 138, 156, *173*
Mayberry, W. R., 202, 217, *229*, *231*
Mazza, G., 152, *175*
Meacock, P. A., 124, 125, *172*
Medoff, G., 91, *102*
Meechan, J. D., 182, *228*
Meers, J. L., 2, *63*
Meissnev, G., 99, *102*
Melchior, N. H., 154, 155, *175*
Mendelson, N. H., 107, 142, 148, 157, 159, 161, 162, *169*, *172*, *174*
Menzel, J., 143, *173*
Mercereau, O., 260, *270*
Merlie, J. P., 190, *230*

Mersel, M., 182, *230*
Mertens, G., 163, *172*
Mertz, J. E., 237, 244, *269*
Messer, W., 132, 133, *170*, *171*
Michniewicz, J. J., 245, *268*
Middleton, J. H., 239, *269*
Miller, J. R., 150, *171*
Miller, R. A., 154, *173*
Milner, L. S., 185, *230*
Mindich, L., 145, *173*, 181, 222, *230*
Minnikin, D. E., 210, *230*
Mirelman, D., 107, 159, *173*
Mitchell, P., 136, 156, *173*
Mitchison, J. M., 117, 126, 133, 136, *173*
Moir, A., 252, *269*
Monacelli, R., 221, *227*
Monty, K. J., 21, 22, 32, 33, *64*, *65*
Moody, R., 265, *267*
Mooney, D. T., 73, 74, *103*
Moore, D. D., 244, 252, *266*, *271*
Moore, R. L., 151, *173*
Morava, J., 127, 155, *175*
Moreno, S., 82, *103*
Morrill, R., 84, *103*
Morrison, A., 239, *269*
Morrison, M., 192, *232*
Morrow, J. F., 249, 257, *269*
Mount, D. W., 163, 164, 165, *171*
Moyle, J., 136, 156, *173*
Mudd, J. B., 217, *228*
Mudd, S., 131, *170*
Muhlradt, P. F., 143, *173*
Mulkins, G. J., 18, 21, *63*
Muller, A., 185, *229*
Munoz, E., 186, *230*
Munson, R. J., 118, *172*
Murakami, S., 165, *173*
Murao, S., 221, *226*
Murphy, M. J., 29, 30, 31, *64*
Murray, E. G. D., 202, *227*
Murray, K., 236, 239, 243, 251, 252, 260, *268*, *269*
Murray, N. E., 215, 252, *269*
Murray, R. G. E., 107, 125, 138, 148, 154, 155, 156, *169*, *173*, *176*
Musumeci, S., 258, *270*
Myers, P. A., 239, 240, *266*, *270*
McAllister, D. J., 202, *230*
McCaman, R. E., 180, 198, *230*
McClements, W., 254, 257, *269*

McConnell, M., 157, *174*
MacDonald-Brown, D. S., 2, 40, 42, 45, *63*
MacDonald, D. W., 11, 16, 17, 19, *63*, *64*
McElhaney, R. N., 220, *233*
McElroy, W. D., 2, 3, 33, 42, *64*
MacFarlane, M. G., 200, 201, 220, *230*
McKenna, C., 19, *64*
McLean, F. I., 118, *172*
McManus, T. T., 217, *228*
McMurray, W. C., 212, 219, 224, *230*
McMurrough, I., 77, 78, *103*
McNair-Scott, D. B., 127, *172*

N

Naaman, S., 120, *170*
Naday, E., 120, *170*
Nagai, K., 162, *173*
Nakaike, S., 219, 220, *230*
Nanni, G., 221, *227*
Nanninga, N., 147, 148, *175*, *176*
Narang, S. A., 244, 245, 246, 247, *266*, *268*, *269*, *270*
Nason, A., 2, 3, 4, 5, 6, 7, 9, 11, 12, 13, 14, 15, 16, 17, 19, 20, 35, 36, 38, 42, 48, 58, 59, *63*, *64*, *65*
Nasu, T., 202, *233*
Nathans, D., 236, 243, 262, *266*, *269*
Nelson, E. T., 217, *230*
Nelson, N., 70, *102*
Nesbitt, J. A., 201, *229*, *230*
Neuberger, A., 191, *228*
Nevers, P., 236, *269*
Newman, C. N., 161, *173*
Newton, A., 108, 125, 150, 152, *169*, *173*, *175*
Newton, J. W., 161, *173*
Ng, M. H., 186, *230*
Nicholas, D. J. D., 2, 3, 7, 10, *64*
Nikerson, W. J., 72, 73, 74, 75, 76, 79, 87, *102*
Nikaido, H., 182, *229*
Niles, W., 246, *270*
Nilson, E. H., 115, 116, *172*
Nishijima, M., 214, 215, 216, 217, 219, 220, *230*
Nishikawa, M., 202, *233*
Nojima, S., 182, 213, 214, 215, 216, 217, 219, 220, 221, 222, 223, *226*, *228*, *229*, *230*

Nomura, M., 92, *102*
Nordstrom, K., 130, 155, *169*, *170*
Norlander, L., 125, *173*
Normark, S., 125, 156, 157, 159, *169*, *173*, *176*
Norris, R. F., 131, *171*
Novick, R., 138, 156, *175*
Nulty, W., 196, *228*
Nunn, D. W., 191, 192, *230*
Nussbaum, A. L., 240, *267*

O

Oakley, C. L., 221, *230*
O'Donnell, J. V., 258, *270*
O'Grady, F., 154, 155, *170*
Ohki, M., 127, 134, *173*, 214, 220, 222, *228*, *230*
Ohsaka, A., 221, *232*
Ohta, A., 205, *230*
Oka, T., 191, *227*
Okada, Y., 150, *170*
Okamoto, N., 214, 220, *226*
Oker-Blom, N., 202, *231*
Okuyama, H., 194, 196, 213, 221, *230*
Old, R. W., 243, *269*
Olden, K., 160, *173*
Olivera, B. M., 260, *269*
Olsen, R. W., 202, *230*
Olson, S. T., 185, 187, *231*
Ono, Y., 216, 221, *230*
Opdenkamp, J. A. F., 185, 187, *226*
Orlowski, M., 80, 90, 92, 93, 94, 96, *103*
Osborn, M. J., 180, 198, *226*
Osley, M. A., 150, *173*
Otnaess, A. B., 221, *230*
Oto, E., 77, *101*
Otsuka, H., 208, *229*
Otten, B. J., 217, *228*
Otten, M. R., 150, *171*
Ottolenghi, S., 258, *269*
Ourebo, S., 149, *172*
Oxman, M. N., 263, *268*

P

Paddock, G. V., 257, 258, *266*, *268*, *269*
Padmanaban, G., 24, *65*
Painter, P. R., 115, 116, 119, 120, *170*, *172*
Pala, V., 221, *227*
Pan, S. S., 6, 7, 13, 14, 15, 16, 19, 35, 36, *64*, *65*

Panasenko, S. M., 255, *266*, *269*
Pardee, A. B., 125, 127, 153, 154, 155, *171*, *175*, *176*
Park, J. T., 151, 154, *170*, *173*
Parker, C. W., 80, *103*
Parkinson, J. S., 251, *269*
Passeron, S., 73, 81, 82, 84, 85, *103*
Pateman, J. A., 2, 16, 17, 18, 19, 20, 21, 40, 41, 42, 44, 45, 46, 48, 60, 61, *63*, *64*, *65*
Patterson, P. H., 180, 198, 205, *230*
Paul, F., 258, *269*
Paulton, R. J. L., 111, 128, 138, *173*
Paveto, C., 73, 81, 82, 84, *103*
Payne, W. J., 2, *65*
Paznokas, J. L., 81, 82, 83, 84, 85, 86, *103*
Pease, P., 146, *169*
Peck, H. D., Jr. 29, *64*
Pelvit, M. C., 107, 159, *170*
Pelzer, H., 166, *176*
Pereira, D. A., 265, *267*
Perkins, H. R., 140, 142, *176*
Perkins, R. L., 154, *173*
Perry, R. P., 127, *173*
Peter, H. W., 185, 186, *231*
Peters, J., 83, 87, 100, *103*
Petersen, M. H., 188, *232*
Petterson, R., 202, *231*
Pictet, R., 258, 259, 260, *271*
Pieringer, R. A., 197, 208, *226*, *231*
Pierucci, O., 112, 120, 128, 129, 130, 150, 152, *170*, *173*
Pieterson, W. A., 212, *226*
Pilipenko, T. W., 208, *226*
Pinphanichakarn, P., 97, *102*
Pizer, L. I., 190, 194, *229*, *230*
Pizzarelli, G., 258, *270*
Plackett, P., 202, 217, *231*
Plempel, M., 99, *103*
Polakis, S. E., 191, *231*
Pollack, R., 263, *268*
Poole, R. K., 126, *173*
Pooley, H. M., 135, 136, 151, *171*, *175*
Poonian, M. S., 240, *267*
Pootrakul, S., 258, *269*
Prensky, W., 258, *267*
Poupard, J. A., 131, *171*
Powell, E. O., 109, 110, 111, 113, 114, 116, 117, *170*, *173*
Powell, G. L., 190, *228*

Previc, E. P., 107, 121, 131, 158, 168, *173*
Primrose, S. B., 151, *176*
Prins, R. A., 205, *232*
Pritchard, J., 258, *269*
Pritchard, R. H., 107, 108, 112, 121, 122, 123, 124, 125, 157, 161, 168, *172*, *173*, *176*
Prodouz, K. N., 25, 36, *63*, *65*
Proulx, P., 202, 203, 213, 219, 221, 222, 223, 224, *226*, *227*, *231*
Prydz, H., 221, *230*
Ptashne, M., 239, 242, 245, *266*, *268*

Q
Quadling, C., 146, *173*

R
Rabbitts, T. H., 257, *269*
Rabinsohn, Y., 182, *232*
Racker, E., 185, *229*, *231*
Radding, C. M., 251, *269*
Raetz, C. R. H., 205, 206, *231*
Rahn, O., 110, 112, *171*, *174*
Rambach, A., 139, 167, *174*, 251, 252, 262, *269*
Ramirez, F., 258, *270*
Rasch, G., 118, *174*
Ratzkin, B., 256, *270*
Ray, P. R., 189, *231*
Ray, T. H., 191, *231*
Ray, T. K., 196, *227*, *231*
Raybin, D. M., 215, 217, *231*
Reanney, D. C., 236, *270*
Redwood, W. R., 186, *231*
Reeve, J. N., 107, 127, 158, 159, 160, 161, 162, 163, *172*, *174*
Rehn, K., 157, 158, *170*
Reiss, J., 201, *229*
Renkonen, O., 202, *227*, *231*
Revelas, E., 112, 120, 129, *170*
Rever, B. M., 16, 17, 18, 19, 20, 21, 40, 41, 42, 48, *63*, *65*
Reyes, E., 70, *102*
Reynolds, J. W., 202, *227*
Ricard, M., 160, 161, 162, 163, 164, 165, *171*, *174*
Richardson, C. C., 244, *270*
Richardson, D., 249, *267*
Richmond, M. H., 116, 117, 119, *169*
Riggs, A. D., 245, 246, 247, *268*, *270*

Roberts, D. B., 16, 19, 40, *65*
Roberts, E. M., 124, *172*
Roberts, R. J., 236, 239, 240, 241, 242, 243, 245, *266*, *267*, *269*, *270*
Roblin, R. O., 262, *266*
Robinow, C. F., 70, *103*
Rodriguez, R. H., 246, 248, *270*
Rodriguez, R. L., 256, *266*
Roeder, W., 91, *103*
Roelofsen, B., 221, *233*
Rogers, H. J., 128, 142, 151, 155, 157, 158, 159, *170*, *171*, *172*, *174*
Rogers, P. J., 72, 84, *103*
Rogers, P. L., 129, *169*
Romeo, D., 185, *231*
Ron, E. Z., 125, 127, 151, *174*
Roselino, E., 85, *103*
Roseman, S., 185, *229*
Rosen, O. M., 82, *103*
Rosenberg, A. J., 219, *226*
Rosenberg, E., 120, *176*
Rosenberg, J., 247, *268*
Rosenberger, R. F., 121, 123, 130, 168, *170*
Rosenthal, D., 30, 31, *64*
Ross, M. J., 245, 250, *266*
Rothfield, L., 185, *228*, *229*, *231*
Rothman, J. E., 183, *231*
Rottem, S., 183, *226*
Rougeon, F., 245, 257, *270*
Rouser, G., 202, *231*
Routledge, V. I., 10, *63*
Rowbury, R. J., 108, 124, 125, 157, 160, 163, 164, 165, *174*, *175*
Roy, P. H., 239, *270*
Rozenhak, S., 127, *174*
Rubin, C. S., 82, *103*
Rudland, P., 80, *103*
Ruettinger, R. T., 185, 187, *231*
Ruiz-Herrera, J., 77, 78, 83, *102*, *103*
Ruska, H., 151, *170*
Russo, G., 258, *270*
Rutherford, E. L., 188, *227*
Rutter, W., 91, *103*
Rutter, W. J., 258, 259, 260, *271*
Ryes, E., 83, *102*
Ryter, A., 107, 127, 139, 140, 143, 146, 147, 148, 149, 151, 159, 160, 162, 163, 167, *169*, *170*, *171*, *174*
Ryu, D. Y., 130, *174*

S

Sadler, J. R., 247, *270*
Saedler, H., 236, *269*
Sager, R. A., 240, *266*
Sagers, R. D., 130, *175*
Sall, T., 131, *170*
Salser, W., 257, 258, *266*, *268*, *269*
Salser, W. A., 243, *270*
Salton, M. R. J., 180, 185, 186, 190, 198, 200, 204, *227*, *230*, *231*
Sambrook, J., 239, *270*
Sandermann, H., Jr., 184, *231*
Sands, J., 188, *232*
Sands, J. A., 188, *231*
Sanzey, B., 260, *270*
Sargent, M. G., 112, 116, 121, 123, 124, 125, 130, 133, 146, 148, 149, 152, 153, 154, 163, 164, 166, 168, *174*
Sarma, P. S., 24, *65*
Sato, T., 157, *172*
Satta, G., 157, *174*
Saverbrey, G., 120, *175*
Scandella, C. J., 188, 213, 214, 215, 216, 217, 219, 223, *231*
Scazzocchio, C., 17, *65*
Schaechter, M., 106, 110, 112, 113, 114, 116, 117, 118, 128, 129, 130, 145, 146, 157, 161, *170*, *174*, 181, *228*
Schafer, R., 181, 188, *232*
Scheller, R. H., 245, 246, 247, *270*
Scherbaum, O., 118, *174*
Schiefer, H. G., 182, *232*
Schindler, H., 188, *231*
Schlessinger, D., 91, *102*
Schloemer, R. H., 37, 38, *65*
Schmidt, J. M., 150, *174*
Schnaitman, C. A., 180, 198, *233*
Schneider, E. G., 196, 197, *232*
Schor, M. T., 186, *230*
Schultz, L., 91, *101*
Schumann, E., 159, *169*
Schumm, J. W., 244, 252, *266*, *271*
Schwartbach, S. D., 257, *269*
Schwarz, U., 107, 132, 139, 143, 146, 155, 157, 158, 159, 167, *170*, *171*, *173*, *174*
Schwinn, G., 208, *232*
Scott, C. C. L., 193, 194, *232*
Seeling, J., 188, *231*
Seiffert, W. E., 80, *103*
Sela, M., 182, *232*
Senff, L. M., 216, *232*
Sergeev, N. S., 19, *63*, *64*
Setlow, R. B., 138, *175*
Sevilla, C. L., 14, *63*
Sgaramella, V., 245, *270*
Shannon, K. P., 108, 124, 125, 157, 160, 164, 165, *174*
Shapiro, B. M., 157, 158, *172*
Shapiro, D., 182, *232*
Shapiro, L., 150, *174*, 261, *270*
Shaposhnikov, G. L., 19, *63*
Shapton, D. A., 263, *270*
Sharp, P. A., 239, *270*
Shaw, E. J., 202, *231*
Shaw, J., 184, *229*
Shaw, M. K., 130, *174*
Sheetz, M. P., 183, *232*
Shehata, T. E., 111, 113, 129, 130, *175*
Sheldon, E. L., 244, 252, *266*
Shen, B. H. P., 134, *175*
Shibuya, L., 205, *230*
Shimada, K., 251, *270*
Shine, J., 247, 258, 259, 260, *268*, *271*
Shively, J. M., 161, *176*, 208, *229*, *232*
Shockman, G. D., 107, 135, 136, 148, 149, 151, 154, 156, *170*, *171*, *175*
Short, S. A., 180, 201, 204, *232*
Shropshire, W. Jr., 99, *102*
Shuman, H., 143, *174*
Sicard, N., 165, *169*
Siccardi, A. G., 152, *175*
Siegel, L. M., 21, 22, 25, 26, 27, 28, 29, 30, 31, 32, 33, *64*, *65*
Sieglin, Y., 184, *227*
Siekevitz, P., 222, 223, *232*
Signer, E. R., 251, *269*, *270*
Signer, F., 95, *102*
Silver, S., 134, *172*
Silverman, M., 255, 261, *270*, *271*
Simon, G., 202, *231*
Simon, M., 255, 261, *270*, *271*
Simon, M. I., 162, *175*
Sinclair, C. G., 130, *175*
Sinensky, M., 196, *232*
Sing, V. O., 77, *103*
Singer, M. F., 262, *266*
Singer, S. J., 183, *232*
Sissenstein, R., 217, *228*
Skalka, A., 261, *270*

Skalka, A. M., 254, 257, *269*
Slater, M., 157, 161, *175*
Slautterback, D. B., 184, *228*
Smith, H., 221, *233*
Smith, H. O., 236, 239, 243, *269*, *270*
Smith, H. S., 127, *175*
Smith, J. A., 116, *175*
Smith, P. F., 202, 217, *229*, *231*, *232*
Smithies, O., 244, 252, *266*
Snider, M. D., 196, *232*
Snipes, W., 188, *232*
Snoek, G. T., 185, 187, *226*
So, M., 246, *270*
Somerville, M., 59, *64*
Sonoki, S., 221, *232*
Sorger, G. J., 12, 17, 18, 21, 37, 47, 50, 51, 52, 53, 54, 55, 56, 57, 58, *63*, *64*
Sorrentino, A. P., 83, *103*
Spaulding, E. H., 263, *268*
Spencer, D. A., 33, *64*
Speth, V., 143, *173*
Spratt, B. G., 154, 157, 163, 164, *174*, *175*
Stacey, K., 153, 165, *175*
Stahl, W. L., 221, *232*
Stanier, R. Y., 150, *174*
Starka, J., 127, 155, *175*
Starr, M. P., 114, *172*
Staugaard, P., 147, 148, *175*
Stawinsky, J., 244, 245, 246, 247, *266*, *269*, *270*
Stechmuller, B., 182, *228*
Steere, N. V., 263, *270*
Stegwee, D., 99, *104*
Steiner, A. L., 80, *103*
Steiner, S., 191, *232*
Steitz, J., 92, *102*
Stellwagen, E., 25, *65*
Sternberg, N., 244, 252, *267*, *270*
Stevens, H. M., 10, *64*
Stewart, F. M., 250, *269*
Stewart, P. R., 72, 84, *103*
Stocker, B. A. D., 255, *270*
Stoeckenius, W., 185, *231*
Stokes, E., 140, 142, 151, 166, *171*
Storck, R., 72, 73, 84, *102*, *103*
Strominger, J. L., 131, 154, 157, *172*, 184, 185, *228*, *231*
Struhl, K., 256, 261, *270*
Subramanian, K. N., 24, 37, 51, 52, 53, 54, 55, 58, *65*

Sud, I. J., 129, 130, *175*
Suda, S., 131, 132, *171*
Sudarsanan, K., 193, *229*
Sugahara, T., 221, *232*
Suganama, A., 131, 132, 138, 156, *171*, *175*
Sugden, B., 239, *270*
Suit, J. C., 127, *172*
Sundman, V., 131, *175*
Sussman, M., 94, *103*
Swarin, R. S., 14, 19, *63*
Swartouw, H. T., 221, *233*
Symons, R. H., 237, 243, 245, *268*
Sypherd, P. S., 73, 74, 76, 79, 80, 81, 82, 83, 84, 85, 86, 87, 90, 92, 93, 94, 96, 100, *102*, *103*
Syvanen, M., 244, *268*
Szer, W., 92, *103*

T

Tabachnik, N. F., 239, *270*
Taber, H. W., 192, *232*
Tahara, Y., 208, *232*
Tait, G. H., 191, *228*
Takahasi, T., 221, *232*
Takanami, M., 239, *270*
Tamori, Y., 219, 220, *230*
Tamura, G., 162, *173*
Tang, C. T., 192, *230*
Tanner, P. J., 151, *171*
Taubeneck, U., 159, *169*
Taylor, C., 158, *174*
Taylor, J. M., 258, *268*, *270*
Taylor, R. C., 10, *64*
Tchurikov, G., 260, *271*
Teather, R. M., 162, *175*
Tecklenburg, M., 247, *270*
Teitelbaum, D., 182, *232*
Tentini, W. C., 116, 117, *172*
Terenzi, H., 72, 75, 85, *103*, *104*
Terenzi, H. F., 72, 83, 85, *103*
Ternynck, T., 260, *270*
Terrana, B., 108, 125, 150, 152, *175*
Terry, C. E., 157, 158, 159, *168*
Terry, D. R., 130, *175*
Thiele, O. W., 208, *232*
Thom, R., 120, *175*
Thomas, C. A., 243, *268*
Thomas, E., 197, *233*
Thomas, M., 240, 251, 252, 253, *270*, *271*

AUTHOR INDEX

Thomas, T., 246, *270*
Thompson, J. S., 136, *175*
Thompson, N., 110, 114, 116, 117, *170*
Thompson, S. T., 25, *65*
Thompson, T. E., 186, *231*
Thornton, M. P., 201, *228*
Thurman, P. F., 158, *174*
Tiemeier, D., 244, 252, *267*, *270*
Tiemier, D., 253, 265, *268*
Tilby, M. J., 142, 161, *175*
Timberlake, W., 93, *103*
Tiollais, P., 251, 252, 262, *269*
Tipper, D. J., 131, *172*
Tischer, E., 258, 259, 260, *271*
Tobias, L., 240, *267*
Todd, D., 258, *268*, *270*
Tomasz, A., 128, 136, 155, *169*, *175*
Tomizawa, J., 127, *168*
Tomsett, A. B., 41, *65*
Topiwala, H. H., 130, *175*
Torregrossa, R. E., 217, *232*
Torres, H. N., 81, *102*
Tove, S. R., 29, *64*
Tropp, B. E., 191, 192, *230*
Truant, J. P., 263, *268*
Tsukagoshi, N., 145, *175*, 188, 203, *232*
Tucker, A. N., 193, 200, 222, 224, *232*, *233*
Tunaitis, E., 204, *232*
Tyring, L., 154, 155, *172*, *175*
Tzagaloff, H., 138, 156, *175*

U

Ulane, R. E., 78, *103*
Ullrich, A., 258, 259, 260, *271*

V

Vagelos, P. R., 180, 188, 190, 194, 196, 197, 198, *226*, *227*, *228*, *231*, *233*
Valwerk, J. J., 212, *226*
Van Alstyne, D., 162, *175*
van Deenen, L. L. M., 182, 185, 187, 212, 213, 220, 221, 223, 224, *226*, *228*, *229*, *231*, *233*
van den Berg, W., 108, 121, 123, 125, 152, *176*
van den Bosch, H., 194, 196, 212, 223, *233*
Van den Ende, H., 99, *103*, *104*

van der Woude, N., 77, *103*
Van de Sanche, J. H., 245, *270*
Van Golde, L. M. G., 205, 220, *232*, *233*
Van Irerson, W., 152, *175*
Vannier, F. S., 127, *127*
Van Tubergen, R. P., 138, *175*
Vapnek, D., 255, *271*
Varmus, H. E., 258, *268*, *270*
Veerkamp, J. H., 202, 217, *228*
Vega, J. M., 22, 23, 24, 25, 26, 27, 28, 29, 30, 31, 32, 33, 60, *65*
Vercellotti, S. V., 7, *62*, *63*
Verger, R., 212, *233*
Vicente, M., 112, 123, 124, 125, 130, 153, *169*
Villarreal, L. P., 261, *271*

W

Wagener, W. S., 216, *232*
Wagner, M., 135, 136, 143, *175*, *176*
Wakil, S. J., 191, 194, 196, *227*, *230*
Walker, J. R., 162, *168*
Wall, R., 257, *268*
Wallace, W., 58, *65*
Walle, J., 265, *269*
Walsh, D. G., 82, *104*
Wang, S., 184, *229*
Ward, C. B., 132, *176*
Ward, C. M., 131, *176*
Ward, J. B., 140, 142, 157, 158, 159, 166, *170*, *176*
Ward, W. H., 110, *170*
Warrack, G. H., 221, *230*
Wassenberg, H. W., 217, *228*
Watson, J., 80, *104*
Watson, J. D., 262, *266*
Weatherall, D. J., 258, *269*
Weber, H. J., 100, *102*
Wechsler, J. A., 163, *176*
Weidel, W., 166, *176*
Weigand, R. A., 161, *176*
Weinberger, M., 150, *169*
Weisberg, J. L., 189, *227*
Weisberg, R., 251, 252, *267*
Weisberg, R. A., 251, *270*
Weisblum, B., 91, *104*
Weissbach, H., 197, *233*
Weissman, S., 262, *266*
Weissman, S. M., 258, *267*, *269*, *271*
Wells, R. D., 241, *266*

Welch, S., 93, *102*
Westling, B., 159, *169*
Westling-Haggstrom, B., 156, *176*
Westmacott, D., 151, *176*
Wherret, J. R., 202, *233*
White, D. A., 180, 198, 215, 219, 220, 221, 222, *226*, *233*
White, D. C., 180, 189, 192, 193, 200, 201, 204, 221, 222, 224, *228*, *230*, *231*, *232*, *233*
Whiteley, H., 91, *104*
Whitfield, J. F., 148, *176*
Wickner, W. T., 207, *228*
Wightman, R. H., 245, *268*
Wilcox, K. W., 239, *270*
Wilkins, A., 153, *176*
Wilkinson, G., 6, *63*
Wilkinson, S. G., 208, 210, *233*
Williams, B. G., 252, *271*
Williams, P. J. L., 208, *228*
Williams, R., 262, *271*
Williams, W. G., 244, 252, *266*
Williamson, J., 139, *172*
Williamson, J. P., 110, 113, 114, 116, 117, *174*
Williamson, R., 258, *269*
Willoughby, E., 154, 157, *172*
Wilson, G., 146, *176*
Wilson, G. A., 254, 265, *267*, *271*
Wilson, J. T., 258, *269*, *271*
Wilson, L. B., 258, *271*
Winblad, B., 156, *176*
Wirtz, K. W. A., 182, *229*, *233*
Wittmann, H., 97, *102*
Wlodarzyk, M., 150, *171*
Woldringh, C. L., 108, 112, 116, 121, 123, 125, 130, 147, 148, 152, 168, *170*, *175*, *176*
Wolff, H., 139, 154, 155, *169*
Wollman, E. L., 251, *268*
Woodward, J. E., 217, *228*
Wright, D. N., 129, *176*
Wu, P. C., 125, 127, 153, *176*
Wu, R., 243, 244, 245, 246, 247, *266*, *267*, *268*, *269*, *270*, *271*
Wyrick, P. B., 157, 158, 159, *170*

Y

Yamada, K., 196, *230*
Yamada, M., 114, 136, *172*, *176*
Yamaguchi, K., 149, *176*
Yamamoto, A., 202, *231*, *233*
Yamamoto, S., 185, 187, *233*
Yameda, Y., 208, *232*
Yano, I., 208, *233*
Yanofsky, C., 248, 261, *268*
Yansura, D. G., 247, *270*
Yokomura, T., 202, *233*
Yoshikawa, H., 149, 165, *173*, *176*
Young, D. C., 10, *64*
Young, F. E., 254, 265, *267*, *271*
Young, H., 91, *104*
Young, R. A., 193, *229*
Yu, V., 260, *271*
Yuan, J. H., 7, *62*

Z

Zanati, E., 128, 155, *175*
Zaritsky, A., 121, 123, 124, 130, 157, 161, 168, *170*, *173*, *176*
Zeeb, D. D., 59, *64*
Zeig, J., 255, *271*
Zelle, M. R., 161, *172*
Zilliken, F., 131, *170*
Zilversmit, D. B., 182, *226*, *229*
Zinder, N. D., 241, 262, *266*, *268*
Zalokov, M., 99, 102
Zorzopulos, J., 72, 75, 83, *103*, *104*
Zuchowski, C., 150, *173*
Zusman, D. R., 120, 125, 162, *176*
Zwaal, R. J. A., 221, *233*

SUBJECT INDEX

A

Aberrations, morphological, in bacteria, 131

Absorption spectra of nitrite reductase from *Neurospora crassa*, 29

Acholeplasma laidlawii, lipid activation of ATPase of, 186

Acholeplasma sp., lipid asymmetry in membrane of, 182

Achromobacter chromococcus, CDP-diglyceride synthesis by, 199

Acinetobacter sp., biosynthesis of phosphatidylserine in, 205
 CDP-diglyceride synthesis in, 198
 occurrence of acylphosphatidylglycerol in, 202
 outer-membrane phospholipase A_1 activity in, 217
 phospholipase C activity produced by, 221
 phospholipid composition of, 193
 turnover of phosphatidylethanolamine in, 222

Actinomyces sp., ornithine-containing lipids in, 208

Actinomycin, as an inhibitor of nitrate assimilation in fungi, 49

Actinomycin D, effect of, on action of ammonia of fungal nitrate reductase, 54

Action of nitrate reductase in *Neurospora crassa*, model for, 35

Activation of nitrate reductase activity in fungi, 59

Active-site probes in nitrate reductase from *Neurospora crassa*, 7

Activity of enzyme in bacteria, and phospholipids, 183

Acylation of phosphatidylglycerol in staphylococci, 202

Acylphosphatidylglycerol, occurrence of, in animal tissues, 202

Acyltransferase, effect of phenethanol on activity of in bacteria, 192

Adenosine triphosphatase, lipid activation of, in bacteria, 185

S-Adenosylmethionine, as a methyl group donor in bacteria, 207
 requirement for, by Type 1 restriction endonuclease, 238
 synthesis of, in *Mucor* sp., 89

Adenyl cyclase, activity of, in *Mucor* spp., 81

Adenylate kinase, thermolabile, from *Escherichia coli* mutants, 196

Adult size in bacteria, concept of, 117

Aerobacter aerogenes, generation times in, 113
 variation in cell size of, 129

Age-classified cells of bacteria, 132

Age distribution in bacterial populations, 109

Agrobacterium sp., occurrence of phosphatidylcholine in, 208
 rod mutants of, 157

Alanine, occurrence of bacterial lipoamino acids in, 201

Alkanes, growth of *Acinetobacter* sp. on, 193

Allosteric effector action of S-adenosylmethionine on Type 1 restriction endonucleases, 238

Allotropy, chemical nature of, in biomembranes, 184

Alpha toxin of *Clostridium perfringens*, 220

Alterations in nitrogen metabolism and dimorphism in fungi, 87

289

SUBJECT INDEX

Alternate electron acceptors in fungal nitrate reduction, 5
Amide-linked fatty acids in bacterial lipids, 208
Amination of α-aminoglutarate, and dimorphism in fungi, 87
Amino-acid starvation, use of, to induce synchrony in bacteria, 127
O-Amino-acyl phosphatidylglycerol biosynthesis in bacteria, 200
Aminophospholipids, reaction of, with sulphanilic acid diazonium chloride, 181
3-Aminopyridine adenine dinucleotide phosphate as an active-site probe for nitrate reductase, 8
Ammonia, assimilation by *Mucor racemosus*, 88
   effect of, on accumulation of nitrate by *Neurospora crassa*, 37
   effect of, on accumulation of nitrate by *Neurospora crassa*, 37
      on activity of nitrate reductase of *Neurospora crassa*, 59
      on nitrate assimilation in fungi, 39
      on synthesis of nitrate-uptake proteins in fungi, 37
      on the product of the *areA* gene in fungi, 44
   reduction of nitrite to, steps involved in, with fungi, 20
   repression, genetic loci involved in, with *Aspergillus nidulans*, 43
   mechanism of, in fungi, 43
   nature of, in fungal metabolism, 39
Ammonia-insensitive synthesis of nitrate reductase in fungi, 45
Amplification of plasmids in *Escherichia coli*, 247
Anabaena spp. as a source of Type II restriction endonucleases, 240
Anaerobic growth of *Mucor* sp., and hexose requirements, 72
Anaerobiosis, effect of, on intracellular levels of cAMP in *Mucor* spp., 81
   effect of, on morphogenesis of *Mucor racemosus*, 74
Anaerovibrio lipolytica, occurrence of phosphatidylserine in, 205
Analysis of bacterial surface extension, genetic approaches, 156

Antibiotic resistance markers, and plasmids in bacteria, 236
Antibodies against nitrate reductase of *Neurospora crassa*, 56
Antibodies, use of, in studies on growth of bacterial surfaces, 135
Antiserum against fungal nitrate reductase, 19
Anucleate cells, of bacteria, growth of, 108
   of *Escherichia coli*, 124
   production of, by bacterial mutants, 163
Apical dome of fungi, enzyme activation in, 78
Apical growth in *Mucor racemosus*, 71
Arginine, effect of, on activity of nitrate reductase from *Neurospora crassa*, 59
   occurrence of, in bacterial lipo-amino acids, 201
Arthrobacter spp., effect of environment on morphology of, 131
Ascospore formation in *Saccharomyces cerevisiae*, and protein synthesis, 92
Asexual life cycle in *Mucor* spp., 69
Asilomar Conference, and genetic engineering, 262
Asparaginase, effect of ammonia on synthesis of, in fungi, 45
Aspergillus nidulans, assimilation of nitrate by, 1
   genes for regulation of nitrate reduction in, 47
   genetic loci involved in ammonia repression in, 43
   genetic loci involved in nitrate assimilation by, 17
   genetics of nitrate assimilation in, 40
   properties of nitrate reductase from, 10
Aspergillus nuclease, use of, in digesting DNA fragments, 247
Assimilation of nitrate in fungi, 1
   regulation of, 38
Assimilatory nitrite reductase, fungal, characterization of, 21
Assimilatory sulphite reductase, fungal, hydroxylamine reductase activity of, 21
Asymmetric division of *Caulobacter* spp., 150

SUBJECT INDEX

Asymmetric growth of bacterial walls, 144
Asymmetry of phospholipids in bacterial membranes, 181
Asymmetry of phospholipids in membranes, 181
Autolysin-deficient mutants of bacteria, 156
Autolysis, of bacterial walls, and growth of, 136
  of *Escherichia coli*, 146
  role of, in action of penicillin on bacteria, 155
Autolytic role of bacterial phospholipase, 224
Autoregulation of nitrate reductase synthesis in *Neurospora crassa*, 48
Average cell size of bacteria, factors determining, 120
Azide, effect of, on fungal nitrate reductase, 2
Azotobacter aquilis, volume of, 119
Azotobacter vinelandii, CDP-diglyceride synthesis by, 199

B

Bacillaceae, phospholipid composition of members of, 180
Bacillus cereus, length extension of, 119
  phospholipase C produced by, 220
Bacillus licheniformis, association of phosphatidylserine with membrane proteins in, 187
  effect of penicillin on, 156
  growth of peptidoglycan in, 140
  lipid activation of penicillinase in, 185
  peptidoglycan chain length in, 142
  phosphatidylglycerol synthesis in, 200
  phosphoglucomutase-deficient mutants of, 157
Bacillus megaterium, asymmetric distribution of lipids in membrane of, 183
  diaminopimelate pool in, 139
  phospholipase A_1 activity of, 215
  varation in cell size of, 129
Bacillus mycoides, generation times in, 113
Basillus spp., as a source of Type II restriction endonucleases, 240
  CDP-diglyceride synthesis in, 198

Bacillus stearothermophilus, ribosome activity in, 92
Bacillus subtilis, advantages over *Escherichia coli* in genetic engineering, 265
  anucleate cells of, 125
  bulging mutants of, 160
  cell-size mutants of, 161
  effect of thymine starvation on surface growth of, 153
  growth of the plasma membrane in, 183
  minicell-producing mutants of, 162
  nuclear division and cell separation in, 123
  osmotically-fragile protoplast mutant of, 224
  production of large anucleate cells by mutants of, 163
  receptor for teichoic acid in, 140
  relationship between cell length and growth rate of, 130
  rod mutants of, 157
  spiral growth mutant of, 142, 161
Bacteria, age distributions in populations of, 109
  Gram-negative, outer membrane of, 142
  rod-shaped, wall growth of, 138
Bacterial phospholipid metabolism, physiology and biochemistry of, 177
Bacterial phospholipids, biosynthesis of, 194
Bacterial surfaces, extension of, 106
  model for extension of, 141
Bacteriochlorophyll content of *Chromatium* sp., effect of light intensity on, 191
Bacteriophage-associated phospholipids, 187
Bacteriophage growth in *Escherichia coli*, restriction-modification systems and, 237
Bacteriophage lambda vectors, in genetic engineering, 251
Bacteriophages, phospholipase A activity of, 216
Bacteriorhodopsin, lipid activation of, in *Halobacterium halobium*, 185
Beta-Lactam antibiotics, effect of, on bacterial morphology, 154

SUBJECT INDEX

Bifidobacterium bifidum, effect of medium composition on morphology of, 131
  occurrence of acylphosphatidylglycerol in, 202
Biochemistry of bacterial phospholipid metabolism, 177
Biochemistry of dimorphism in *Mucor* spp., 68
Biohazards of genetic engineering, possible nature of, 236
Biological characteristics of bacterial phospholipids, 178
Biological containment, and genetic engineering, 264
Biosynthesis of bacterial phospholipids, 180
Biosynthesis of microbial phospholipids, 194
Blastocladiella emersonii, zoosporangia production and protein synthesis in, 93
Blebs in the outer membrane of Gram-negative bacteria, 146
Block points in bacterial growth, 127
Bordetella pertussis, ornithine-containing lipids in, 208
Branching in bacteria, effect of environment on, 131
Branching of hyphae in *Mucor racemosus*, 71
Breakage and rejoining of DNA molecules as a basis of genetic engineering, 236
Brucella melitensis, ornithine-containing lipids in, 208
Brucella sp., occurrence of phosphatidylcholine in, 208
Budding bacteria, growth of, 150
Bulging mutants of bacteria, 160

C

Cadaverine synthesis by *Mucor* sp., 89
Calcium ions, effect of, on bacterial morphology, 132
Calcium requirement for bacterial phospholipase A_1, 215
Candida lipolytica, CDP-diglyceride synthesis by, 199
Carbamoyl phosphate, effect of, on stimulation of nitrate reductase activity in *Neurospora crassa*, 59

Carbon dioxide, and morphogenesis in *Mucor* sp., 72
  effect of, on dimorphism in *Mucor* spp., 71
Carbon metabolism associated with dimorphism in fungi, 84
Carbon monoxide complex, formation of, with nitrite reductase, 29
Carbon monoxide, effect of, on fungal nitrite reductase, 27
Carbon source, effect of, on dimorphism in *Mucor* spp., 71
  effect of, on morphology of *Arthrobacter* spp., 131
Carboxyl ester hydrolases in bacteria, 210
Cardiolipin, biosynthesis by bacteria, 203
  hydrolysis by bacterial phospholipase D, 222
  hydrolysis of, in bacteria, 213
  phospholipase A_1, presence of in *Acinetobacter* sp., 217
  specificity of bacterial, 217
  see diphosphatidylglycerol
  specific phospholipase D of *Haemophilus parainfluenzae*, role for, 224
  synthetase in bacteria, activity of, 204
Casamino acids, effect on synthesis of nitrate-uptake systems in *Neurospora crassa*, 37
Catabolism of glucose, and dimorphism in fungi, 84
Catabolism of microbial lipids, 210
Catalase synthesis by fungi, effect of nitrite on, 24
Catalytic efficiencies of fungal nitrite and nitrate reductases, 36
Caulobacter crescentus, ribosome activity in, 92
Caulobacter spp., growth in, 150
  septum formation in, 125
Cell cycle, and surface extension, in prokaryotes, 105
  nuclear division in the bacterial, 149
Cell density of bacteria, 126
Cell division, and bacterial nuclear division, 149
  of bacteria, synchronous, agents that cause, 126
Cell elongation, and penicillin-binding proteins in bacteria, 154

SUBJECT INDEX

Cell length of bacteria and surface area, 121
Cell separation and nuclear division in *Bacillus subtilis*, 123
Cell size, at birth of bacteria, 115
  effect of temperature on, for bacteria, 130
  of bacteria, average, factors determining, 120
  mutants affecting, 160
Cell surface, bacterial, topography of, 134
Cell wall, morphogenesis in *Mucor* sp., 90
  structure, and morphogenesis in *Mucor* sp., 75
Cell width, bacterial, mutations affecting, 160
Cellulose utilization by *Mucor* spp., explanation of, 83
Cellular dimensions of bacteria, environmental effects on, 128
Cellular level, dimorphism in *Mucor* spp. at the, 68
Cellular phospholipases, action of, in bacteria, 212
Central formation of septa in bacteria, 106
Centripetal growth of bacterial walls, 138
Cephaloridine, effects of, on peptidoglycan biosynthesis, 154
Cephalothin, effect of peptidoglycan biosynthesis, 154
Ceramide, phosphorylation of, by bacteria, 197
Charon phage λ, role of, in end-to-end joining of DNA molecules, 244
Chelating agents, effect of, on morphogenesis of *Mucor* sp., 75
Chicken DNA, cloning of DNA from, in bacteria, 257
Chitin, in the walls of *Mucor rouxii*, 76
  synthesis of, in *Mucor rouxii*, 77
  synthetase activity in *Mucor rouxii*, 77
Chitosan in walls of *Mucor rouxii*, 76
Chlamydomonas spp. as a source of Type II restriction endonucleases, 240
Chloramphenicol, and development in *Mucor* sp., 72
  effect of, on dimorphism in *Mucor* spp., 71

use of, to amplify plasmids in *Escherichia coli*, 247
Chlorpromazine, effect of, on bacterial growth, 161
Cholesterol, asymmetric distribution of, in mycoplasma membranes, 183
Chromatium sp., effect of light intensity on phospholipid composition of, 191
Chromophore, haem-like, in fungal nitrite reductase, 28
Chromosomal clocks in bacteria, 107
Chromosomal segregation and surface extension in bacteria, 149
Chromosome-initiation mutants of bacteria, 127
Chromosome replication in bacteria, and surface growth, 152
Chromosome separation in bacteria, 148
Cloned DNA, screening methods for, 260
Cloning, of higher eukaryotic DNA in bacteria, 257
  of prokaryotic genes in genetic engineering, 254
  of recombinant DNA in genetic engineering, 254
  of yeast DNA in bacteria, 256
Clostridium butyricum, occurrence of phosphatidyl-N-methylethanolamine in, 208
  phosphatidic acid biosynthesis in, 197
Clostridium perfringens, α toxin of, 220
  inhibition of peptidoglycan biosynthesis in, 154
  occurrence of lipo-amino acids in, 200
Clostridium welchii, lysylphosphatidylglycerol biosynthesis in, 201
Cocci, bacterial, topography of surface of, 135
Cohesive ends of DNA molecules, in genetic engineering, 244
Cold lability of enzymes, and allotropy, 184
Collins and Richmond equations, and growth law for bacteria, 119
  for growth of bacteria, 116
Competitive inhibition of nitrite reductase activity by hydroxylamine, 33
Complementation of nitrate reductase activity of *Aspergillus nidulans*, 16
Composition of medium and growth rate of bacteria, 128

Concentrations of nitrate and nitrite reductases in fungal mycelia, 35
Conditional yeast mutants of *Mucor* spp., 89
Conjugative plasmids, use of, in genetic engineering, 248
Conservation of phospholipids in bacterial growth, 145
Control development of gene expression in *Mucor* sp., 94
Control of length extension in bacteria, 124
Corynebacterium humiferum, ability of restriction endonucleases to recognize DNA sequences in, 241
Coulter counter, use of, in determining volume distributions in bacterial populations, 119
Culture age, effect of, on bacterial phospholipid composition, 190
Cyanide, effect of, on fungal nitrate reductase, 2
  effect of, on nitrite uptake by fungi, 38
Cyanogen bromide-activated dextran, use of, in study of membrane asymmetry, 182
Cyclic AMP-dependent protein kinase of *Mucor* spp., 82
Cyclic AMP, effect of, on inducibility of maltose permease in bacteria, 143
  phosphodiesterase activity of *Mucor* spp., 82
  role of, in relief of catabolite repression in microbes, 79
Cyclic GMP, role of, in *Mucor* spp., 80
Cyclic nucleotides, role for, in fungal dimorphism, 79
Cycloheximide, as an inhibitor of nitrate assimilation in fungi, 49
  effect of, on pyruvate kinase synthesis by *Mucor* sp., 85
  effect on action of ammonia on fungal nitrate reductase, 54
Cyclopropane fatty-acid residues in bacterial ornithine-containing lipids, 208
Cytidine diphosphate diglyceride biosynthesis by bacteria, 198
Cytidine triphosphate, effect of, on lipid metabolism of *Escherichia coli*, 190

Cytidine triphosphate: phosphatidic acid cytidyl transferase, activity in bacteria, 198
  properties of bacterial, 198
Cytochrome b-557 in fungal nitrate reductase, 3
Cytochrome c, reduction of, with nitrate reductase, 3
  site of action of, in nitrite reductase of *Neurospora crassa*, 35
Cytochrome initiation in bacteria, 115
Cytochromes, association of, with nitrate reductase from *Aspergillus nidulans*, 11
  of bacteria, transmission of, to progeny, 145

D

Decadeoxyribonucleotides, use of, in insertion of DNA sequences, 245
Decarboxylation of phosphatidylserine in bacteria, 204
Defective nitrate reductase mutants of *Aspergillus nidulans*, 13
Density of bacterial cells, 126
Deoxyribonucleic acid molecules end-to-end joining of, 243
  use of restriction endonucleases in analysis of, 236
Deoxyribonucleic acid polymerase I, role of, in end-to-end joining of DNA molecules, 244
Deoxyribonucleic acid synthesis, and length extension in bacteria, 21
  inhibition of, and surface extension of bacteria, 151
Deoxyribonucleic yeast, cloning of, in bacteria, 256
Deoxynucleotidyl transferase, role of, in end-to-end joining of DNA molecules, 243
Derepressible nature of nitrate reductase activity, 49
Desulfovibrio gigas, CDP-diglyceride synthesis by, 199
Detergent requirement for *in vitro* synthesis of CDP-diglyceride, 198
Detergent sensitivity of bacterial phospholipases, 214

SUBJECT INDEX

Development control of gene expression in *Mucor* sp., 94
Development studies in micro-organisms, 68
Diacyllysocardiolipin, occurrence of, in *Acinetobacter* sp., 217
Diaminopimelic acid-requiring auxotrophs of *Escherichia coli*, 138
Diaphorase activity, and fungal nitrate reduction, 5
  association of, with nitrite reductase, 33
Diaphorase, nitrite reductase as a, 32
Dibutyryl-cAMP in dimorphism in *Mucor* sp., 79
Dibutyryl-3'5'-cyclic adenosine monophosphate, effect of, on dimorphism in *Mucor* spp., 72
Dictyostelium discoideum, UDP-glucose synthesis in, 94
Differentiation in micro-organisms, interest in, 68
Diglyceride, phosphorylation of, by bacteria, 197
Dimensions, cellular, bacterial, environmental effects on, 128
Dimorphic cycle in *Mucor racemosus*, 69
Dimorphic transitions in *Mucor racemosus*, 71
Dimorphism, carbon and energy metabolism associated with, in fungi, 84
  fungal, role for cyclic nucleotides in, 79
  in *Mucor* spp., biochemistry of, 67
2,4-dinitrophenol, effect of, on nitrite uptake by fungi, 38
Diphosphatidylglycerol, biosynthesis in bacteria, 203
  structure of, 179
Diplococcus pneumoniae, topography of surface of, 135
Dispersion of generation times in bacteria, 112
Distribution of ages in bacterial populations, 109
Distribution of sizes in exponentially growing bacteria, 117
Dithionite reductase, induction of, by nitrite in *Neurospora crassa*, 32
Dithionite reduction by nitrite reductase, 32

Dithionite, use of, to protect fungal nitrite reductase, 25
Division, cellular, synchronous, agents that cause, 126
  of bacteria, time-course of, 110
Dominance of ammonia repression over nitrate reduction in fungi, 45
Dormancy of zygospores in *Mucor* sp., 99
Double-strand cleavage of DNA, and restriction enzymes, 238
Drosophila melanogaster-Escherichia coli DNA recombinant plasmids, 250
Drug-resistance plasmids, and association with phospholipid, 188
Dye reduction by nitrite reductase, 32

E

Egg phosphatidylcholine as an activator of isoprenoid alcohol phosphokinase in bacteria, 185
Electron donors to fungal nitrate reductase, 2
Electron paramagnetic resonance spectra, low temperature, of nitrate reductase, 10
Electron-transfer reactions catalysed by nitrite reductase, 31
Electon-transfer reactions of nitrate reductase from *Neurospora crassa*, 7
Electron-transport chain in *Haemophilus parainfluenzae*, 192
Embden-Meyerhof pathway, and dimorphism in fungi, 84
Endonucleases, restriction, use of, in genetic engineering, 237
End-to-end joining of DNA molecules, 243
Energy metabolism associated with dimorphism in fungi, 84
Energy requirement in transport of nitrite and nitrate by fungi, 37
Engineering, genetic, in micro-organisms, 235
Enteric bacteria, formation of filaments by, 152
Enterobacteriaceae, phospholipid composition of members of, 180
Environmental conditions, effect of, on dimorphism in *Mucor* spp., 71

Environmental effects on bacterial cellular dimensions, 128
Environmental factors, effect of, on dimorphism in *Mucor* spp., 68
Enzyme activity in bacteria, and phospholipids, 183
Enzymology of nitrate assimilation by fungi, 2
Equatorial location of septa in bacteria, 134
Equatorial zone growth in bacteria, 139
Escherichia coli, biosynthesis of *bis*-phosphatidic acid by, 202
  cardiolipin biosynthesis in, 204
  CDP-diglyceride synthesis in, 198
  cell density of, 126
  cellular phospholipase C activity of, 221
  chemical nature of lipopolysaccharide of, 184
  Drosophila melanogaster DNA recombinant plasmids, 250
  effect of phenethanol on phospholipid composition of, 191
  fatty-acyl residue turnover in, 224
  generations, times in, 113
  growth of, in relation to content of R factors in, 130
  growth rate and size of, 123
  inhibition of peptidoglycan biosynthesis in, 154
  length distribution in, 112
  lipid activation, of murein lipoprotein in 185
  of pyruvate oxidase in, 185
  lysophospholipase activity in membrane of, 220
  minicell-producing mutants of, 162
  mutant of defective, in phosphatidylserine decarboxylase, 207
  mutants of defective, in phosphatidylserine synthetase, 206
  outer-membrane phospholipase A_2 activity in, 219
  phosphatidylglycerol synthesis in, 200
  phosphatidylserine decarboxylase of, 207
  phosphatidylserine synthetase in, 205
  phospholipase A_2 activity in, 219
  phospholipase A_1 of, 213
  phospholipase D activity of, 221
  phospholipid metabolism in, 188
  phosphorylation of diglyceride by, 197
  polyglycerophosphatide cycle in, 223
  production of large anucleate cells by, 163
  properties of glycero-3-phosphate acyltransferase from, 196
  rod mutants of, 157
  sirohaem from, 30
  sphaeroplasts, role of, in end-to-end joining of DNA molecules, 244
  surface extension of, 143
  turnover of phosphatidylethanolamine in, 222
  unsaturated fatty-acid auxotroph of, 191
  use of plasmids of, in genetic engineering, 247
  volume of, 119
Escherichia spp. as a source of Type II restriction endonucleases, 240
Ethylemine diamine tetraacetate, effect of, on permeability in fungal mycelia, 52
Ethylene diamine tetraacetic acid, effect of, on morphogenesis of *Mucor* sp., 75
Euglena gracilis, cloning of DNA from, in bacteria, 257
Eukaryotic DNA sequences, cloning of, 246
Eukaryotic, higher, DNA, cloning of, in bacteria, 257
Excretion of cAMP in *Mucor* spp., 81
Exponential-phase rod-shaped bacteria, size of, 117
Extension of bacterial surfaces, model for, 141
Extension of the surface and the cell cycle in prokaryotes, 105
Extracellular phospholipase activity in bacterial cultures, 220

F

Fatty-acid exchange reactions in mammalian tissues, 223
Fatty-acid synthetase complex of *Saccharomyces cerevisiae*, 197
Fatty acids, incorporation of, by bacteria, 189

SUBJECT INDEX

Fatty-acyl donors in bacterial synthesis of phosphatidic acid, 194
Fatty-acyl phosphatidylglycerol biosynthesis in bacteria, 201
Fatty-acyl residue turnover in *Escherichia coli*, 224
Feedback inhibition of nitrate reductase activity in fungi, 59
Fermentable hexoses, and morphogenesis in *Mucor* sp., 72
  effect of, on dimorphism in *Mucor* spp., 72
Fermentation rates, relationship of, to dimorphism in fungi, 84
Ferredoxin, lack of, in fungi, 22
Ferric ions, need for, in fungal nitrite reductase, 22
Ferrobacillus sp., occurrence of phosphatidylcholine in, 208
Filament formation, by *Caulobacter* sp., 150
  effect of temperature on, with bacteria, 130
Filament production by anucleate cell mutants of bacteria, 164
Filamentous mutants of *Escherichia coli*, 125
Filaments, formation of, by enteric bacteria, 152
Fixed size of bacteria and chromosomal origin, 120
Flagella, antigens in *Salmonella* spp., 255
  segregation of, in bacteria, 146
Flavin adenine dinucleotide, protection of fungal nitrite reductase by, 25
Fluorescamine, use of, in study of membrane asymmetry, 181
Folded chromosome structure in bacteria, 150
Frequency distributions of generation times in bacteria, 110
Fucose in walls of *Mucor rouxii*, 76
Fungi, assimilation of nitrate in, 1

G

Galactosyl transferase, lipid activation of, in *Salmonella typhimurium*, 185
Gallus domesticus, cloning of DNA from, in bacteria, 257

Gene cloning, journal devoted to studies on, 236
Gene expression in nitrate assimilation by fungi, 40
*are*A Genes, role of, in ammonia repression in fungi, 44
*are*A Gene in *Aspergillus nidulans*, 17
cnx Genes in *Aspergillus nidulans*, 17
*gdh*A Gene in *Aspergillus nidulans*, 17
*gdh*A Genes, and ammonia repression in fungi, 45
*nia*A Gene in *Aspergillus nidulans*, 17
nit Genes in *Neurospora crassa*, 17
*tam*A Gene in *Aspergillus nidulans*, 17
*tam*A Gene products, and ammonia repression in fungi, 45
Genes, prokaryotic, cloning of, in genetic engineering, 254
Generation times, in bacteria, dispersion of, 112
  of bacteria, 109
    relation of, to initiation of chromosome synthesis, 115
  of mothers and daughters in bacteria, 114
Genetic approaches to analysis of bacterial surface extension, 156
Genetic engineering in micro-organisms, 235
Genetic loci involved in nitrate assimilation in fungi, 17
Genetics, of *Mucor* sp., 99
  of nitrate assimilation in fungi, 1
Genome segregation in bacteria, 147
Germ-tube formation following germination of sporangiophores in *Mucor* spp., 70
Globin, rabbit, biosynthesis of, 91
Gluconobacter cerinus, ornithine-containing lipids in, 208
Gluconobacter sp., occurrence of phosphatidylcholine in, 208
Glucose repression in *Mucor bacilliformis*, 84
Glucosidases of *Mucor racemosus*, effect of cAMP on, 83
Glucosyl transferases in bacteria, lipid activation of, 185
Glucosylation of teichoic acids, inability of mutant bacteria to, 157

SUBJECT INDEX

Glutamate, ability of fungi to grow on, 43
  effect of, on fungal dimorphism, 87
  on nitrate reduction in *Neurospora crassa*, 46
Glutamate dehydrogenase activity, in *Neurospora crassa*, and ammonia repression, 46
  of *Mucor* spp., 84, 86
Glutamate dehydrogenase, repression of, by glucose, 88
Glutamine, effect of, on nitrate reductase activity of *Neurospora crassa*, 59
  synthase, and ammonia assimilation by *Mucor racemosus*, 88
Glycan chains, shape of, in spiral-growth bacterial mutants, 161
Glycerol auxotrophs of bacteria, use of, 189
Glycero-3-phosphate, acyltransferase activity of bacteria, 194
  auxotrophs of *Escherichia coli*, 196
sn-Glycerophosphate, sequential acylation of, in bacteria, 194
Glycine, effect of, on activity of nitrate reductase in *Neurospora crassa*, 59
Gram-negative bacteria, outer membrane of, 142
  phospholipid composition of, 180
Gram-positive bacteria, phospholipid composition of, 180
Growth law, and bacterial growth, 117
Growth model, symmetric, for bacteria, 147
Growth rate, and medium composition of bacteria, 128
  of bacteria, and cell length, 121
Growth relative to internal markers in bacteria, 146
Guanosine-3′,5′-bis(diphosphate), role of, in bacterial lipid metabolism, 190
Guidelines for research Involving recombinant DNA, 262

H

Haem, as a prosthetic group in nitrite reductase of *Neurospora crassa*, 26
  component of fungal nitrate reductase, 3
Haemophili as a source of Type II restriction endonucleases, 239

Haemophilis aegyptius, as a source of Type II restriction endonuclease, 239
Haemophilus parainfluenzae, CDP-diglyceride synthesis by, 199
  phospholipase D activity of, 221
  phospholipid metabolism in, 192
  turnover of phosphatidylethanolamine in, 222
Haemophilus spp., existence of isoschizomers in, 241
Halobacterium halobium, lipid activation of bacteriorhodopsin in, 185
Heat, effect of, on fungal nitrate reductase, 5
  shock, use of, to induce synchrony in bacteria, 127
Heavy metals, effect of, on morphogenesis of *Mucor* sp., 75
Hexadecane, growth of *Acinetobacter* sp. on, 193
Hexoses, effect of, on dimorphism in *Mucor* spp., 71
Higher eukaryotic DNA, cloning of, in bacteria, 257
Histidine, effect of, on activity of nitrate reductase in *Neurospora crassa*, 59
Histoplasma sp., multiple forms of RNA polymerase in, 91
Hydrogen peroxide, inactivation of fungal nitrite reductase by, 23
Hydroxylamine, as a competitive inhibitor of nitrite reductase activity, 33
  reductase activity in fungi, regulation of, 50
  reductase activity of fungal nitrite reductase, 21
  reductase, induction of, by nitrite in *Neurospora crassa*, 32
  reductase isoenzymes in *Neurospora crassa*, 32
  reduction catalysed by nitrite reductase, 31
Hydroxylase, lipid activation of, in *Pseudomonas putida*, 185
Hyphal cells, production of, following germination of sporangiophores in *Mucor* spp., 70
Hyphal growth in *Mucor racemosus*, 71
Hyphal tip growth in fungi, 93
Hyphal-yeast dimorphism in *Mucor* spp., 68

SUBJECT INDEX

Hyphomycrobium sp., growth of, 150
  occurrence of phosphatidylcholine in, 208
Hypoxanthine, synthesis of nitrate reductase by fungi grown on, 18

I

Idealized age distributions in bacterial populations, 109
Immunochemical techniques, use of, to study membrane asymmetry, 181
Immunofluorescence studies, use of, in bacterial wall growth, 138
Immunogenicity of phospholipids, 182
Immunological identification of pyruvate kinases in *Mucor* sp., 85
Inactivation of fungal nitrate reductase, by ammonia, 53
  by proteases, 58
Inactivity of tungstan-containing fungal nitrate reductase, 55
Inducible subunits of nitrate reductase of *Neurospora crassa*, 14
Induction of nitrite reductase activity in mutants of *Aspergillus nidulans*, 48
Induction of nitrate reductase, relation to induction of cytochrome c reductase in fungi, 42
Inhibition of nitrate reductase, from *Aspergillus nidulans*, 11
  from *Neurospora crassa*, 8
Inhibition of protein synthesis, and bacterial growth, 151
Initiation mutants of bacteria, 165
Initiation of cytochrome synthesis in bacteria, 115
Inner membrane location of amino-acyl esters of phosphatidylglycerol in *Acholeplasma* sp., 182
Insertion of specific DNA sequences in genetic engineering, 244
Insulin, rat, insertion of genes for, into bacteria, 259
Intercalation of new material into bacterial wall peptidoglycans, 140
Internal markers relative to growth in bacteria, 146
Interpretations of size distributions of bacteria, 118
Intracellular concentrations of cAMP in *Mucor racemosus* 80

Intracellular inclusion, membrane-bound, in *Acinetobacter* sp., presence of, 194
Intracytoplasmic membranes in bacteria and occurrence of phosphatidylcholine, 208
Iron, haem, in nitrate reductase from *Neurospora crassa*, 9
Isoprenoid phosphokinase, lipid dependence of, in bacteria, 184
Isoschizomers, definition of the term, 241

J

Joining, end-to-end, of DNA molecules, 243
Jumping of chromosomes in bacteria, 149

K

Kanamycin-resistance plasmids, use of, in genetic engineering, 248
Kinase, diglyceride, activity of, in bacteria, 197
Klebsiella sp., rod mutants of, 157

L

Labelling, chemical, of membrane phospholipids, 181
Lability, of fungal nitrite reductase, 21
  of mRNA for nitrate reductase in fungi, effect of ammonia on, 53
Beta Lactamase, cloning staphylococcal of, in *Escherichia coli*, 254
Lactobacillus acidophilus, occurrence of lipo-amino acids in, 201
Lactoperoxidase-catalysed iodination, use of, to study membrane asymmetry, 182
Lambda bacteriophage vectors in genetic engineering, 251
Lampropedia hyalina, synchronous division of, 114
Length, average, of bacteria, 106
  extension, and growth rate in bacteria, 123

SUBJECT INDEX

Length, extension—*continued*
  in rod-shaped bacteria, 106
  of bacteria, 117
  processes in bacteria, 107
  of bacteria, rate of increase of, 119
Life cycle of *Mucor rouxii*, chemical differentiation in wall of, during, 76
Life lengths of sisters of bacteria, 114
Ligation of DNA fragments, in genetic engineering, 244
Light intensity, effect of, on phospholipid content of *Chromatium* sp., 191
Lipid dependency of bacterial enzyme activity, 184
Lipid synthesis in bacterial membranes, and growth rate, 133
Lipids, bacterial, role of, in cellular physiology, 178
  microbial, catabolism of, 210
Lipolytic enzymes in bacteria, 210
Lipopolysaccharide, of *Escherichia coli*, 184
  synthesis in Gram-negative bacteria, 143
Listeria monocytogenes, occurrence of acylphosphatidylglycerol in, 202
Loci, genetic, involved in nitrate assimilation in fungi, 17
Low-temperature electron paramagnetic resonance spectra of nitrate reductase, 9
Lysine, decarboxylation of, by *Mucor* sp., 90
  occurrence of, in bacterial lipo-amino acids, 201
Lyso-*bis*-phosphatidic acid, occurrence of, in baby hamster kidney cells, 202
Lysogens, in genetic engineering, 251
Lysophosphatidic acid acyltransferase activity in bacteria, 194
Lysophosphatidic acid, biosynthesis of, in bacteria, 194
Lysophospholipase activity, in *Escherichia coli*, 213
  of bacteria, 219
Lysophospholipids, bacterial, hydrolysis of, 219
  in bacteria, possible significance of, 225
Lysylphosphatidylglycerol, synthesis of, 201

M

Macromolecular synthesis in bacteria, exponential rate of, 133
Magnesium-activated ATPase activity in bacteria, lipid activation of, 187
Maize-root protease, properties of, 58
Malate-vitamin K reductase, lipid activation of, in *Mycobacterium phlei*, 185
Maltose permease, identity of, with phage receptor in bacteria, 143
Maltose utilization by *Mucor* spp., explanation of, 83
Mannose contents of walls of *Mucor rouxii*, 76
Maturation of nitrate reductase in fungi, effect of nitrate on, 55
Maximal activities of fungal nitrite and nitrate reductases, 36
Mechanism of nitrate reductase from *Neurospora crassa*, 9
Mecillinam, effect of, on peptidoglycan biosynthesis, 154
Medium composition and growth rate of bacteria, 128
Megasphaera elsdenii, occurrence of phosphatidylserine in, 205
Melanin in walls of *Mucor rouxii*, 76
Membrane-associated proteins in bacterial membranes, and phospholipid biosynthesis, 181
Membrane-bound enzymes in bacteria, 184
Membrane, plasma, growth of, in bacteria, 144
  protein synthesis, and thymine starvation in bacteria, 154
Membranes, asymmetry of phospholipids in, 181
Messenger RNA, for nitrate reductase in fungi, 53
  synthesis, effect of nitrate on, 52
Metabolic equivalence of mRNAs in micro-organisms, 95
Metabolic responses to morphogenic signals in fungi, 79
Metabolism of phospholipids by bacteria, 177
Metal analysis of fungal nitrate reductase, 9

SUBJECT INDEX

Metal-binding agents, nature of inhibitory effect of, on nitrate reductase, 6
Metalloflavoprotein nature of fungal nitrate reductase, 2
Metals, heavy, effect of, on morphogenesis of *Mucor* sp., 75
  in nitrate reductase from *Aspergillus nidulans*, 11
Methanol stimulation of phospholipase activity in bacteria, 214
Methionine, requirement of, by *coy* mutants of *Mucor* sp., 89
Methodology, importance of, in studies on bacterial phospholipid metabolism, 178
Methylase, modification, as a restriction enzyme in bacteria, 238
Methylated forms of phosphatidylethanolamine in bacteria, 207
Methylation of phosphatidylethanolamine by bacteria, 207
6-Methylpurine, effect of induction of the nitrite-uptake system in fungi, 38
Michaelis constants for nitrate reductase from *Aspergillus nidulans*, 11
Microbial lipids, catabolism of, 210
Microbial phospholipids, structural, functional and biological characteristics of, 178
Microbial viruses, phospholipids and, 187
Micrococci, growth of walls of, 136
Micrococcus flavus, wall growth of, 138
Micrococcus lysodeikticus, asymmetric distribution of lipids in membrane of, 183
  biosynthesis of cardiolipin in, 203
  CDP-diglyceride synthesis in, 198
  effect of culture age on phospholipid composition of, 190
  lipid activation of ATPase of, 186
Micrococcus spp., synchronous division of, 114
Micro-organisms, genetic engineering in, 235
Minicell-producing bacterial mutants, 162
Minicell-producing strains of *Escherichia coli*, membrane growth in, 146

Mitomycin C, effect of, on surface extension of bacteria, 152
Model for gene action in regulation of synthesis of nitrate and nitrite reductases in *Aspergillus nidulans*, 42
Model for nitrate reductase in fungi, 34
Modification of mRNAs in micro-organisms, 92
Modification methylase, as a restriction endonuclease, 238
Molecular weight of fungal nitrite reductase, 22
Molecular weight of nitrate reductase from *Aspergillus nidulans*, 11
Molydbate, use of, in subunit aggregation of nitrate reductase, 14
Molybdenum cofactor in fungal nitrate reductase, nature of, 19
Molybdenum-containing component of fungal nitrate reductase, nature of, 13
Molybdenum, effect on synthesis of nitrate reductase in fungi, 55
  in fungal nitrate reductases, 3
  use of, in aggregation of subunits of nitrate reductase, 14
Moraxella lwoffi, CDP-diglyceride synthesis by, 199
Moraxella spp. as a source of restriction endonucleases, 240
Morphogenesis, control of, and protein synthesis in fungi, 90
  in micro-organisms, studies on, 68
Morphogenic signals, metabolic responses to, in fungi, 79
Morphological changes in bacterial growth, 106
Morphological effects produced by nutritional factors in bacteria, 131
Morphologies of *rod* mutants of bacteria, 157
Morphology of bacteria, and inhibition of wall synthesis, 156
Morphology of *Mucor racemosus*, effect of cGMP on, 80
Morphology mutants, of bacteria, 108
Mucor sp., 89
Morphopoeitic agents, and *Mucor* development, 72
  effect of, on *Mucor* spp., 71

SUBJECT INDEX

Mosaicism in bacterial membranes, evidence for, 192
Mother-daughter correlations in bacterial growth, 113
Mucor bacilliformis, respiratory-deficient mutant of, 84
Mucor genevensis, dimorphism in, 71
  effect of chloramphenicol on, 72
  respiratory capacity in, 84
Mucor racemosus, dimorphic cycle in, 69
  effect of anaerobiosis on morphogenesis of, 73
  effect of cyclic AMP on morphogenesis of, 73
  electrophoresis of ribosomal proteins of, 98
  glutamate dehydrogenase activity of, 83
  glutamate dehydrogenases in, 87
  hyphal growth of, 71
  intracellular concentration of cAMP in, 80
  peptide chain elongation in, 94
  pyruvate kinases of, 83
  structure of ribosomes of, 96
  translation in, 92
Mucor rouxii, adenyl cyclase of, 81
  cell-wall composition of, in relation to development, 76
  dimorphism in, 71
  morphogenesis of, 73
  multiple forms of RNA polymerase in, 91
  pyruvate kinase isoenzymes in, 85
  response of, to cAMP, 80
Mucor sp., genetics of, 99
  glutamate dehydrogenase activity of, 86
Mucor spp., biochemistry of dimorphism in, 67
  role of cyclic AMP in dimorphism of, 79
Mucoran in walls of *Mucor rouxii*, 76
Multimeric nature of fungal nitrate reductases, 16
Murein lipoprotein in *Escherichia coli*, lipid activation of, 185
Mutants, affecting initiation in bacteria, 165
  bacterial, affecting cell size, 160
  that produce large anucleate cells, 163

morphological, of *Mucor* spp., 89
  properties of nitrate reductases from, 12
nit Mutants of *Neurospora crassa*, 12
Mutations, bacterial, affecting cell width, 160
Mycelia, fungal, concentrations of nitrite and nitrate reductases in, 35
Mycelium-yeast dimorphism in *Mucor* spp., 68
Mycobacteria, CDP-diglyceride synthesis by, 199
Mycobacterium bovis, ornithine-containing lipids in, 208
Mycobacterium phlei, lipid activation of malate-vitamin K reductase in, 185
  phospholipase A_1 activity of, 216
Mycoplasma capricolum, asymmetric distribution of cholesterol in membrane of, 183
Mycoplasma laidlawii, membrane-bound lysophospholipase activity in, 220
Mycoplasmas, occurrence of acyl-phosphatidylglycerol in, 202
Myxococcus spp. as a source of Type II restriction endonucleases, 240
Myxococcus xanthus, cell length of, 120

N

Nalidixic acid, effect of, on surface extension of bacteria, 152
Neisseria gonnohoeare, effect of benzylpenicillin on morphology of, 156
  phospholipase A activity in, 216
Neurospora crassa, absorption spectra of nitrite reductase from, 29
  assimilation of nitrate by, 1
  concentration of nitrite and nitrate reductases in, 36
  electron-transfer reactions of nitrate reductase from, 7
  genetic loci involved in nitrate assimilation by, 17
  model for action of nitrite reductase in, 35
  nitrate reductase regulator genes in, 47
  properties of nitrate reductase from, 2
  properties of nitrate reductase from mutants of, 12
  spectra of nitrite reductase of, 27

SUBJECT INDEX

subunits of nitrate reductase of, 14
turnover of nitrate reductase in, 56
uptake of nitrite and nitrate by, 37
Nicotinate, synthesis of nitrate reductase by fungi grown on, 18
nit genes, function of, 47
Nitrate, as an inducer of transcription of mRNA for nitrate reductase, 55
  assimilation in fungi, 1
    regulation of, 38
  effect of, on ornithine utilization in fungi, 43
    on synthesis of protease in fungi, 58
  reductase, activity in fungi, effect of NADPH on, 60
    from *Aspergillus nidulans*, properties of, 10
    from *Neurospora crassa*, electron-transfer reactions of, 7
    mechanism of action of, 9
    in *Aspergillus nidulans*, regulatory role of, in its own synthesis, 41
    in fungi, regulation of activity of, 58
    instability of, 50
    mRNA, effect of ammonia ion, in fungi, 53
    of fungi, properties of, 2
    of *Neurospora crassa*, subunits of, 14
  uptake, by *Neurospora crassa*, 37
    effect of, on catalase synthesis by fungi, 24
    on nitrate assimilation by fungi, 24
    distinction of, from nitrite uptake in *Neurospora crassa*, 37
Nitrite, competition by, for fungal nitrite reductase, and carbon monoxide, 29
  interaction of, with sirohaem in action of nitrite reductase, 35
  reductase, activity, inhibition of, by carbon monoxide, 29
    chromophore, relation of, to sirohaem, 30
    electron-transfer reactions catalysed by, 31
    fungal, interaction with carbon monoxide, 28
    lability of, 21
    properties of, 20, 21
    sensitivity of, to oxidation, 24
    sirohaem as a prosthetic group for, 26

subunit structure of, 25
in fungi, model for, 34
of *Neurospora crassa*, spectra of, 27
uptake, by *Neurospora crassa*, 37
mutants of *Neurospora crassa*, 38
Nutrition for nitrogenous compounds, and assimilation of nitrate, 49
Nutritional factors, morphological effects produced by, in bacteria, 131
Nitrogen flow rate, effect of, on morphogenesis of *Mucor racemosus*, 74
Nitrogen metabolism, alteration in and dimorphism in fungi, 87
Nitrogen source, effect of, on formation of molybdenum-containing cofactor of fungal nitrate reductase, 18
Nitrogenase, molybdenum cofactor of, 19
Nitrosocystis sp., occurrence of phosphatidylcholine in, 208
Nitroxyl reductase of fungi, possible role of, 20
Nomenclature of bacterial phospholipase, 211
Nuclear material, and surface extension of bacteria, 148
Nuclei, attachment of, to surface of bacteria, 148
  average number of, in bacteria, 111
Nucleotides, cyclic, role for, in fungal dimorphism, 79

O

Origin of size distributions in bacterial populations, 118
Ornithine, ability of fungi to grow on, 43
  containing lipids, bacterial, structures of, 209
  in bacteria, 208
  occurrence of, in bacterial lipo-amino acids, 201
Oscillations in cell density of bacteria, 126
Osmotic remedial mutants of bacteria, 157
Osmotic shock, use of, to induce synchrony in bacterial growth, 127

Osmotically-fragile protoplast mutant of *Bacillus subtilis*, 224
Outer membrane, of *Desulfovibrio gigas*, ornithine-containing lipids in, 210
of Gram-negative bacteria, 142
phospholipase A_2 activity in *Escherichia coli*, 219
Oxidation, sensitivity of fungal nitrite reductase to, 24
Oxidized *versus* reduced difference spectra of nitrate reductase from *Neurospora crassa*, 4
Oxygen, and morphogenesis in *Mucor* sp., 72

P

Paracolonobacter sp., CDP-diglyceride synthesis by, 199
Pasteur, studies on *Mucor* spp. by, 68
Penicillin-binding proteins in bacteria, 154
Penicillin, effect of, on length extension in bacteria, 154
induced filament formation in bacteria, 127
Penicillinase, lipid activation of, in *Bacillus licheniformis*, 185
lipid activation of, in bacteria, 187
Pentose metabolism in *Aspergillus nidulans*, 60
Pentose phosphate cycle, effect of nitrate on synthesis of enzymes on, in *Aspergillus nidulans*, 60
Peptide bridges in bacterial wall polymers, 142
Peptidoglycan biosynthesis, and length extension in bacteria, 154
Peptidoglycan growth in bacteria, 124
Peptidoglycan layer in bacteria, growth of, 138
Peptidoglycan synthesis, and thymine starvation in bacteria, 153
in bacterial *rod* mutants, 158
Peripheral wall growth in bacteria, 107, 139
Peripheral wall synthesis and septation in bacteria, 160
Permease activity in bacteria, and growth rate, 134

Peroxide, inactivation of fungal nitrite reductase by, 23
Phage attachment in *Bacillus subtilis*, 140
Phenethanol, effect of, on phospholipid metabolism by bacteria, 191
Phenethyl alcohol, and morphogenesis in *Mucor* sp., 72
effect of, on dimorphism in *Mucor* spp., 71
Phosphate, in walls of *Mucor rouxii*, 76
limitation, effect of, on *Bacillus subtilis*, 140
-limited *Pseudomonas fluorescens*, ornithine-containing lipid in, 210
Phosphatidic acid biosynthesis in bacteria, 194
bis-Phosphatidic acid, structure of, 179
bis-Phosphatidic acids, biosynthesis of, in bacteria, 201
Phosphatidylcholine biosynthesis in bacteria, 207
Phosphatidylcholine choline-phosphohydrolase activity produced by bacteria, 220
Phosphatidylcholine, structure of, 179
Phosphatidylethanolamine, activation of glucosyl transferases in bacteria, 186
as an acyl donor in bacterial lipid metabolism, 203
biosynthesis in bacteria, 204, 207
hydrolysis by bacteria, 213
structure of, 179
turnover in *Haemophilus parainfluenzae*, 193
turnover of, in bacteria, 190
Phosphatidylglycerol, activation of ATPase in *Acholeplasma laidlawii*, 186
as a substrate in bacterial cardiolipin biosynthesis, 203
as a substrate in biosynthesis of acylphosphatidylglycerol by bacteria, 203
association of, with bacteriophage, 187
hydrolysis by bacteria, 214
phosphatase, role of, in bacteria, 200
phosphate, occurrence of, in bacteria, 199
synthetase activity in bacilli, 200
reaction of, with sulphanilic diazonium chloride, 181

SUBJECT INDEX

specific phospholipase activity in bacteria, 215
structure of, 179
synthesis by bacteria, 199
Phosphatidylglycerophosphate, synthesis of, in bacteria, 199
Phosphatidyl-N-methylethanolamine biosynthesis in bacteria, 207
Phosphatidylserine, association of, with membrane proteins in *Bacillus licheniformis*, 187
biosynthesis in bacteria, 204
decarboxylase activity in bacteria, 207
structure of, 179
synthetase activity in bacteria, 204
Phosphodiesterase activity in *Mucor* spp., 81
Phosphoglucomutase-deficient mutants of *Bacillus licheniformis*, 157
Phospholipase A_2 activity in bacteria, 213
of *Bacillus subtilis*, intracellular activation of, 224
Phospholipase A_2 activity in bacteria, 219
Phospholipase C activity in bacteria, 212, 220
Phospholipase D, activity in bacteria, 212, 221
Phospholipase digestion of membranes, 181
Phospholipases A, presence of, in bacteria, 212
Phospholipases, bacterial, nomenclature of, 211
bacterial, possible physiological roles of, 222
cellular, action of, in bacteria, 212
in bacteria, 210
Phospholipid biosynthesis in bacteria, pathways for, 195
Phospholipid composition, of bacteria, regulation of, 188
of *Thermus aquaticus*, 189
Phospholipid exchange proteins, use of, in study of membrane asymmetry, 181
Phospholipid metabolism, bacterial, physiology and biochemistry of, 177
Phospholipid synthesis and membrane growth in bacteria, 144

Phospholipid turnover in bacteria, 222
Phospholipids, and microbial viruses, 187
as antigens, 182
asymmetric distribution of, in bacterial membranes, 181
in enzyme activity, 183
inhibitory effects of cardiolipin biosynthesis in bacteria, 204
Phospholipoprotein nature of penicillinase, 187
Phosphoric diester hydrolases in bacteria, 210
Phosphorylation, of C_{55} isoprenoid alcohols by bacteria, 184
of ribosomal proteins in *Mucor racemosus*, 97
Photosynthetic organisms, difference in nitrite reductase from that in fungi, 22
Phycomyces blakesleeanus, zygospore dormancy in, 99
Physical containment, and genetic engineering hazards, 263
Physiological studies on bacterial growth, 151
Physiology of bacterial phospholipid metabolism, 177
Plasma membrane, growth in bacteria, 144
synthesis in bacteria, growth of, 133
Plasmid vectors in genetic engineering, 247
Plasmids, amplification of, in *Escherichia coli*, 247
effect of, on growth of bacteria, 130
in *Escherichia coli*, and production of phosphatidylserine synthetase, 206
Plasticity of spore wall in *Mucor* spp., 70
Pleiotropic nature of thermolabile adenylate kinase mutants of *Escherichia coli*, 196
Polar blebs in bacteria, location of, 148
Polar caps of Gram-negative bacteria, growth of, 143
Polar carboxyl groups in sirohaems, 30
Polyadenylylation of mRNA, 91
Polyamines, and fidelity of translation in *Mucor racemosus*, 97
synthesis of, in *Mucor* sp., 89

Polyglycerophosphatide cycle, presence of, in *Escherichia coli*, 223
Polynucleotide recognition sequences in DNA, and restriction endonucleases, 242
Polyoxin D, effect of, on *Mucor rouxii*, 77
Polypeptide chain elongation in *Mucor racemosus*, 94
Polyuronides in walls of *Mucor rouxii*, 76
Pool of diaminopimelate in bacteria, existence of, 139
Pool sizes in bacteria, growth rate and, 134
Populations of bacteria, age distributions in, 109
Positional specificity of bacterial phospholipase A_1, 214
Positive regulatory role of the *nir*A gene in synthesis of nitrate and nitrite reductases in fungi, 42
Potential biohazards of genetic engineering, 262
Potential for synthesis of nitrate reductase in fungi, 54
Probes for active sites in fungal nitrate reductases, 7
Products of genes involved in nitrate assimilation in fungi, 40
Prokaryotes, surface extension and the cell cycle in, 105
Prokaryotic genes, cloning of, in genetic engineering, 254
Promiscuous plasmids, and genetic engineering, 236
Prosthetic group, of fungal nitrite reductase, sirohaem as, 26
  sirohaem as a, with nitrite reductase from *Neurospora crassa*, 30
Protease, action of, on nitrate reductase *in vivo*, 58
Protein kinases, effect of cAMP on, in *Mucor* spp., 82
Protein, membrane, synthesis, and thymine starvation in bacteria, 154
  protein interactions in bacterial membranes, 145
  synthesis, and dimorphism in fungi, 90
  and membrane growth in bacteria, 144
  burst of, in morphogenesis, 93

inhibition of, and bacterial growth, 151
  rate of, in *Mucor racemosus*, 92
Proteolytic activation of chitin synthetase in *Mucor rouxii*, 78
Proteus sp., *rod* mutants of, 157
Pseudomonas aeruginosa, generation times in, 113
Pseudomonas aureofaciens, phospholipase C activity of, 221
Pseudomonas putida, lipid activation of hydroxylase in, 185
Pseudomonas rubescens, ornithine-containing lipids in, 208
Pseudomonas schuykilliensis, phospholipase C activity produced by, 221
Pseudomonas sp., occurrence of acylphosphatidylglycerol in, 203
Purification of fungal nitrite reductase, problems in, 25
Purine catabolism, effect of ammonia on, in fungi, 45
Puromycin, effect of, on induction of nitrite-uptake system in fungi, 38
  effect of, on ribosomes from *Mucor racemosus*, 96
Putrescine synthesis by *coy* mutants of *Mucor* sp., 89
Pyridine haemochromogen derivative of fungal nitrate reductase, 3
Pyruvate kinase isozymes in *Mucor rouxii*, 85
Pyruvate kinases of *Mucor racemosus*, 83
Pyruvate oxidase, lipid activation of, in *Escherichia coli*, 185

R

Rabbit globin, biosynthesis of, 91
Rat insulin, insertion of genes for, into bacteria, 259
Real-age distributions in bacterial populations, 118
Receptor for teichoic acid in *Bacillus subtilis*, 140
Recognition sites for bacterial restriction endonucleases, 241
Recombinant DNA, in genetic engineering of micro-organisms, 235

SUBJECT INDEX

Redox state of fungal nitrate reductase, and sensitivity to inhibition, 6
Reduction of nitrate by fungi, 1
Reduction of nitrite to ammonia, steps involved in, with fungi, 20
Regulation, of activity of chitin synthetase in *Mucor rouxii*, 77
  nitrate assimilation in fungi, 38
  of nitrate reductase activity in fungi, 58
  of phospholipid composition in bacteria, 188
  of phospholipid metabolism in bacteria, 222
Regulator action of *nir*A gene action in *Aspergillus nidulans*, 41
Regulatory mechanisms in nitrate assimilation by fungi, 39
Replication of DNA in bacteria, velocity of, 124
Resistance factors in bacteria, and growth rate, 130
Resistance of bacterial phospholipases to detergents, 214
Respiratory-deficient mutant of *Mucor bacilliformis*, 84
Responses, metabolic, to morphogenic signals in fungi, 79
Restriction endonucleases, use of, in genetic engineering, 237
Restriction-modification systems in bacteria, nature of, 237
Rhizopus stolonifer, dormancy of zygospores of, 99
Rhodopseudomonas palustris, growth of, 150
Rhodopseudomonas sp., occurrence of phosphatidylcholine in, 208
Rhodopseudomonas spheroides, ornithine-containing lipids in, 208
  phospholipid composition of, 191
Rhodospirillum rubrum, ornithine-containing lipids in, 208
Ribonucleic acid polymerases, and morphogenesis in *Mucor* sp., 91
Ribonucleic acid synthesis, and thymine starvation in bacteria, 153
Ribosome, biosynthesis in *Escherichia coli*, 92
  of *Mucor racemosus*, structure of, 96
Ribosomal proteins of *Mucor racemosus*, electrophoresis of, 98
Rod forms of bacteria, growth of, 137
Rod mutants of bacteria, 157
Rod-shaped bacteria, from exponential-phase cultures, size of, 117
  growth of, 106
  wall growth of, 138

S

Saccharomyces cerevisiae, cloning of DNA from, in bacteria, 256
  glycero-3-phosphate acyltransferase of, 197
  multiple forms of RNA polymerase in, 91
Salmonella spp., bacteria-phase variation in, 255
  flagella antigens in, 255
Salmonella typhimurium, CDP-diglyceride synthesis in, 198
  cell-width mutants of, 160
  effect of temperature on growth rate of, 130
  generation times in, 113
  growth rate of, 128
  inhibition of peptidoglycan biosynthesis in, 154
  lipid activation of galactosyl transferase in, 185
  occurrence of acylphosphatidylglycerol in, 202
  production of large anucleate cells by, 163
  turnover of phosphatidylethanolamine in, 193, 222
  wall growth in, 108
Saturated fatty-acyl residues in bacterial membranes, 189
Screening methods for cloned DNA, 260
Sea urchin sperm DNA, cloning of, 246
Sedimentation coefficient of fungal nitrate reductase, 5
Segregation, of bacterial membranes, 145
  of flagella in bacteria, 146
  of genomes in bacteria, 147
Selenomonas ruminantium, occurrence of phosphatidylserine in, 205

SUBJECT INDEX

Self assembly of subunits of nitrate reductase of *Neurospora crassa*, 16
Sensitivity of DNA to restriction endonucleases, 248
Separation, of chromosomes in bacteria, 148
  of daughter cells of bacteria, 110
Septa of bacteria, formation of, 106
Septal site, choice of position of, in anucleate bacterial mutants, 163
Septal wall formation in bacteria, 135
Septation, and penicillin-binding proteins in bacteria, 154
  mutants of bacteria, 161
  start of, in bacteria, 127
Septum formation, in bacteria, 107
  in bacterial *rod* mutants, 159
Sequences of DNA, specific, insertion of, in genetic engineering, 244
Sequences recognized by restriction endonucleases in DNA, 240
Serratia marcescens, CDP-diglyceride synthesis, by, 199
Serratia spp., phospholipase C activity produced by, 221
Sex in members of the Mucorales, 99
Shape maintenance, and penicillin-binding proteins in bacteria, 154
Sirohaem, as a prosthetic group of fungal nitrite reductase, 26
  nitrite as the prosthetic group of nitrite reductase in *Neurospora crassa*, 30
  location of, in action of nitrite reductase, 35
  structure of, 31
Sister cells of bacteria, differences in growth rate of, 116
Sister-sister correlations in bacterial growth, 113
Sites of growth in bacteria, location of, 134
Size, bacterial cell, mutants affecting, 160
  distribution in exponentially growing bacteria, 117
  distributions of bacteria, interpretation of, 118
  of bacterial cells, average, factors determining, 120
  of exponential-phase rod-shaped bacteria, 117

Snake venoms as a source of phospholipases, 212
Sodium fluoride, effect of, on adenylcyclase of *Mucor* spp., 82
Specific activities of nitrate reductase from *Neurospora crassa*, 7
Spectra, absorption, of nitrite reductase from *Neurospora crassa*, 29
  of nitrite reductase of *Neurospora crassa*, 27
Spermidine synthesis in *coy* mutants of *Mucor* sp., 89
Sphaeroplasts of *Escherichia coli*, role of, in end-to-end joining of DNA molecules, 244
Sphingomyelin, hydrolysis of, by bacterial phospholipase C, 221
Spiral-growth mutants, of *Bacillus subtilis*, 142
  of bacteria, 161
Spirillum serpens, outer membrane growth of, 143
Spiroplasma citri, phospholipid composition of, 217
Sporangiospore germination in *Mucor* spp., 68
Sporangiospore formation in *Mucor* spp., 68
Sporangiospores of *Mucor rouxii*, wall composition of, 76
Spore germination in *Mucor* spp., 70
Spores, of *Bacillus megaterium*, phospholipase A_1 activity of, 215
  of *Mucor rouxii*, wall composition of, 76
Stability, *in vivo*, of nitrate reductase in *Neurospora crassa*, 56
Stabilizing effect of nitrate on nitrate reductase in *Neurospora crassa*, 57
Staphylococci, inhibition of wall synthesis in, 156
Staphylococcus aureus, cardiolipin biosynthesis in, 204
  growth of wall of, 136
  lipid activation of isoprenoid alcohol phosphokinase in, 165
  occurrence of lipo-amino acids in, 200
  phospholipid metabolism in, 192
Staphylococcus spp., synchronous division of, 114
Starvation for thymine, effect of, on surface growth of bacteria, 153

SUBJECT INDEX

Stochastic element in division of bacteria, 116
Stochastic fluctuations in the bacterial growth cycle, 112
Streptococci, wall growth in, 107
Streptococcus faecalis, cell volume of, 136
  effect of mitomycin C on, 154
  generation times in, 113
  lipid activation of ATPase of, 186
  occurrence of lipo-amino acids in, 201
  septum formation in, 135
Streptococcus lactis, inhibition of wall synthesis in, 156
Streptococcus pyogenes, topography of surface of, 135
Streptomyces stoyaensis, ornithine-containing lipids in, 208
Streptomyces spp. as a source of restriction endonucleases, 240
Structural characteristics of bacterial phospholipids, 178
Structure of sirohaem, 31
Subunit structure, of nitrate reductase from *Aspergillus nidulans*, 12
  of nitrate reductase from *Neurospora crassa*, 7
Subunits of nitrate reductase of *Neurospora crassa*, 14
Sulphanilic acid diazonium chloride, use of, to study membrane asymmetry, 181
Sulphide, effect of, on fungal nitrate reductase, 6
Sulphite, as a competitive inhibitor of nitrite reductase, 34
  inhibition of nitrite reductase of fungi by, 23
Sulphydryl-binding agents, effect of, on fungal nitrate reductase, 2
Sulphydryl groups, role for, in fungal nitrate reductase, 8
  role of, in nitrate reductase from *Neurospora crassa*, 7
Surface area, increase, effect of protein synthesis inhibition on, in bacteria, 151
  of bacteria, and cell length, 121
Surface, cell, bacterial, topography of, 134
  extension, and the cell cycle in prokaryotes, 105

in bacteria and chromosomal segregation, 149
of bacteria, genetic approaches to analysis of, 156
Symmetric growth model for bacteria, 147
Symmetry of septum formation in bacteria, 107
Synchronous cell division of bacteria, agents that cause, 126
Synchronous cultures and age-classified cells of bacteria, 132
Synchronous populations of bacteria, 111
Synchrony of division in bacteria, 114
Synthesis of cell walls, and morphogenesis of *Mucor* sp., 75

T

Taurocholate, requirement for, by bacterial phospholipase, 215
Teichoic acid synthesis, and thymine starvation in bacteria, 153
Teichuronic acid, inability of mutant bacteria to synthesize, 157
Temperature, effect of, on growth of bacteria, 130
  effect of, on inactivation of nitrate reductase from *Aspergillus nidulans*, 60
  on phospholipid composition of bacteria, 189
Temperature-sensitive flagella production by bacteria, 146
Temperature-sensitive *rod* mutants of bacteria, 158
Terminus attachment of the bacterial chromosome, 149
Tetrahydroporphyrin nature of sirohaem, 29
Theophylline, effect of, on adenylcyclase of *Mucor* spp., 82
Thermolabile glycero-3-phosphate acyltransferase from *Escherichia coli*, 196
Thermus aquaticus, phopholipid composition of, 189
Thickness of bacterial walls, cause for, 140
Thickness of walls in bacteria, effect of environment on, 131

Thiobacillus thiooxidans, ornithine-containing lipids in, 208

Thiourea, effect of, on fungal nitrate reductase, 2

Threonine starvation, effect of, on wall formation in bacteria, 135

Thymine-requiring strain of *Escherichia coli*, velocity of chromosome replication in, 124

Thymine starvation, and surface extension of bacteria, 152

Timing of division in bacteria, 106

Topography of the bacterial cell surface, 134

Transacylating enzymes in bacteria, 223

Transcription, of inserted DNA sequences, 246

of mRNA for nitrate reductase, effect of nitrate on, 55

Transcriptional control in *Mucor* sp., 95

Transcriptional events in morphogenesis in *Mucor* sp., 91

Transfer RNAs, role of, in biosynthesis of lysylphosphatidylglycerol, 201

Transition group metals, and morphogenesis in *Mucor* sp., 75

Translation, of inserted DNA sequences, 246

of mRNA for nitrate reductase, effect of nitrate on, 55

in fungi, 54

Transmission of bacterial cytochromes to progeny, 145

Transpeptidation in bacterial wall synthesis, 142

Transport, of nitrate and nitrite in *Neurospora crassa*, 37

systems, conservation of, in bacterial growth, 145

Triacyllysocardiolipin, presence of, in *Acinetobacter* sp., 217

Trigger for cell division of bacteria, 116

Triglyceride lipase activity of bacteria, 215

Triton X-100, inhibition of acyltransferases by, 196

Tryptophan, effect of, on activity of nitrate reductase in *Neurospora crassa*, 59

Tungstate as an antagonist of nitrate reductase in fungi, 49

Tungsten, incorporation of, into nitrate reductase from *Neurospora crassa*, 15

Turgor as a factor in bacterial wall growth, 107

Turnover number, of fungal nitrite and nitrate reductases, 36

of fungal nitrite reductase, 25

Turnover, of lipids in bacteria, 210

of nitrate reductase in *Neurospora crassa*, 56

of peptidoglycan in bacteria, 139

of wall components in bacteria, 107

Type I restriction endonucleases in bacteria, 238

Type II restriction endonucleases from bacteria, 239

Types of bacterial restriction endonucleases, 238

U

Ubiquinone-6, activation of pyruvate oxidase, 186

Unsaturated fatty-acyl residues in bacterial membranes, 189

Uptake of nitrite and nitrate by *Neurospora crassa*, 37

Urea transport in fungi, effect of ammonia on, 45

Uric acid, synthesis of nitrate reductase by fungi grown on, 18

Uridine diphosphate-glucose synthesis in *Dictyostelium discoideum*, 94

Ustilago maydis, nitrate reductase in, 49

V

Valency changes undergone by molybdenum in nitrate reductase, 10

Vanadium, incorporation of, into nitrate reductase from *Neurospora crassa*, 15

Variation in size of bacteria, 106

Vectors, plasmid, in genetic engineering, 247

Vegetative growth in *Mucor* spp., 68

Veillonella parvula, occurrence of phosphatidylserine in, 205

Velocity of DNA replication in bacteria, 124

SUBJECT INDEX

Viral phospholipids, role of, 188
Viruses, microbial, and phospholipids, 187
Volatile factors, and morphogenesis in *Mucor racemosus*, 73
Volume increase in bacteria, 132
Volumes of bacteria, and thymine starvation, 153

W

Wall, association of, with bacterial plasma membranes, 145
  degradation in bacteria, 126
  morphogenesis in *Mucor* sp., 90
Walls, bacterial, asymmetric growth of, 144
  bacterial, centripetal growth of, 138
Water content of bacteria, 126
Width, bacterial cell, mutations affecting, 160
  of bacteria, 106

X

Xanthomonas spp. as a source of Type II restriction endonucleases, 240
Xylose, and morphogenesis in *Mucor* sp., 72
  effect of on hyphal branching in *Mucor* spp., 1

Y

Yeast development in *Mucor rouxii*, 73
Yeast DNA, insertion of, in genetic engineering, 254
Yeast growth of *Mucor racemosus*, 71
Yeast-hyphal transformation, and protein synthesis, 94
Yeast-mycelium dimorphism in *Mucor* sp., 68

Z

Zygospore dormancy in *Mucor* sp., 99

Surface Extension in the Cell Cycle in Prokaryotes
M. G. SARGENT

Note added in proof

Koppes, Woldringh and Nanninga (*Journal of Bacteriology* **134**, 423, 1978) have recently provided evidence for an exponential growth pattern in *E. coli*, using the Collins-Richmond principle (p. 118). However, Cullum and Vicente (*Journal of Bacteriology* **134**, 330, 1978) using a simpler method have concluded that the rate of length extension is initially constant followed by a period in mid-cycle during which a doubling in rate occurs, followed by another period of constant rate. Unfortunately, the statistical significance of this conclusion is less clear (p. 118) than for the report of Koppes *et al.* The coefficient of variation of cell length at initiation found by Koppes *et al.* is greater than that found at birth and similar to that at division. The authors regard this as evidence that the timing of cell division is strictly deterministic after initiation (p. 116). Previously Koch (*Advances in Microbial Physiology* **16**, 49, 1977) had analysed published data for *E. coli* which suggested that the variation in cell size at initiation was greater than at division and therefore that the timing of division may not be solely regulated by the chromosomal clock.

Begg and Donachie (*Journal of Bacteriology* **133**, 452, 1978) have suggested that the increased width of *E. coli* 15*thy*⁻ grown at low thymine concentrations (p. 124) is due to an unrecognized imbalance in growth possibly caused by the activation of the prophage carried by this strain. In three other strains they find that the increased mass per cell, found at low thymine concentrations is entirely due to an increase in cell length. The authors consider that this observation eliminates the possibility that the control of length extension is dependent on termination (p. 121). However, they do not appear to have considered that in these strains, the D period may be lengthened. In the absence of data for DNA content, or C and D times, these conclusions seem premature.

An autoradiographic study of the segregation of labelled wall in an autolytic-deficient mutant of *B. subtilis* has recently provided evidence that length extension occurs from a small number of growth sites per cell (Pooley, Schaeppi and Karamata, *Nature*, London **274**, 263, 1978). Koppes, Overbeeke and Nanninga (*Journal of Bacteriology* **133**, 1053, 1978) have confirmed the high resolution autoradiographic observations on *E. coli* pulse-labelled with ³H-diaminopimelic acid published by Ryter, Hirota and Schwarz (*Journal of Molecular Biology* **78**, 185, 1973) (p. 139) but have not investigated the segregation process. This remains the most critical experiment in determining the pattern of length extension.

72628